T0137107

eThekwini's Green and Ecological Infrastructure
Policy Landscape

Richard Meissner

eThekwini's Green and Ecological Infrastructure Policy Landscape

Towards a Deeper Understanding

 Springer

Richard Meissner
Council for Scientific and Industrial
Research (CSIR)
Pretoria, South Africa

Centre for Water Resources Research
University of KwaZulu-Natal
Scottsville, South Africa

ISBN 978-3-030-53053-2   ISBN 978-3-030-53051-8   (eBook)
https://doi.org/10.1007/978-3-030-53051-8

This Springer imprint is published by the registered company Springer Nature Switzerland AG
The registered company address is: Gewerbestrasse 11, 6330 Cham, Switzerland

To Colleen and Johann

Acknowledgements

My heartfelt thanks to my wife, Colleen, and our son, Johann, for being patient with me while researching this book. They always endure the inconvenience of my absence while I am away on field trips. Then there are my CSIR colleagues, some of whose ideas are contained in the pages of this book. I would like to acknowledge the input of Inga Jacobs-Mata, who has always been a sparring partner with whom I can discuss my thoughts. A special word of thanks goes to Marc Pienaar who, with his computer wizardry, produced the digital version of PULSE^3. A big thank you goes to Luanita Snyman-Van der Walt, who produced the map of eThekwini, and to Krishi Krishna, who patiently trawled the Internet in search of articles written by eThekwini officials.

As usual, a word of appreciation goes out to my friend and colleague, Jeroen Warner, from the Wageningen University in the Netherlands. He has provided me with valuable insights that have enriched my analysis. Sverre van Klaveren, a former student at the Wageningen University, also deserves a mention. He conducted research on eThekwini's green infrastructure policy landscape for his Master's thesis under Jeroen's supervision and my co-supervision.

Johan Marais, from the African Snake Bite Institute, shared the picture of the black mamba, and Douw Steyn, the Sustainability Director of Plastics SA, provided snapshots of the waste pollution in the port of Durban and on the city's beaches. Duncan Hay, Executive Director of the Institute of Natural Resources, contributed photographs of the uMngeni Vlei. I am grateful for all these photos, since one picture can speak a thousand words. I would like to thank Sharon Rees for diligently proofreading the manuscript.

Numerous individuals willingly sacrificed their time so that I could conduct interviews with them. These interviews have supplied the bulk of the information that I have analysed and reported on in this book. Although I assured them of their anonymity before the interview commenced, I would like to express my gratitude to the following: Sabine Stuart-Hill, Andre Mather, Bronwyn Goble, Duncan Hay, Cathy Sutherland, Fatima Alli, Pearl Gola, Geoff Tooley, Ismail Banoo, Jim Taylor, Joanne Douwes, Sean O'Donnoghue, Lee D'Eathe, Magash Naidoo, Martin Clement and Kate Pringle. To ensure their anonymity, I will refer to them as numbered respondents in numerical order throughout the manuscript, although their numbers will be inconsistent with the list of names.

Finally, I would like to acknowledge the funding provided by the South African National Research Foundation for this research endeavour. Although numerous individuals have read various parts of this manuscript, have made suggestions and given their valuable input, I bear full responsibility for any errors that may be contained therein.

July 2020   Richard Meissner

Contents

1 Green and Ecological Infrastructures and Water Security at a Municipal Level: An Overview 1
  1.1 Introduction 1
  1.2 The Conceptualisation of Green and Ecological Infrastructures 3
    1.2.1 Green Infrastructures 4
    1.2.2 Ecological Infrastructures 8
  1.3 Misinterpreted Multi-functionality 10
  1.4 Green and Ecological Infrastructures as Instrumental Policies 11
  1.5 Water Security 15
  1.6 South African Municipalities and Water Resources Management 17
  1.7 Conclusion 19
  References 20

2 The Use of Paradigms and Theories in the PULSE³ Analytical Framework 25
  2.1 Introduction 26
  2.2 Conceptual Building Blocks 27
    2.2.1 Paradigms 28
    2.2.2 Theory 30
    2.2.3 Productive Utility 32
  2.3 Paradigms and Theories as Multiple Inquiry Systems 35
    2.3.1 Epistemological and Practical Value 35
    2.3.2 Actions and Situations 35
    2.3.3 Configurations 36
    2.3.4 Insightful Consequences 37
    2.3.5 Science Speaking Truth to Power 39

2.4 Pulse[3]   41
    2.4.1 Rationale...   41
    2.4.2 Components   42
2.5 Conclusion...   71
References   73

3 eThekwini's Green and Ecological Infrastructure Policy
   Landscape..   81
3.1 Introduction   81
3.2 The eThekwini Metropolitan Municipality   82
3.3 Methodology   85
    3.3.1 Data Collection   86
    3.3.2 Data Coding   87
3.4 Analysis...   87
    3.4.1 Paradigm Profile   87
    3.4.2 Theories 112
    3.4.3 Causal Mechanisms 175
    3.4.4 The Problem-Solving and Critical Theory
       Perspectives... 196
3.5 Practice Theory as a Direction-Finding Beacon 199
    3.5.1 Conceptualising Practice 200
    3.5.2 The Practice Perspective............................... 200
3.6 Conclusion... 219
References 221

4 Discussion 235
4.1 Introduction 235
4.2 Self-Reflection and Exclusivity 236
4.3 Reality.. 239
    4.3.1 Empiricism 239
    4.3.2 Systemic Constructivism 240
4.4 Group-Think 240
    4.4.1 Ecological Dominance 241
    4.4.2 International Relations of eThekwini as an Autonomous
       Actor 243
4.5 Conclusion... 244
References 246

Appendix... 249

Index 281

Abbreviations

ACT	African Conservation Trust
AIMS	Agential, ideational, material and structural
ALCOSAN	Allegheny County Sanitary Authority
ANC	African National Congress
BII	Biodiversity Intactness Index
CBD	Convention on Biological Diversity
CMA	Catchment Management Agency
COP	Conference of Parties
COVID-19	Coronavirus disease 2019
CRP	Community Reforestation Programme
CSI	Corporate Social Investment
CSIR	Council for Scientific and Industrial Research
CSR	Corporate Social Responsibility
CWRR	Centre for Water Resources Research
D'MOSS	Durban Metropolitan Open Space System
DA	Democratic Alliance
DAC	Durban Adaptation Charter
DBSA	Development Bank of Southern Africa
DCCS	Durban Climate Change Strategy
DRAP	Durban Research Action Partnership
DUCT	Duzi Umngeni Conservation Trust
DUT	Durban University of Technology
DWS	Department of Water and Sanitation
EFF	Economic Freedom Fighters
EPA	Environmental Protection Agency
EPCPD	Environmental Planning and Climate Protection Department
FIFA	*Fédération Internationale de Football Association*
GDP	Gross Domestic Product
GIZ	*Gesellschaft für Internationale Zusammenarbeit*
GWP	Global Water Partnership

IFP	Inkatha Freedom Party
INR	Institute of Natural Resources
IPCC	Intergovernmental Panel on Climate Change
IR	International Relations
KZN	KwaZulu-Natal
MOU	Memorandum of Understanding
MPI	Migration Policy Institute
NGO	Non-governmental Organisation
NRF	National Research Foundation
ORI	Oceanographic Research Institute
PDF	Portable Document Format
ProEcoServ	Project on Ecosystem Services
PRRP	Palmiet River Rehabilitation Project
PRVC	Palmiet River Valley Community
PULSE	People Understanding and Living in a Sustained Environment
PWR	Palmiet River Watch
RIPE	Regulatory International Political Economy
SANBI	South African National Biodiversity Institute
SAPPI	South African Pulp and Paper Industries
SASA	South African Sugar Association
SDG	Sustainable Development Goals
SEA	Strategic Environmental Assessment
SMME	Small, Medium and Micro Enterprise
SuDS	Sustainable Drainage System
TEU	Twenty-foot Equivalent Unit
TKZN	Tourism KwaZulu-Natal
TNC	The Nature Conservancy
TPA	Transnet Port Authority
UEIP	uMngeni Ecological Infrastructure Partnership
UKZN	University of KwaZulu-Natal
UN	United Nations
UNESCO-IHP	United Nations Educational, Scientific and Cultural Organisation-International Hydrological Programme
UNFCC	United Nations Framework Convention on Climate Change
UN-Water	United Nations Water
USA	United States of America
WEF	Water-Energy-Food
WESSA	Wildlife and Environment Society of Southern Africa
WfE	Working for Ecosystems
WHO	World Health Organisation
WIOMSA	Western Indian Ocean Marine Science Association
WRC	Water Research Commission
WWF	World Wide Fund for Nature

List of Figures

Fig. 1.1   Map of the eThekwini Metropolitan Municipality.   3
Fig. 1.2   Municipal solid waste washed up on Durban's beaches,
            April 2019   4
Fig. 1.3   Plastic pollution on Durban's beaches in the aftermath
            of the April 2019 storm that devastated the city.   5
Fig. 1.4   The port of Durban.   6
Fig. 1.5   Volunteers removing municipal solid waste from the port
            of Durban, April 2019   7
Fig. 1.6   Tons of municipal solid waste and other debris floating
            in the port of Durban, April 2019.   8
Fig. 3.1   The Port of Durban's oil storage facilities   83
Fig. 3.2   The Durban Marina   83
Fig. 3.3   Research worldview assessment   88
Fig. 3.4   eThekwini's roof-top garden initiative   94
Fig. 3.5   The Palmiet River.   95
Fig. 3.6   A thunderstorm over the Magaliesberg silhouetted
            by lightening at Hartbeespoort   96
Fig. 3.7   Snowfall over Hartbeespoort, with the Magaliesberg
            in the background, August 2012.   97
Fig. 3.8   The Dome Pools of the Magaliesberg near Mooinooi,
            September 2019   98
Fig. 3.9   Durban Botanical Gardens   99
Fig. 3.10   The Drakensberg from the Monk's Cowl Nature Reserve,
            with a wetland in the foreground 100
Fig. 3.11   Howick and the Howick Falls from the air, with Midmar Dam
            on the uMngeni River in the background 102
Fig. 3.12   The headwaters of the uMngeni near Nottingham Road
            in the Natal Midlands. 103
Fig. 3.13   The Inanda Dam supplies water to the eThekwini Metropolitan
            Municipality. 104
Fig. 3.14   The Msunduzi River, a tributary of the uMngeni River 105

Fig. 3.15   The River Horse wetland 106
Fig. 3.16   Perspectives present in the analysed interviews.............. 113
Fig. 3.17   Businesses close to the River Horse wetland 118
Fig. 3.18   Municipal workers cleaning municipal solid waste
             in the wake of the devastating April 2019 storm 119
Fig. 3.19   View across the uMngeni Vlei—conceptualised as an
             ecological infrastructure 119
Fig. 3.20   uMngeni Vlei Note the snow-capped hills and Drakensberg
             Mountains in the background 120
Fig. 3.21   Midmar Dam spill way................................. 120
Fig. 3.22   oHlanga Estuary, at Umhlanga Rocks, north of Durban 121
Fig. 3.23   Umkomaas Estuary, south of Durban 128
Fig. 3.24   An egret hunting fishes in the oHlanga Estuary 128
Fig. 3.25   Luxury apartments on the beach front at Umhlanga Rocks..... 153
Fig. 3.26   Small fish in the oHlanga Estuary north of Durban These fish
             are Glassies (*Ambassis natalensis*), which is one of few fish
             species that completes its life-cycle in South African estuaries,
             whereas most are species that spawn at sea, recruit into the
             estuaries as young juveniles and use the systems
             as nurseries....................................... 154
Fig. 3.27   An angler collecting sand prawn *Callichirus kraussi*
             for bait from the oHlanga Estuary at low tide............... 155
Fig. 3.28   An angler with collected sand prawn *Callichirus kraussi*
             from the oHlanga Estuary.............................. 156
Fig. 3.29   Alien vegetation and erosion near Durban 175
Fig. 3.30   Anglers at the Umkomaas Estuary 176
Fig. 3.31   A rat (*Rattus*) along the banks of the uMngeni River near
             the Quarry Road informal settlement 177
Fig. 3.32   A black mamba (*Dendroaspis polylepis*) 177
Fig. 3.33   The sum of causal mechanisms in the analysed interviews..... 178
Fig. 3.34   Respondent 1's identified causal mechanisms................ 179
Fig. 3.35   Respondent 2's identified causal mechanisms................ 180
Fig. 3.36   Respondent 3's identified causal mechanisms................ 181
Fig. 3.37   Respondent 4's identified causal mechanisms................ 182
Fig. 3.38   Respondent 5's identified causal mechanisms................ 184
Fig. 3.39   Respondent 6's identified causal mechanisms................ 185
Fig. 3.40   Respondent 7's identified causal mechanisms................ 186
Fig. 3.41   Respondent 8's identified causal mechanisms................ 187
Fig. 3.42   Respondent 9's identified causal mechanisms................ 188
Fig. 3.43   Respondent 10's identified causal mechanisms............... 189
Fig. 3.44   Respondent 11's identified causal mechanisms............... 190
Fig. 3.45   Respondent 12's identified causal mechanisms............... 191
Fig. 3.46   Responding 13's identified causal mechanisms............... 192

Fig. 3.47   Respondent 14's identified causal mechanisms   193
Fig. 3.48   Respondent 15's identified causal mechanisms   194
Fig. 3.49   Respondent 16's identified causal mechanisms   196
Fig. 3.50   Critical and problem-solving theory types   197

List of Tables

Table 2.1   Paradigm assessment index 45

Table 2.2   Repertoire of theories. 67

Table 3.1   The elements of green infrastructure conceptualisations 101

Table 3.2   The elements of ecological infrastructure conceptualisations . . . 101

Table 3.3   The hydrosocial contract theory 123

Table 3.4   Liberal Institutionalism and eThekwini's green and ecological
infrastructure policy landscape 134

Table 3.5   Social constructivism and eThekwini's green and ecological
infrastructure policy landscape 161

Table 3.6   Risks identified in the interviews 168

Table 3.7   Scientific publications produced by eThekwini officials 204

Chapter 1
Green and Ecological Infrastructures and Water Security at a Municipal Level: An Overview

Abstract Humankind has increased water security in highly-industrialised states by supplying sizeable populations with potable water and sanitation services. However, to a large extent, in developing countries like South Africa, cities like Durban still face water insecurity concerns. In the context of such concerns, green and ecological infrastructures play a role in water security and in other normative objectives, such as biodiversity protection, climate change adaptation and mitigation, as well as in sustainable development. Water security is linked to the notion of environmental and human security. It is not only at international level that water security has become a prominent objective, concept and practice; the local governments in South Africa also have a constitutional obligation to provide water services to their citizens. The government apparatus that is closest to the people are the municipalities, and their services to residents include water purification and reticulation, wastewater treatment and storm-water discharge. This chapter explains the political context of green and ecological infrastructures, coupled with water security, in the eThekwini Metropolitan Municipality.

Keywords eThekwini Metropolitan Municipality · Green infrastructure · Ecological infrastructure · Pollution · Flooding · Water security

1.1 Introduction

A paramount and relentless environmental challenge facing humankind in the twenty-first century is water resources management (Pahl-Wostl 2015, 2017) and the use of multiple practices at various levels. At least in industrialised countries, humans have increased water security (Cook and Bakker 2012) for all, and this has resulted in severe environmental consequences due to technological innovations (Pahl-Wostl 2015), use and disposal. This is the concealed irony of water security: while securing water for the human population, technological advances are damaging the ecosystems that provide the water and protect the settlements against disaster. However, it is not only technology that plays an unfavourable and inauspicious role. According to Ghosh and Kansal (2019), '[t]he dynamic interaction between society and nature is

R. Meissner, *eThekwini's Green and Ecological Infrastructure Policy Landscape*,
https://doi.org/10.1007/978-3-030-53051-8_1

influenced by the prevailing normative, cognitive and regulative societal systems, which guide the relationship between society and nature …' Our decisions, laws and policies are, in many instances, aberrations, whether they be based on empirical or interpretivist scientific results.

During these environment-society associations, green and ecological infrastructures become an important element for offsetting or at least mitigating, the negative consequences of regulative and technological improvements, such as flooding and pollution, which, in turn, influence the water systems unfavourably. This shows an exceedingly positive association between the infrastructure types and their influence on development-related negative consequences. Taking the regulative advances into consideration, green and ecological infrastructures contain inherent normative and cognitive structures that influence such initiatives. Regulatory structures do not only come about through empirical scientific innovations, but they are also constituents of a wider social process, where inter-subjective elements, like ideology (Schidler 2020a, b), play a central role. This means that these infrastructure types are not immune to the vagaries of the invisible elements relating to cognitive processes.

To explore these cognitive, development and regulatory associations further, let us consider Durban, which is located in the eThekwini Metropolitan Municipality (Fig. 1.1). Here, flooding and pollution (Figs. 1.2 and 1.3) are major concerns, not only for residents in low-lying areas, but also for the transport infrastructure, in or near the port of Durban (Fig. 1.4). In October 2017, Durban was hit by a violent storm, which resulted in eight fatalities (Wolhuter 2017) and the Durban South roads being closed, due to local flooding (Singh 2017). In April 2019, another enormous rainstorm ravished the city, leaving 85 dead and infrastructure damage costing millions of rands (Somdyala 2019). According to Head (2019), more than 200 mm of precipitation fell overnight in some parts of KwaZulu-Natal during this tempest. He went on to say that several portions of Durban received more than 100 mm of rain in 24 h, while the average April rainfall for the city, under normal conditions, is about 71 mm (Head 2019).

What became apparent in the aftermath of both of the abovementioned events was that tons of municipal solid waste (Figs. 1.5 and 1.6) were washed up on the city's beaches and in the port (Somdyala 2019). According to the news reports, the Transnet Port Authority (TPA), which operates the harbour, could only open the facility after the flood had subsided, but large volumes of solid waste continued to affect the movement of ships (Singh 2019), which had a knock-on effect on the port's daily operations and on the costs by shipping companies operating large vessels in and out the port.

With this context in mind, I will review the nature and role of green and ecological infrastructures in this chapter. I will begin by outlining how scholars define them, before reviewing the role of municipalities in implementing such enterprises. Thereafter, I will discuss the nature and extent of water security, as a concept, as well as the role of local governments in water resources management, bearing in mind their constitutional and democratic mandates. In conclusion, I will outline how the rest of the book will unfold.

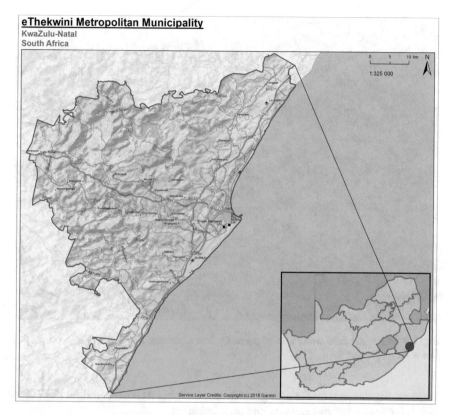

Fig. 1.1 Map of the eThekwini Metropolitan Municipality. Map produced by Luanita Snyman-van der Walt

1.2 The Conceptualisation of Green and Ecological Infrastructures

Built-up areas have distinguishable biophysical characteristics, compared to rural areas, due to their altered surface cover (Gill et al. 2007), like paved roads and sidewalks. These distinctive features are manifested, for instance, in altered energy exchanges, which create urban heat islands and hydrology changes that generate increased water runoff (Gill et al. 2007).

Because of this, green and ecological infrastructure provisioning varies in built-up and rural settings. Green infrastructures seem to be more applicable to metropolitan environments than ecological infrastructures, which are used in rural locations. The place and purpose of both infrastructure types depends, to a certain degree, on the nature and extent of civil engineering development, or its absence. Even so, the eThekwini Municipality is not exclusively reliant on green infrastructures to mitigate flooding and pollution, but also on ecological infrastructures, since large parts of the municipality are peri-urban and rural (Meissner et al. 2018a, 2019). The Durban

Fig. 1.2 Municipal solid waste washed up on Durban's beaches, April 2019. Photo courtesy of Douw Steyn, Plastics SA

municipality depends on the uMngeni River, which has its source in rural KwaZulu-Natal, for sustaining its economy and population.

With this background in mind, the concept of green and ecological infrastructures is, according to Cameron et al. (2012), ill-defined. This creates problems for city planners regarding the type of infrastructure that is required to get maximum benefit, under certain conditions (Cameron et al. 2012). These difficulties are compounded by the issue of time, since green infrastructure plans do not deliver benefits overnight. A distinction exists between both infrastructure types in scientific literature.

1.2.1　Green Infrastructures

Benedict and McMahon (2002) first defined the concept of a 'green infrastructure', and since then, scholars have coined various other definitions (Roe and Mell 2013). For Williamson (2003 cited in Roe and Mell 2013), a green infrastructure is 'an interconnected network of protected land and water that supports native species, maintains natural ecological processes, sustains air and water resources and contributes to the health and quality of life for … communities and people'.

Naumann et al. (2011: 1) broadly defined a green infrastructure as:

Fig. 1.3 Plastic pollution on Durban's beaches in the aftermath of the April 2019 storm that devastated the city. Photo courtesy of Douw Steyn, Plastics SA

…the network of natural and semi-natural areas, features and green spaces in rural and urban, and terrestrial, freshwater, coastal and marine areas, which together enhance ecosystem health and resilience, contribute to biodiversity conservation and benefit human populations through the maintenance and enhancement of ecosystem services. Green infrastructure can be strengthened through strategic and coordinated initiatives that focus on maintaining, restoring, improving and connecting existing areas and features, as well as creating new areas and features.

For Naumann et al. (2011), a green infrastructure is not only common in urbanised settings, but also in rural areas. They also define this infrastructure through the lens of several practices, giving effect to the notion of a network of interconnected spaces and rehabilitating existing and dilapidated areas. These practices are usual within engineering and give us the first hint of the epistemological and ontological features of green infrastructures.

Sandström (2002) noted that, before the concept 'green infrastructure' was coined, scholars spoke of a 'green space', when referring to natural areas, such as urban parks and nature reserves, with recreation being the main function. To move away from this unilateral function, Sandström (2002) introduced the concept of a 'green infrastructure' to highlight the multipurpose role of a green space. He also elevates green infrastructures to the same level and dignity as technological infrastructures in traditional town planning.

Fig. 1.4 The port of Durban

For Tzoulas et al. (2007) a green infrastructure consists of '… all, natural, semi-natural and artificial networks of multifunctional ecological systems within, around and between urban areas, at all spatial scales'. For them, green infrastructures highlight both the quality and quantity aspects of green spaces, together with their multifunctional role and their interconnectedness between various habitats.

According to these descriptions, a green infrastructure is not a single component, but a holistic system, and it has the multifunctional ability to contribute to improved water security as well as the broader human wellbeing in urban, peri-urban and surrounding rural areas. By so doing, a green infrastructure could 'fulfil the needs of a variety of stakeholders' (Madureira and Andresen 2014: 38).

Wachsmuth and Angelo (2018) talked about 'green urban nature', which signifies the return of nature to the city in the form of street trees, urban gardens, as well as the greening of post-industrial landscapes. Their definition differs from those that are promoted by Benedict and McMahon (2002), Roe and Mell (2013), Williamson (2003), Sandström (2002), Tzoulas et al. (2007) and Naumann et al. (2011), in that they lump green and ecological infrastructures together, within a wide range of human and natural settings.

The benefit of green infrastructures is related directly to their environmental functions with regard to increased urban biodiversity, the provision of ecosystem services, including water and food resources, limited climatic regulation, atmospheric purification and storm-water management, coupled with flood alleviation (Madureira and

Fig. 1.5 Volunteers removing municipal solid waste from the port of Durban, April 2019. Photo courtesy of Douw Steyn, Plastics SA

Andresen 2014). Urban planners, therefore, inextricably link water resources to the objectives of green infrastructures (Naumann et al. 2011), with water insecurity mitigation being advanced through such interventions.

Although these definitions seem straightforward, green infrastructures means different things to people (Benedict and McMahon 2002; Roe and Mell 2013), and they are coupled with various functions. For one stakeholder group, like a rural community, a green infrastructure could be a food source, while for urban dwellers, the same infrastructure could act as a buffer against flood surges, or as a climate change adaptation and mitigation mechanism. This suggests that urban planners and decision-makers consider green infrastructure policies and projects from varying perspectives and that they expect different benefits from them. It also implies an association between green and engineered infrastructures. By so doing, we can conclude that a green infrastructure is epistemologically rooted in the natural sciences, with the aim of solving various problems, which is also the case with an ecological infrastructure.

Fig. 1.6 Tons of municipal solid waste and other debris floating in the port of Durban, April 2019. Photo courtesy of Douw Steyn, Plastics SA

1.2.2 Ecological Infrastructures

Whereas green infrastructures are common in urban and peri-urban areas, Sigwela et al. (2017) relates ecological infrastructures to a sustainable rural livelihood. Whereas Sigwela et al. (2017) defines an ecological infrastructure functionally in rural areas, Li et al. (2017: 12) speak of an 'urban ecological infrastructure'. They define this type of infrastructure as '… an organic integration of blue (water-based), green (vegetated), and grey (non-living) landscapes, combined with exits (outflows, treatment or recycling) and arteries (corridors), at an ecosystem scale'.

In their definitions of an ecological infrastructure, Pringle et al. (2015) and Jewitt et al. (2015) also observed it in relation to the ecosystem. They view it as 'a naturally-functioning ecosystem that produces and delivers valuable services to people' and that is 'the nature-based equivalent of a built infrastructure'. Their definition was borrowed from the South African National Biodiversity Institute (SANBI), which notes that an '[e]cological infrastructure refers to a naturally-functioning ecosystem that delivers valuable services to people, such as water and climate regulation, soil formation and disaster risk reduction. It is the nature-based equivalent of a built or hard infrastructure and can be just as important for providing services and underpinning socio-economic development' (SANBI 2019). This means that these definitions, like Sandström's (2002) conceptualisation of a green infrastructure, elevate an ecological infrastructure to the same status, and with the same function, as technological infrastructures.

The research of Pringle et al. (2015) and Jewitt et al. (2015) focused on ecological infrastructures of the uMngeni River Catchment. The uMngeni provides ecosystem services in the form of water, climate regulation, soil formation and disaster risk reduction (Kotzé 2013). Since large parts of this catchment are rural, Kotzé (2013) argued that such areas contain the vital ecological infrastructure for maintaining and restoring the river. These essentials, including indigenous forests and wetlands, contribute directly to rural livelihoods. Kotzé's (2013) article in *The Water Wheel*, a publication of the Water Research Commission (WRC), reported on the Project on Ecosystem Services (ProEcoServ) of the Council for Scientific and Industrial Research (CSIR) in the following way: 'ProEcoServ aims to enhance the integration of ecological infrastructures and ecosystem services into national development planning …' (SAPECS, no date). The ProEcoServ researchers have adopted SANBI's definition as their conceptualisation of an ecological infrastructure.

Therefore, as emphasised by Li et al. (2017), an interdependent link exists between ecological and built or hard infrastructure. Pringle et al. (2015) argue that an engineered infrastructure can enhance the benefits that we derive from nature, but that continued development compromises the ability of the environment to deliver numerous services. Here we deal with a compromising situation, namely, that an ecological infrastructure, together with a built infrastructure, must be sustainably managed and restored (Pringle et al. 2015) to deliver such services.

To bring this notion closer to the subject at hand, in a press release, dated 28 February 2013, after a workshop to establish the uMngeni Ecological Infrastructure Partnership (UEIP), officials from the eThekwini's Water and Sanitation and Environmental Planning and Climate Protection departments defined ecological infrastructures as follows: 'Ecological (or natural) infrastructures refer to functioning ecosystems that produce and deliver services that are of value to society, such as fresh water, climate regulation, soil formation and disaster risk reduction' (eThekwini Metropolitan Municipality 2013). This definition is very similar to that of the SANBI description. In fact, SANBI played a major role in the formation of the UEIP during the workshop (Pringle 2020; Stuart-Hill 2020).

eThekwini convened the workshop 'to catalyse a new partnership initiative that was aimed at unlocking the potential of natural ecosystems in the water security equation' (eThekwini Metropolitan Municipality 2013). During the workshop, eThekwini officials indicated that an ecological infrastructure has a dual function, namely, to improve the quality of the uMngeni's water resources and to create jobs by restoring and maintaining it. In the light of the establishment of this partnership, Neil McCleod, the former head of Durban's Water and Sanitation Department, said that the engineering solutions for securing water for Durban had not worked and that a 'novel approach is needed for improving the quality and the quantity of water in the uMngeni through the restoration and maintenance of the catchment's [sic.] natural infrastructure i.e. its ecosystems' (eThekwini Metropolitan Municipality 2013).

eThekwini defines an ecological infrastructure as the water security improvements that are linked to water quality and quantity. In the light of this, eThekwini juxtaposes two ontologies (realities), namely, an ecological and a grey infrastructure, with the ecological infrastructure supplementing the grey infrastructure.

Kotzé (2013), Pringle et al. (2015) and Jewitt et al.'s (2015) definitions of an ecological infrastructure are in line with eThekwini's conceptualisation. This is no coincidence, since both Pringle, a former researcher at the INR and now an independent consultant, and Jewitt, formerly with the UKZN and now with IHE Delft Institute for Water Education, the Netherlands, have produced extensive research on the uMngeni River Basin and its ecosystems.

When one looks at the definitions that are conceptualised by other researchers and those that are closely linked to the uMngeni River Basin, one can see that an ecological infrastructure, in the global and Durban contexts, has a much narrower function than that of a green infrastructure. Globally, researchers emphasise the ecosystem service functioning capabilities of an ecological infrastructure, while, in the case of eThekwini, this infrastructure has a closer relationship with water security and is defined in terms of water quality and quantity. This brings me to the 'misinterpreted multi-functionality' of green and ecological infrastructures that has been highlighted by research over the years.

1.3 Misinterpreted Multi-functionality

To reiterate the previous findings outlined above, green and ecological infrastructures, coupled with built infrastructures, have inter-sectoral benefits that observers perceive as transformative technologies, to mitigate the risks and enhance the opportunities of ecosystem services. Municipalities promote green infrastructure projects because of their multi-functionality, since they are attractive as a '… means of achievement of several environmental, social, cultural and economic urban policy aims' (Madureira and Andresen 2014: 38).

Nevertheless, there is growing concern and criticism over the applicability of green infrastructures (Madureira and Andresen 2014). They are typically sited by local governments, based on their specific benefits, such as storm-water reduction or climate change adaptation, '… rather than a suite of socio-economic and environmental benefits' (Meerow and Newell 2017: 62). A point of contention revolves around this alleged misinterpretation of the multi-functionality of a green infrastructure, as it narrows the perceptions of the planners regarding its use (Meerow and Newell 2017). This 'misinterpretation' grows, in part, from '… the lack of stakeholder-informed, city-scale approaches to systematically identify ecosystem service tradeoffs, synergies and "hotspots" associated with green infrastructure and its siting' (Meerow and Newell 2017: 62).

If we consider the criticism of green infrastructures by Madureira and Andresen (2014) and Meerow and Newell (2017), which is also applicable to an ecological infrastructure because of its interdependent functionality with the former, it is evident that such projects could contribute to inclusive socio-economic development. As such, green and ecological infrastructures could not only serve urban or inner-city environments, but also peri-urban and rural areas around large cities (Madureira and Andresen 2014); this is the discursive drive behind the promotion of these infrastructure types. In this vein, the infrastructure is geared towards encompassing socio-economic development. However, what is not considered by observers is the normativity around the promotion and functionality of both infrastructures, coupled with the wider publicly-informed and encompassing stakeholder perceptions of their performance. By this, I mean that, since both infrastructure types have a specific ecosystem functionality, it is possible that green and ecological infrastructure proponents could focus exclusively on the empiricism that is coupled with both infrastructure types, while ignoring the interpretive and normative sides of the practicality coin (more on this later).

In this book, I will report on an investigation that not only studied the multi-functionality of green and ecological infrastructure projects, but also their interpretive and normative sides and, more specifically, to what extent paradigms and theories influence the development and application of green and ecological infrastructures. It is important to not only understand the green and ecological initiatives from an instrumental-policy perspective, but also from a worldview and theoretical angle. This is to gain a deeper understanding of the motives behind the promotion of such infrastructures by local governments and the hidden implications for the environment and the broader public. To do this, I need to briefly review the policy outlook of both infrastructure types.

1.4 Green and Ecological Infrastructures as Instrumental Policies

Following on from the misinterpreted multi-functionality argument above, Dunn (2010) argues that local authorities usually do not implement green infrastructures in the poorer neighbourhoods of urban environments. However, applying such an infrastructure can also benefit the more needy communities in terms of ecosystem services; by improving the quality of urban water, reducing urban air pollution, improving public health, enhancing urban aesthetics and safety, as well as generating green collar jobs and urban food security. Even so, according to Dunn (2010), the fact that green infrastructures assist the urban poor, is not always the issue. The concern is that both legal and instrumental policy barriers to green infrastructure implementation are inhibiting its benefits for the poor. Dunn (2010) advocates the removal of legal and policy obstacles, an increase in public financing and the raising of awareness of the benefits of green infrastructures among policy-makers and the wider public, which will facilitate an awareness of the their benefits among those that are deprived.

Schäffler and Swilling (2013) investigated the importance of robust planning for green infrastructure implementation in Johannesburg. For them, the important considerations are the public value of a green infrastructure and the opportunity cost of not investing in it. By linking social-ecological resilience with the urban infrastructure transition perspectives, Schäffler and Swilling (2013) show how a green infrastructure can be a response to challenging urban environments. They argue that since many South African metropolitan municipalities are experiencing service delivery protests, the risk exists that municipal officials might overlook green infrastructure investment during their planning and that they might focus exclusively on service provision. Despite this negative political landscape that has hampered the acceptance of a green infrastructure, Schäffler and Swilling (2013) recommend several policy interventions for change. For public officials to properly understand what a green infrastructure is and what its benefits are, they need to mix their thinking on social-ecological systems with infrastructure dynamics and a transition theory. There is a need to evaluate ecosystem services at a local government level, as there is a paucity of knowledge about them. Lastly, such an appraisal needs to take into consideration the unique circumstances and conditions of a municipality (Schäffler and Swilling 2013). An interesting observation was made regarding research on environmental issues at a city level:

> Writing from a developing country, we are sensitive to the particular dynamics that characterise Southern African cities, where research into, and planning around, environmental issues in general are both in their infancy. In these cities, environmental concerns are largely absent from academic, policy and even civil society activist discourse, being dwarfed by the seemingly more pressing matters of service delivery deficits, economic exclusion and poverty (Schäffler and Swilling 2013: 2).

This argument emphasises the role of increased knowledge as a facilitative tool for policy change, as it enhances the benefits of a green infrastructure for all, through policy change and improvement.

In related research, Lennon and Scott (2014) proposed the re-scoping of planning policies and processes and that they be incorporated into green infrastructure thinking and practice. For them, the ecology, ecosystem services and environmental risks need to become the central concerns in local government planning practices. They also advocate the identification of ecological principles by local government officials, in order to inform spatial planning strategies and practices. Furthermore, they argue that the green infrastructure approach gives local government planners an effective way of operationalising the ecosystem service concept when conducting spatial planning. Facilitating the latter does not merely entail a recalibration of the current modes of spatial planning, thinking and practice, but it also involves the transformation of spatial planning structures and how spatial planners view the world in which they operate and act (Lennon and Scott 2014).

Regarding the latter recommendation, Mell (2014) identified the localised interpretation of the green infrastructure concept within a variety of delivery approaches of urban planning. This adds to the level of epistemological complexity when developing green infrastructure policy instruments and investments.

Similarly, Finewood (2016) noted that the intention of green and ecological infrastructure approaches is to expand water governance practices and public participation in city planning enterprises. Nevertheless, he argued that local government officials frame green infrastructure methods through a dominant 'grey epistemological approach' (Finewood 2016: 1000, 1001). He used the concept '*grey epistemology* to indicate a specific way of knowing water that focuses on the technical, abiotic aspects of a water system and the possibilities created by engineering as an objective problem-solving tool (as opposed to a particular form of knowledge)' (emphasis in the original) (Finewood 2016: 1001). Regarding the meaning of the concept 'grey epistemology', he goes on to say that:

> Specifically, grey infrastructure (largely impervious surfaces made of asphalt and concrete, such as pipes and sewers) provides a standard practice for capturing urban stormwater and treating wastewater as quickly as possible. It provides an attributable set of public health benefits as well as a calculable number of gallons that can be managed under certain conditions. Presumably these measurable characteristics provide an apolitical strategy for meeting urban water challenges as they isolate water and manage it largely separate from social systems and the messiness of day-to-day human interactions. (Finewood 2016: 1001–1002)

Finewood (2016) makes no excuses when he argues that such a grey epistemology was behind the unsuccessful US$2–4 billion Allegheny County Sanitary Authority (ALCOSAN) proposal that was submitted to the United States' Environmental Protection Agency (EPA) to mitigate storm-water overload in the ALCOSAN sewer system. That said, how we view the world through specific paradigms and theories influences our practices. For Finewood (2016), the proposal was unsuccessful because it rested largely on grey infrastructures and the thinking of an epistemological community that consisted of conservative and risk-averse water utilities and engineering companies. Based on this argument, Finewood (2016: 1000) went on to argue that alternative and more creative forms of greening a city, as seen through a grey epistemology, 'may not necessarily represent a more democratic process, but instead reproduce uneven urban landscapes under greener cover'. With this, Finewood (2016) inferred that a green infrastructure may appear more democratic because it involves a broader set of stakeholders, but if green infrastructures are promoted by the same institutions, it only conceals the supremacy of prevailing management arrangements and conventional practices. Furthermore, this signifies that, although green and ecological infrastructures could bring about a host of positive consequences for both the environment and the people, city planners do not consider the potential negative normative implications, such as the undemocratic practices that do not consider public involvement and how the public views such projects. There is, therefore, a dominant theory operating in such thinking, which results in a one-sided rational focus on the engineering costs and benefits, while ignoring the broader political processes.

Cousins (2017) reframes Finewood's (2016) argument by saying that a preference for science and data-driven green and ecological infrastructure approaches might be instituted to reinforce the dominant and powerful planning 'paradigms', instead of engaging in more experimental environmental governance arrangements. This entails that green infrastructure approaches that are maintained by a grey epistemology, may impede democratic processes because of their focus on the technicalities of water governance and management (Cousins 2017).

Wachsmuth and Angelo (2018) argued along a similar line when they say that there is an ideological tension between green and grey infrastructures, which makes

green infrastructure policies the realm of public participation and grey infrastructure strategies the domain of technocratic expertise. Because of this strain, municipal officials and other champions could take green infrastructures 'at face value
as inherently sustainable', which Wachsmuth and Angelo (2018: 1052) called 'the
realistic illusion', while grey infrastructure implementation suffers from 'the illusion
of transparency' that is characterised by considering a problem to be technological
optimisation.

In this respect, Finewood et al. (2019: 910) expressed a further argument by stating
that, 'The impacts of hidden or obscured values and normative biases embedded in
[green] infrastructure suggest that it is important to be attentive to who designs them
and for what purpose'. In this book, I intend to uncover how green and ecological
infrastructure policies and initiatives are designed, and by whom? Finewood et al.
(2019: 910) went on to say that their research has discovered that a green infrastructure is the 'purview of engineers'. These professionals design and implement such
systems by utilising technocratic decision-making processes for the protection of
urban properties and for investment opportunities, by complying with regulations, as
well as by stimulating and promoting urban development. Even so, it would appear
that engineers are not the only ones that 'dominate' in the equation and that the wider
epistemic community are also involved.

In a study on the role of the epistemic community in shaping the local government policy in Fukuoka City, Japan, Mabon et al. (2019), observed that 'Locally-
situated scholars … have, for several decades, sought to shape local responses to
environmental change by influencing the policy for the built environment and the
green space.' In this sense, Mabon et al. (2019) argued that the role of the epistemic
community in Fukuoka tends to lean towards 'technocratic' governance and that
scholars could act 'as a higher authority for setting an appropriate environmental
policy'. Furthermore, Boelens and Vos (2012) argued that municipal officials inter-
nalise the expression of the concepts and processes that are expressed by 'experts'.
When considering the conclusion of Mabon et al. (2019), as well as the assertion of
Boelens and Vos (2012), Reus-Smit and Snidal (2008) expressed the view that these
interpretations by city planners entail the empirical side (p. 28) of the theoretical coin
of green and ecological infrastructures. What about the normative side? In this book,
I will explore the theoretical tension between the empirical and normative stances in
eThekwini's green and ecological policy landscape, and I will indicate how the wider
epistemic community has contributed to the thinking of eThekwini city planners in
conceptualising and implementing green and ecological infrastructure enterprises.

All in all, this brief literature review indicates that not all local government officials
understand a green infrastructure in the same way; they have divergent considerations of the concept, of the operating environment of the infrastructure, and of the
ways and means of its implementation. What the examination also tells us is that a
one-size-fits-all approach to green and ecological infrastructure policy development
and implementation would most probably not work, because unique socio-economic
and ecological circumstances influence the green and ecological infrastructure policy
environment of local governments. Furthermore, the examination of literature indicates that green and ecological infrastructure approaches are not merely technocratic

practices that ameliorate the problems around water security and water governance, in general, but that they have implications for political and, by default, also power relations within a city's boundaries, and possibly beyond.

In the next section of this chapter, I will discuss water security and how it is linked to the instrumental policy arrangements of green and ecological infrastructures.

1.5   Water Security

Since we tend to focus on the concept of 'security', the notion of 'water security' often conjures up images of military-like action to keep water resources 'safe'. There is, however, more to the notion than meets the proverbial eye (Steyn et al. 2019). A single definition of water security does not exist (Siwar and Ahmed 2014; Steyn et al. 2019), and there are a variety of conceptualisations, depending on the human and environmental needs (Cook and Bakker 2012; Siwar and Ahmed 2014), as well as how people interact with one another and the environment to realise their water requirements (Meissner et al. 2018a, b, 2019). This shows that the debate on water security is complex, both locally and internationally.

The complexity of the conceptions and the discussions surrounding water security are manifested in their origins, intent and use. By way of illustration, at the international level, and after the end of the Cold War, observers of world politics revised the concept 'security' because of the changing geo-political landscape; they expanded it to include not only military or physical security, but also economic, environmental and human security (Bocardi et al. 2016; Funke et al. 2019). Not only did a new meaning surface for the word 'security', and particularly 'water security', but avenues for researching water security politics were also unlocked (Warner and Zeitoun 2008).

Remaining on a political level, 'proponents use the concept to raise water-related concerns to the level of a political priority' (Funke et al. 2019: 51), in a bid to express the urgency of water crises, at all levels, and to establish immediate action to prevent potentially violent situations or extreme human stress (Pahl-Wostl et al.2016; Funke et al. 2019). This is also the case with South African academics who write about water security; they seek 'to help elevate water-related challenges to the political agenda' (Funke et al. 2019: 57). In this regard, theories and concepts are not always neutral because they benefit actors in the political arena (Cox 1996; du Plessis 2000; Warner and Zeitoun 2008). That said, the concept 'security' constitutes a psychological component that is used as a potential trigger for action on the part of the epistemic and policy communities.

According to the Global Water Partnership (GWP) (2000: 12):

> Water security, at any level from the household to the global, means that every person has access to enough safe water at affordable cost to lead a clean, healthy and productive life, while ensuring that the natural environment is protected and enhanced... The term "water security" aims to capture the complex concept of holistic water management and the balance

between resource protection and resource use. Water security needs to be considered at local, national and regional levels.

Water Aid (2012) defined water security as the reliable access to an adequate quantity and quality of water for basic human needs, small-scale livelihoods and local ecosystem services, coupled with the well-managed risk of water-related disasters. The United Nations Educational, Scientific and Cultural Organisation—International Hydrological Programme (UNESCO-IHP) (2012) noted that water security is the capacity of a population to safeguard access to adequate quantities of acceptable quality water for sustaining livelihoods, human well-being and socioeconomic development, for ensuring protection against water-borne pollution and water-related disasters and for preserving ecosystems in a climate of peace and political stability. This conceptualisation of UNESCO-IHP (2012) applies at a watershed level. According to Grey and Sadoff (2007), water security is the reliable availability of an acceptable quantity and quality of water for health, livelihoods and production, coupled with a satisfactory level of water-related risks to people, environments and economies. Water security, in this regard, acts as a risk minimiser. The United Nations Water (UN-Water) (2013, cited in Steyn et al. 2019: 3), defines water security in a similar way to Grey and Sadoff (2007), stating that it is:

> The capacity of a population to safeguard sustainable access to adequate quantities of acceptable quality water for sustaining livelihoods, human well-being, and socio-economic development, for ensuring protection against water-borne pollution and water-related disasters, and for preserving ecosystems in a climate of peace and stability.

The above definitions relate to water quantity and quality, as well as the notion of security being a safeguard against the risk of political instability (UNESCO-IHP 2012), water-borne diseases and water-related disasters, such as droughts and flooding. These explanations also refer to the economic means of production and welfare creation (Water Aid 2012; UNESCO-IHP 2012; Grey and Sadoff 2007). If these definitions are phrased differently, water security relates either to the stability of political systems against the backdrop of the non-appearance of violence and instability, or to the absence of biophysical risks originating in the natural environment, such as water-borne diseases. Water security as a concept, therefore, plays several roles, either as a risk minimiser, a socio-economic developer or a political stabiliser.

These water security notions reflect an empiricism that is geared towards the creation of wealth and the absence of physical and psychological harm. What we also need is a normative sense of what water security means and how it could promote the human condition, as well as the natural environment. It is here where governance and management are important to consider with respect to their empirical and normative practices.

Since specialists delineate the concept of water security in the above-mentioned manner, it is an aspirational subject and an all-encompassing notion in socio-economic development, from an individual level to a global level (Meissner 2016). Indeed, water security, which is briefly defined as the safe and secure access to an adequate quantity and quality of water for human and environmental use, is a multifaceted undertaking that involves a plethora of stakeholders and management practices. A more elaborate definition of water security notes that it 'is a state of mind based on context-specific (i.e. localised and individualised) perceptions and practices held by individuals of water-related threats and/or opportunities and how it influences them, their surroundings and their interactions with others' (Meissner et al. 2018a: 112). This perception of water security gives us a bottom-up view of individuals, other than engineers and policy practitioners, who play a central role in achieving the aspirations of water security (Meissner 2016). Water security is not an end-goal, but a strategic activity for achieving sustainable socio-economic development, as well as psychological well-being. The idea of a green and ecological infrastructure is part of achieving water security by means of an environment-centric governance and management regime. The goal is for water security to find its expression at local government level, where the South African Constitution (Act 108 of 1996) (RSA 1996) obliges municipalities to practice responsible water resources management for the benefit of all its residents.

1.6 South African Municipalities and Water Resources Management

Observers of municipalities consider these entities to be the sphere of government that is closest to its citizens (Songer 1984; Zybrands 2011; Meissner 2015; Meissner et al. 2018a, b, 2019). It is at local government level, therefore, that populations come into direct contact with the state and its governing apparatus. Any policy, initiative, regulation or decision taken at a local level has the potential to directly influence people's lives, for better or for worse (Meissner et al. 2018a). This is one of the reasons why the South African Constitution (Act 108 of 1996) specifies that water service provisioning and management are the responsibility of the local government (RSA 1996; van Koppen and Schreiner 2014; Meissner 2015; Meissner et al. 2018b).

The Constitution provides the framework that divides the functions and powers between the three governmental spheres (municipal, provincial and national) (de Visser 2005). On this separation of functions and powers, Pieterse (2019: 55) notes that: '[S]ection 156(1) of the Constitution determines that municipalities have "executive authority in respect of" and "the right to administer" functional areas listed in Schedules 4B and 5B of the Constitution'. These Schedules '…relate to issues of everyday urban operation (such as public works, storm water management, municipal roads and parking) and delivery of essential services (such as municipal health services, water and sanitation services and municipal public transport), though also extend to areas more determinative of life in the city (such as management of public spaces) and to the cities' developmental trajectory (such as municipal planning)' (Pieterse 2019: 55).

Because of these functions and powers stipulated in the Constitution, du Plessis (2010: 265) argues that, 'Local government is no longer [particularly after the constitutional transformation of 1996] regarded as mere functionary or an agent of national and provincial government and its constitutional mandate centres, in the main, on the idea of "developmental local government"'. Local governments therefore enjoy considerable independence when deciding what to do and how to do it within the constitutional confines. They also play a developmental role that is expressed by means of inclusive service delivery to all their residents.

With regard to these functions and operational areas, previous research conducted on the public administration of South African local governments has indicated that there are a plethora of problems and opportunities (Beall et al. 2000; Parnell et al. 2002; Barichievy et al. 2005; Koma 2010; Meissner et al. 2018a, 2019). These relate not only to the regulatory landscape in which municipalities operate, but also to the socio-economic, financial and political surroundings. For instance, Koma (2010) identified the historical and socio-economic structural conditions that hamper local governments from achieving the aspirations of a developmental state. Because of the differences between the historical and socio-economic progression trajectories of local governments, municipalities often find it difficult to implement strategies for meaningful service delivery, as defined by their mandatory mission (Koma 2010; Meissner et al. 2018a). Other socio-economic circumstances that are prevalent, at a local government level, include the high unemployment rate, accompanied by elevated poverty levels, housing backlogs and informal settlements, which are issues that are partly constituted by limited budgets. Experts link these problems directly to the apartheid legacy of separate development. Coming to grips with, and mitigating, these challenges requires well-developed and measured policies (Beall et al. 2000), by-laws and standards (Meissner et al. 2018a).

As already mentioned, in terms of water resources management, and particularly their ambitions with regard to water security, municipalities have legislative powers and duties regarding water and sanitation services to the communities within the bounds of their jurisdiction, whether they are urban, peri-urban or rural. These legislative powers and duties have a direct bearing on the water security of the people. Even so, water security, at local government level, does not only depend on the correct interpretation and implementation of various Acts. As stated previously, other social, political and ideological considerations are at play when it comes to water security services (Meissner et al. 2018a).

Relating to this argument, the Minister for Provincial and Local Government in 2002, Sydney Mufamadi (cited in Parnell et al. 2002, and Meissner et al. 2018a), noted the following:

> Municipal government is at the heart of the democratic system that has emerged in South Africa since the political transition. Democracy makes sense when it connects with and gives expression to, the everyday challenges of ordinary South Africans. This means that municipal government must be at the forefront of involving citizens in all aspects of governance and development by providing them with practical and effective opportunities for participation.

Mufumadi's quote also applies to the overly technocratic approach to green and ecological infrastructures that seemingly excludes the citizenry from the planning and implementation of such initiatives. Because of the local governments' proximity to its citizens, Barichievy et al. (2005: 370) argue that 'local government is required to govern in ways more democratic than the provincial and national spheres'. Proper contact between the citizens and local government is not only a theoretical exercise relating to green and ecological infrastructures, coupled with water security, but it is also enshrined in legislative arrangements and has a practical value and implications (Meissner et al. 2018a). In this sense, the acceptance of green and ecological infrastructures by citizens and the water security aspirations of local government relate not only to inadequate legislative structures, but also to the democratic ambitions that connect with the everyday challenges and expectations of its citizens.

1.7   Conclusion

The devastating floods that pounded eThekwini and large parts of KwaZulu-Natal in 2017 and 2019 showed the importance of the natural environment in providing not only ecosystem services, but also in generating human hardship. In the eThekwini context, green and ecological infrastructures are not only linked to water security aspirations, but also to one of the world's most popularised environmental concerns, namely, long-term anthropogenic climate change (eThekwini Municipality 2019). If one adds this issue to the water security mix, it becomes clear how complicated water resources management at local government level can become. The problem relates not only to environmental issues, such as climate change playing out on a global level, but also to how city planners and scholars define and implement green and ecological infrastructures at municipal level.

   This literature review has shown us that green infrastructure initiatives are common in urban and rural settings. The extent to which municipalities can implement these infrastructure types depends not only on their financial resources, but also on their technical capabilities and paradigmatic framing, in relation to traditional engineering matters, such as storm-water and sewer system construction. Ecological infrastructures, on the other hand, seem to be more common in rural areas. Large parts of eThekwini are rural and, the ecological infrastructure in these areas plays an important part in the livelihoods of most people, since they rely on agriculture for their income. From a constitutional and political perspective, South African municipalities have considerable independence when it comes to deciding where to implement such infrastructures. However, it is clear that local governments have an obligation to not only provide the necessary services for the improved well-being of their citizens, but what they do must also be in line with the wishes of the wider public. In other words, green and ecological infrastructure initiatives and water security aspirations are not the realm of the engineers and a select few stakeholders only.

   With regard to local decisions, from an environmental perspective, a river does not stop providing services to people as soon as it enters an urban environment. In the case of the uMngeni River, it provides vital services to people from its source in the KwaZulu-Natal Midlands, to its mouth in Durban. We can say the same for the many estuaries along the Durban coastline.

Photos of the aftermath of the devastating storm in April 2019 show that Durban's rivers and streams are not only a blessing, in terms of eco-system services, but that they can also be a means by which the municipal solid waste can be transported (Naidoo et al. 2015). It must be noted that it is not the rivers and streams that are at fault, but that it is human behaviour, such as littering, that is predominantly responsible. It is here where we see the direct relationship between the functioning of green and ecological infrastructures, as ecosystem services, and water security, as an aspiration and association between the environment and humans.

When one considers the normative function of positioning water security on the agenda, it is possible that the issue of municipal solid waste washing up on Durban's beaches and in the port could become a political priority. At this stage, it is unknown how municipal solid waste relates to water security. However, it is possible that, should proponents frame the issue, in terms of environmental, human and water security, it could become more urgent on the political agenda of the municipality. The scientific community could also play an important part in this, together with interest groups like Plastics SA, and the public. A topic of investigation for the epistemic community could be how green and ecological infrastructures can play a role in alleviating the concern regarding municipal solid waste.

In Chap. 2, I will discuss the nature and function of research paradigms and theories on which I have based the PULSE[3] analytical framework. Using this analytical framework, I will analyse the content of 16 semi-structured and face-to-face interviews that I conducted with stakeholders in Durban and Pietermaritzburg in August 2018. These stakeholders are involved in eThekwini's green and ecological infrastructure policy landscape. I will present the results of my analysis in Chap. 3, before ending with a discussion and conclusion in Chap. 4.

References

Barichievy K, Piper L, Parker B (2005) Assessing 'participatory governance' in local government: a case study of two South African cities. Politeia 24(3):370–393

Beall J, Crankshaw O, Parnell S (2000) Local government, poverty reduction and inequality in Johannesburg. Environ Urban 12(1):107–122

Benedict MA, McMahon ET (2002) Green infrastructure: smart conservation for the 21st century. Sprawl Watch Clearinghouse, Washington D.C.

Bocardi J, Spring UO, Brauch HG (2016) Water security: past, present and future of a controversial concept. In: Pahl-Wostl C, Bhaduri A, Gupta J (eds) Handbook on water security. Edward Elgar, Northampton, MA

Boelens R, Vos J (2012) The danger of naturalizing water policy concepts: water productivity and efficiency discourses from field irrigation to virtual water trade. Agric Water Manag 108:16–26

Cameron RW, Blanuša T, Taylor JE, Salisbury A, Halstead AJ, Henricot B, Thopmpson K (2012) The domestic garden—its contribution to urban green infrastructure. Urban Forest Urban Green 11:129–137

Cook C, Bakker K (2012) Water security: debating and emerging paradigm. Glob Environ Change 22(1):94–102

Cousins JJ (2017) Structuring hydro-social relations in urban water governance. Ann Am Assoc Geogr 107(5):1144–1161

Cox RW (1996) Social forces, states and world orders. In: Cox RW, Sinclair TJ (eds) Approaches to world order. Cambridge University Press, Cambridge

De Visser J (2005) Developmental local government: a case study of South Africa. Stellenbosch Law Rev 21(1):90–115

Dunn AD (2010) Siting green infrastructure: legal and policy solutions to alleviate urban poverty and promote healthy communities. Pace Law Faculty Publications Paper 559

Du Plessis A (2000) Charting the course of the water discourse through the fog of international relations theory. In: Solomon H, Turton A (eds) Water wars: enduring myth or impending reality, Africa dialogue series, vol 2. ACCORD/Green Cross International and the African Water Issues Research Unit, Durham/Pretoria

Du Plessis A (2010) 'Local environmental governance' and the role of local government in realising Section 24 of the South African Constitution. Stellenbosch Law Rev 21(2):265–296

eThekwini Metropolitan Municipality (2013) New solutions explored for water security and service delivery: investments in ecological infrastructure in the greater uMngeni Catchment. Joint Media release, 28 Feb 2013

eThekwini Metropolitan Municipality (2019) Durban climate change strategy. Available at: https://www.durban.gov.za/City_Services/energyoffice/Pages/DurbanClimateChangeSt rategy.aspx. Accessed on: 6 June 2019

Finewood MH (2016) Green infrastructure, grey epistemologies, and the urban political ecology of Pittsburgh's water governance. Antipode 48(4):1000–1021

Finewood MH, Matsler AM, Zivkovich J (2019) Green infrastructure and the hidden politics of urban storm-water governance in a post-industrial city. Ann Am Assoc Geogr 109(3):909–925

Funke N, Nortje K, Meissner R, Steyn M, Ntombela C (2019) An analysis of international and South African discourses and perspectives on water security. In: Meissner R, Funke N, Nortje K, Steyn M (eds) Understanding water security at local government level in South Africa. Palgrave Macmillan, London

Ghosh R, Kansal A (2019) Anthropology of changing paradigms of urban water systems. Water Hist 11(1–2):59–73

Gill SE, Handley JF, Ennos AR, Pauliet S (2007) Adapting cities for climate change: the role of green infrastructure. Built Environ 33(1):115–133

Global Water Partnership (GWP) (2000) Towards water security: a framework for action. Global Water Partnership, Stockholm

Grey D, Sadoff CW (2007) Sink or swim? Water security for growth and development. Water Policy 9(6):545–571

Head T (2019) Durban floods: here's how much rain has fallen since Monday. The South African, 23 April. Accessed at: https://www.thesouthafrican.com/weathersa/durban-floods-how-much-rai nfall-kzn-23-april/. Accessed on: 23 Mar 2020

Jewitt G, Zunckel K, Dini J, Hughes C, de Winnaar G, Mander M, Hay D, Pringle C, McCosh J, Bredin I (2015) Investing in ecological infrastructure to enhance water security in the uMngeni River catchment. Department of Environmental Affairs, Pretoria

Koma SB (2010) The state of local government in South Africa: issues, trends and options. J Pub Adm 1:111–120

Kotzé P (2013) Investing in ecological infrastructure for our future water security. The Water Wheel, pp 24–27

Lennon M, Scott M (2014) Delivering ecosystems services via spatial planning: reviewing the possibilities and implications of a green infrastructure approach. Town Plan Rev 85(5):563–587

Li F, Liu X, Zhang X, Zhao D, Liu H, Zhou C, Wang R (2017) Urban ecological infrastructure: an integrated network for ecosystem services and sustainable urban systems. J Clean Prod 163(supplement 1):S12-18

Mabon L, Shih W-Y, Kondo K, Kanekiyo H, Hayabuchi Y (2019) What is the role of epistemic communities in shaping local environmental policy? Managing environmental change through planning and greenspace in Fukuoka City, Japan. Geoforum 104:158–169

Madureira H, Andresen T (2014) Planning for multifunctional urban green infrastructures: promises and challenges. Urban Des Int 19(1):38–49

Meerow S, Newell JP (2017) Spatial planning for multifunctional green infrastructure: growing resilience in Detroit. Landscape Urban Plan 159:62–75

Meissner R (2015) The governance of urban wastewater treatment infrastructure in the Greater Sekhukhune District Municipality and the application of analytic eclecticism. Int J Water Gov 2:79–110

Meissner R (2016) Water security in Southern Africa: discourses securitising water and the implications for water governance and politics. In: Pahl-Wostl C, Bhaduri A, Gupta J (eds) Handbook on water security. Edward Elgar, Northampton, MA

Meissner R, Steyn M, Moyo E, Shadung J, Masangane W, Jacobs-Mata I (2018a) South African local government perceptions of the state of water security. Environ Sci Policy 87:112–127

Meissner R, Funke N, Nortje K, Jacobs-Mata I, Moyo E, Steyn M, Shadung J, Masangane W, Nohayi N (2018b) Water security at local government level in South Africa: a qualitative interview-based analysis. Lancet Planet Health 2(supplement 1):S17

Meissner R, Steyn M, Jacobs-Mata I, Moyo E, Shadung J, Nohayi N, Mngadi T (2019) The perceived state of water security in the Sekhukhune District Municipality and the eThekwini Metropolitan Municipality. In: Meissner R, Funke N, Nortje K, Steyn M (eds) Understanding water security at local government level in South Africa. Palgrave Macmillan, London

Mell IC (2014) Aligning fragmented planning structures through a green infrastructure approach to urban development in the UK and USA. Urban Forest Urban Green 13(4):612–620

Mufamadi S (2002) Foreword. In: Parnell S, Pieterse E, Swilling M, Wooldridge D (eds) Democratising local government: the South African experiment. University of Cape Town Press, Cape Town

Naidoo T, Glassom D, Smit AJ (2015) Plastic pollution in five urban estuaries of KwaZulu-Natal, South Africa. Mar Pollut Bull 101:473–480

Naumann S, Davis M, Kaphengst T, Pieterse M, Rayment M (2011) Design, implementation and cost elements of green infrastructure projects. Ecologic Institute and GHK Consulting, Berlin

Pahl-Wostl C (2015) Water governance in the face of global change: from understanding to transformation. Springer, Cham, Switzerland

Pahl-Wostl C (2017) An evolutionary perspective on water governance: from understanding to transformation. Water Resour Manage 31(10):2917–2932

Pahl-Wostl C, Gupta J, Bhaduri A (2016) Water security: a popular but contested concept. In: Pahl-Wostl C, Bhaduri A, Gupta J (eds) Handbook on Water Security. Edward Elgar Publishing Limited, Cheltenham

Parnell S, Pieterse E, Swilling M, Wooldridge D (2002) Democratising local government: the South African experiment. University of Cape Town Press, Cape Town

Pieterse M (2019) A year of living dangerously? Urban assertiveness, cooperative governance and the first year of three coalition-led metropolitan municipalities in South Africa. Politikon 46(1):51–70

Pringle K (2020) Personal communication. Independent consultant, Pietermaritzburg, South Africa, 7 Apr 2020

Pringle C, Bredin I, McCosh J, Dini J, Zunckel K, Jewitt G, Hughes C, de Winnaar G, Mander M (2015) An investment plan for securing ecological infrastructure to enhance water security in the uMngeni River Catchment. Development Bank of Southern Africa, Midrand

Republic of South Africa (RSA) (1996) Constitution of the Republic of South Africa No. 108 of 1996. Government Printer, Pretoria

Reus-Smit C, Snidal D (2008) Between utopia and reality: the practical discourse of international relations. In: Reus-Smit C, Snidal, D (eds) The Oxford Handbook of international relations. Oxford University Press, Oxford

Roe M, Mell I (2013) Negotiating value and priorities: evaluating the demands of green infrastructure development. J Environ Planning Manage 56(5):650–673

Sandström UG (2002) Green infrastructure planning in urban Sweden. Plan Pract Res 17(4):373–385

Schäffler A, Swilling M (2013) Valuing green infrastructure in an urban environment under pressure—the Johannesburg case. Ecol Econ 86:246–257

Schindler S (2020a) Microanalysis as ideology critique: the critical potential of 'zooming in' on everyday social practices. In: Martill B, Schindler S (eds) Theory as ideology in international relations: the politics of knowledge. Routledge, London

Schidler S (2020b) The task of critique in times of post-truth politics. Rev Int Stud 1–19

Sigwela A, Elbakidze M, Powell M, Angelstam P (2017) Defining core areas of ecological infrastructure to secure rural livelihoods in South Africa. Ecosyst Serv 27:272–280

Singh K (2017) Massive storm, floods hit Durban, residents urged to stay indoors. News24, 10 Oct. Accessed at: https://www.news24.com/news24/southafrica/news/massive-storm-floods-hit-durban-residents-urged-to-stay-indoors-20171010. Accessed on: 23 Mar 2020

Singh K (2019) KZN Floods: movement of ships in Durban harbour limited by mass debris pile-up. News24, 26 Apr. Accessed at: https://www.news24.com/SouthAfrica/News/kzn-floods-movement-of-ships-in-durban-harbour-limited-by-mass-debris-pile-up-20190426. Accessed on: 23 Mar 2020

Siwar C, Ahmed F (2014) Concepts, dimensions and elements of water security. Pak J Nutr 13(5):281–286

Somdyala K (2019) PICS: Durban harbour, beaches awash with plastic pollution after floods. News24, 24 Apr. Accessed at: https://www.news24.com/SouthAfrica/News/pics-durban-harbour-beaches-awash-with-plastic-pollution-after-floods-20190424. Accessed on: 21 Mar 2020

Songer DR (1984) Government closest to the people: constituent knowledge in state and national politics. Polity 17(2):387–395

South African National Biodiversity Institute (SANBI) (2019) Ecological infrastructure. South African National Biodiversity Institute, Pretoria. Accessed at: https://www.sanbi.org/biodiversity/science-into-policy-action/mainstreaming-biodiversity/ecological-infrastructure/. Accessed on: 5 Mar 2019

Southern African Program on Ecosystem Change and Society (SAPECS) (No date) ProEcoServ: strengthening the science policy interface of biodiversity and ecosystem services. Council for Scientific and Industrial Research, Pretoria. Accessed at: https://sapecs.org/wp-content/uploads/2013/05/ProEcoServ_2013.pdf. Accessed on: 5 Mar 2020

Steyn M, Meissner R, Nortje K, Funke N, Petersen V (2019) Water security and South Africa. In: Meissner R, Funke N, Nortje K, Steyn M (eds) Understanding water security at local government level in South Africa. Palgrave Macmillan, London

Stuart-Hill S (2020) Personal communication centre for water resources research, University of KwaZulu-Natal, Pietermaritzburg, South Africa, 7 Apr 2020

Tzoulas K, Korpela K, Venn S, Yli-Pelkonen V, Kaźmierczak A, Niemela J, James P (2007) Promoting ecosystem and human health in urban areas using Green Infrastructure: a literature review. Landscape Urban Plann 81(3):167–178

UNESCO-IHP (2012) Strategic plan of the eight phase of IHP (IHP-VIII, 2014–2012). UNESCO-IHP, Paris, France

United Nations-Water (2013) What is water security? Infographic. [Online]. Available at: https://www.unwater.org/publications/water-security-infographic/. Accessed on: 10 June 2019

Van Koppen B, Schreiner B (2014) Moving beyond integrated water resource management: developmental water management in South Africa. Int J Water Resour Dev 30(3):543–558

Wachsmuth D, Angelo H (2018) Green and grey: new ideologies of nature in urban sustainability policy. Ann Am Assoc Geogr 108(4):1038–1056

Warner JF, Zeitoun M (2008) International relations theory and water do mix: a response to Furlong's troubled waters, hydro-hegemony and international water relations. Polit Geogr 27:802–810

Water Aid (2012) Sanitation and water for poor urban communities: a manifesto. Water Aid, London

Williamson KS (2003) Growing with green infrastructure. Heritage Conservancy, Doylestown, PA

Wolhuter B (2017) #DurbanStorm—at least 8 dead. IOL, 11 Oct. Accessed at: https://www.iol.co.za/news/south-africa/kwazulu-natal/durbanstorm-at-least-8-dead-11540781. Accessed on: 4 Mar 2020

Zybrands W (2011) Local government. In: Venter A, Landsberg C (eds) Government and politics in South Africa, 4th edn. Van Schaik Publishers, Pretoria

Chapter 2
The Use of Paradigms and Theories in the PULSE^3 Analytical Framework

Abstract In Chap. 1, I indicated how paradigms and theories influence and enable green and ecological infrastructure projects, particularly within the framing of a grey epistemology. In this chapter, I will expand on the role and functioning of paradigms and theories from the philosophy of the natural and social sciences. Paradigms and theories are vital in generating knowledge, which, in turn, constitutes agency through policy development, recommendations and implementation. It is significant to note that the way we think of the world and surrounding situations, influences our actions. Paradigms and theories are part of an interdependent and nonlinear causal chain. Based on the role and importance of these cognitive instruments in the policy process, I have developed an analytical framework by utilising the meta-theoretical assumptions of five identified research paradigms, several theoretical perspectives, four causal mechanism types, and I also use the distinction between the grand and problem-solving theories. I have called this framework PULSE^3, which stands for People Understanding and Living in a Sustained Environment. The cube (^{3}) denotes three forces, namely, thinking, acting and change, the latter of which is omnipresent in society and the natural environment. I base PULSE^3 on the argument that positivism, when employed as the sole paradigm by investigators, has difficulty investigating and explaining fundamental social processes that are manifested in ambiguities, paradoxes, uncertainties and contradictions. The purpose of PULSE^3 is to generate a healthier appreciation of issues that are faced by policymakers and practitioners. All in all, PULSE^3 is an analytical tool that can be used to assist researchers and practitioners to investigate policies, plans, programmes and strategies and to gain a deeper understanding of issues and practices, based on their paradigmatic and theoretical foundations.

Keywords Paradigm · Theory · Causation · Practice · Analytic eclecticism · Causal mechanism · PULSE^3

R. Meissner, *eThekwini's Green and Ecological Infrastructure Policy Landscape*, https://doi.org/10.1007/978-3-030-53051-8_2

2.1   Introduction

In this chapter, I will elaborate on the role and function of research paradigms (here-after called paradigms) and theories that inform the eThekwini green and ecological policy landscape. My intention is to show how paradigms (worldviews) and theories (perspectives) play a role in how we view the world and react to it in the policy and political domains. When talking about paradigms and theories, it is inevitable that normative elements, such as beliefs, culture, ideas, ideologies and social values, enter the fray. Researchers and practitioners alike often ignore these elements, espe-cially when dealing with seemingly technical problems, such as green and ecolog-ical infrastructures and water security. This means that practitioners and researchers often develop policies that are based on incomplete, or one-sided, thinking (Meissner 2017). In a previous publication, I showed how such thinking and understanding operate within the context of several policies, and I claimed that, with a fuller appre-ciation of paradigms and theories, science can play a more meaningful role in the service of society (Meissner 2017).

To highlight the normative elements further and to bring the rest of the study into the domain of cognitive elements, I will elaborate on how these aspects influence policy paradigms. I need to add that there is a difference between policy paradigms and research paradigms, with the former focusing on policy changes that are infused by changing ideas within a fluid context, and the latter informing the research practices in both the sciences.

In the policy domain, scholars often use the concept 'policy paradigm' to indicate normativity, and more particularly, the ideas that influence the policy process. By way of example, Menahem (1998) described the policy changes within Israel's water management realm, since its independence in 1948. Through her research on the role of policy networks in the emergence, persistence and changes in policy, she identified two paradigms. The earlier policy paradigm focused on the expansion of water resources and agricultural production, whereas the second concentrated its attention on the priority of agricultural expansion over water conservation. For her, policy paradigms are systems of ideas and standards that specify policy goals, the types of instruments used to attain such objectives and the nature of problems that they are supposed to confront (Menahem 1998). She argues that the social context is influential in defining Israel's interests (Menahem 2001).

In another study from a completely different domain, namely, economic policy-making in Britain, Hall (1993) asserted that, 'In order to understand how social learning takes place, we also need a more complete account of the role that ideas play in the policy process. After all, the concept of social learning implies that ideas are central to policy making …' Still on the topic of British economic policy-making, Carstensen and Matthijs (2018) investigated how ruling policy paradigms persist and why they often undergo internal ideational changes. Hall's (1993) ideas around policy paradigms are influential in their research and they show how policy-makers use their institutional and ideational power to redefine and reinterpret a dominant neoliberal understanding of the economy, so that it matches their own ideas and policy priorities (Carstensen and Matthijs 2018).

Although policy paradigms will not be covered in this book, the three examples above show that, when it comes to paradigms and theories, the ideational elements play an important role. My concern will be to investigate and report on the role of research paradigms and social theories in defining eThekwini's green and ecological infrastructure policy landscape. To be sure, from the brief review above, we notice that paradigms and theories are about a variety of ideational elements that influence the policy process, and that the context in which they evolve is of central importance.

The chapter begins with a description of paradigms and theories and how researchers use them to describe the world and to inform policy, be it at a local or national level, or in world politics. In the second section, the importance of paradigms and theories will be outlined. I will focus on the significance of these cognitive processes in the discipline of International Relations (IR), since I am most familiar with this discipline. This will be followed by a discussion on the importance of causation and why we need to take it more seriously. With regard to the importance of paradigms, theories and causality, I will discuss why researchers often negatively perceive paradigms and theories, before ending with my conclusion.

2.2 Conceptual Building Blocks

As I have already mentioned, I will consider two conceptual building blocks from which the rest of the investigation follow, namely, paradigms and theories. In short, these intangible cognitive tools play an important role in causal chains in that concrete activities within policy processes are often not causes in themselves, but they are the consequence of paradigms and theories. To this end, according to Guba (2002), Ridley (2003), Lhabitant (2004), Andreaotti (2006) and Komisarczuk and Welch (2006), a Chinese proverb states that 'Theory without practice is idle, practice without theory is blind' (Andreotti 2006). Gay (1979: 178) cites Zais's (1976: 87) version of this statement in the following way: 'Theory without practice is idle speculation while practice without theory constitutes little more than blind or random groping'. This is possibly what Immanuel Kant referred to when he indicated that 'Thoughts without contents are empty; intuitions without concepts are blind' (Lloyd 2017: 1). In fact, Lloyd (2017) paraphrases Kant's dictum in the Chinese proverb, when he says that: 'Neither of these qualities [i.e. concepts versus intuitions or precepts] is preferable to the other. Without sensibility[,] objects would not be given to us, and without understanding[,] they would not be thought by us'.

These quotes are particularly relevant when considering the interdependent relationship that underpins paradigms and theories, on the one hand, and practice, on the other.

2.2.1   Paradigms

Linguists introduced the word 'paradigm' to the English language in the late fifteenth century through the Late Latin form of the Greek word *paradeknunai*, which means to 'show side by side'. A paradigm can mean 'a typical example or pattern of something; a pattern or model' (OALD 2013, cited in Meissner 2017). Paradigms have different meanings, both across, and within, scientific disciplines, yet they all relate to scientific enquiry.

In this sense, a paradigm is a worldview that underlies the theories and methodologies of scientific subjects. For Guba and Lincoln (1994: 105, cited in Christians 2018), a paradigm is 'the basic belief system or worldview that guides the investigator, not only in choices of method but in ontologically and epistemologically fundamental ways'. Sociologists, Burrell and Morgan (1979) treated paradigms as perspectives that bind the work of theorists, since they are sets of ontological and epistemological assumptions (Schultz and Hatch 1996).

For Warner et al. (2002: 9) 'Paradigms are sets of references that frame the way in which science, management and people understand and act upon the world around them'. From an art education perspective, Pearse (1983) indicated that a paradigm is an internally consistent orientation that serves as a platform for constructing a conceptual and operational approach to functioning in the world.

I follow Sil (2000a), Sil and Katzenstein (2010) and Schultz and Hatch (1996), who used the word 'paradigm' as a research tradition that helps to organise and guide science, to bind the work of a number of theorists and that sets ontological, epistemological and methodological assumptions about the practice of science. From these conceptualisations, we notice that epistemology, methodology and ontology are central features of paradigms, and that their meta-theoretical assumptions are central to their utility in the sciences.

2.2.1.1   Scientific Practice

Concerning science as a practice, it is the view of Crewe and Forsyth (2003) that paradigms assist practitioners to understand their own work and how their efforts relate to society and the natural environment. This implies that the central characteristics of a paradigm play a relational role between the researchers and their investigations. The epistemology, methodology and ontology of one or more paradigms, inform scholars and practitioners on how to view, relate to and investigate reality.

That said, paradigms constitute various practices, not only among academia, but also in the world of work. From a methodology perspective, Guba (1990: 17) argued that a paradigm is 'a basic set of beliefs that guides action, whether of the everyday garden variety or action taken in connection with a disciplined inquiry'. Cloete and de Coning (2018: 34) advanced a notion that is similar to Guba's when they argued that: 'A *paradigm* is a collection or pattern of commonly held assumptions, values, beliefs, concepts, models and/or theories constituting a general prescriptive, intellectual framework or approach to scientific activities (e.g. ideologies like liberalism, Marxism, nationalism, apartheid, fascism, feminism, globalism, environmentalism and Darwinism)'.

As noted earlier, scholars describe the substance, role and function of paradigms in various ways. Cloete and de Coning (2018) equated them with ideologies and add positivism, relativism, developmentalism, capitalism, socialism, liberalism and conservatism to their list of examples. It is not surprising that they do this because they later discuss the nature of a 'policy paradigm', which, for them, '… is a prescriptive approach to public policy that is either promoted or discouraged on normative grounds (e.g. evidence-informed versus opinion-informed policy making)' (Cloete and de Coning 2018: 34).

For Cloete and de Coning (2018), ideologies play an important part when 'thinking about policy challenges' (Béland and Cox 2013: 193). In the context of 'thinking' and 'policy challenges', Blyth (2013: 211) stated that: 'Mere facts will (sometimes) not be allowed to get in the way of a good ideology' where 'the "truth" about the [economic] crisis and the ideas that made it possible really does depend upon what the most powerful members of a group (or society) consent to *believe*' (emphasis added).

From the various definitions, we can see that there are several elements that delimit a 'paradigm'. We also notice that the ontological and epistemological elements reflect how we see reality and produce knowledge about actuality. The philosophies that guide actions happen in a specific setting, that is, in either a research or scientific environment. The beliefs of scientists guide their actions through epistemological, methodological and ontological assumptions.

A paradigm deals with the way in which researchers generate knowledge, by guiding their investigative action from a cognitive perspective that is constituted by their ontological, epistemological and methodological notions. These elements are invisible in a research-informed policy (Meissner 2017). If we could make the invisible visible, we would be able to understand how science influences policy processes.

This does not mean that it is only researchers or scientists who produce, acquire and utilise paradigms. I would go so far as to say that paradigms are omnipresent in the lives of most individuals, either implicitly or explicitly, when assessing a situation or issue. As Guba (1990) and Warner et al. (2002) argued, this means that paradigms guide our management practices and, taking it a step further, they are the building blocks of our practices (Meissner 2017). The central characteristic of a research paradigm is about having a basic belief about the various aspects when generating knowledge.

Although we can identify numerous policy paradigms over a certain period and within a specific context (p. 22), as far as I know, there are at present only five research paradigms available to natural and social scientists. I use Lincoln et al.'s (2011, 2018) paradigm typology, namely, positivism, post-positivism, interpretivism/constructivism, critical theory and the participatory paradigm, as the foundation for PULSE[3] (Lincoln et al. 2011, 2018; Meissner 2017). These paradigms each have specific notions of: (a) the way of understanding reality (ontology), (b) the relationship between the researcher and the thing being researched (epistemology), (c) the process of research (the method), (d) the theory of truth, (e) what the data measure (validity) and (f) the reproduction of research results (reliability) (Guba 1990; Weber 2004; Meissner 2017). Again, a basic set of epistemological, methodological and ontological beliefs guide our research practices (Guba 1990). These are not general principles, but the attitude with which researchers should conduct their investigations. Scientists adhering to the positivist paradigm, for instance, believe that the researcher and reality are separate, and for the adherent of the participatory paradigm, a participative reality exists, which links the objective and subjective realities (Lincoln et al. 2011, 2018; Meissner 2017).

2.2.2   Theory

Researchers should not confuse the concept of 'theory' with that of 'paradigm' (Meissner 2017); theories are part of paradigms that act as knowledge systems, which makes theories operational. In the late sixteenth century, linguists introduced the word 'theory' to the English language through Late Latin from the Greek word *theōria*, which means 'contemplation or speculation'. This meaning is, in turn, derived from the expression *theōros*, which means 'spectator'. A lexical definition states that a theory is 'a supposition or a system of ideas to explain something, especially one based on general principles that are independent of the thing to be explained' (OALD 2013, cited in Meissner 2017). Another explanation notes that theory is 'a set of principles on which the practice of an activity is based' or 'an idea used to account for a situation or justify a course of action' (OALD 2013, cited in Meissner 2017). Consequently, theories explain and provide action-orientating principles. Having said that, the difference between a paradigm and theory is that a paradigm is a more holistic foundation for research than a theory; it is more specific, or to the point, and it is a simplified explanation of reality.

Theories are representations of reality, which help to make sense of a complex world. In simple terms, when referring to a 'theory', I mean a supposition, a set of principles or a system of ideas that is used to explain, understand and justify an action. Theories: (a) explain a phenomenon in society and the natural environment, (b) are ideas that account for, or give an understanding of, a situation, (c) are sets of principles on which we base our practices, (d) are a framework or system of concepts and propositions that provide causal explanations, through (e) simplified explanations and understandings, for (f) making sense of reality.

2.2.2.1   Comprehension

From a scholarly perspective, Scheiner and Willig (2008: 882, cited in Owusu 2016) asserted that a theory is 'a framework or system of concepts and propositions that provide causal explanations of phenomena within a particular domain'. This gives an idea of the elements that constitute a theory and how they relate to practices. A theory is a simplified picture of reality and, as such, theories help us to understand how the world works in specific domains. Since we experience the world as complex and difficult to understand, with limited knowledge at our disposal, we make sense of it through theories. In order to gain understanding, we must decide which factors are more important during research and we must discard those that are not. In so doing, the world becomes more understandable (Mearsheimer and Walt 2013; Meissner 2017) as we break it down into easily intelligible features (Meissner 2017) or basic explanations (Koslowski et al. 1989). We can compare theories to a map that represents reality in a simplified form. Whereas a map contains sketches and symbols, theories explain reality through the causal narratives (Mearsheimer and Walt 2013) of the actors, as well as the ideas, material resources and social structures that guide their conduct (Meissner 2017). In this sense, theories declare that researchers can explain a phenomenon with a single factor or an assembly of elements (Mearsheimer and Walt 2013; Meissner 2017).

General principles, ideas and explanations are variables or concepts that constitute a theory, which helps researchers to define key concepts or variables. This entails the making of assumptions about the important actors that are affected by, and that affect, the issues or phenomena. The central purpose of a theory is to identify how independent, intervening and dependent variables correspond. The interdependence of variables enables scholars to conclude testable hypotheses, or to determine how they anticipate concepts to 'co-vary' (Mearsheimer and Walt 2013; Meissner 2017). To make an inference is to decide, or reach an opinion, that something is true, based on the gathered information (Mearsheimer and Walt 2013), although what constitutes the truth is debatable (Meissner 2017). According to Mearsheimer and Walt (2013: 432, cited in Meissner 2017), what is important, is that '…a theory explains why a particular proposition should be true, by identifying the causal mechanisms that produce the expected outcome(s). Those mechanisms—that are often unobservable are supposed to reflect what is actually happening in the real world'. Causal mechanisms could act as dependent, independent and interceding variables to explain causal relationships. According to Koslowski et al. (1989), the mechanism equates to the theory, with causal mechanisms being crucial to understand causation (Falleti and Lynch 2009; Kurki 2008) (causation and causal mechanisms will be discussed on p. 55).

In the social sciences, scholars use theories that they can either explain or understand. Explanation occurs through causal analysis, by investigating the general patterns in social processes, while understanding happens by inquiring into the formation of the meaning and the reason for the actions. This implies that explanation is about discovering the cause with understanding interested into the meaningful context of an action. Both these functions of a theory are legitimate and can be combined (Kurki 2008) to produce deeper comprehension. This is linked to the empirical and normative roles of a theory (p. 28). According to Hayes and James (2014: 401):

...theories actually embody logics or particular modes of thought that are reflected in the
thought processes of [practitioners] in the real world. The theories can therefore be understood not just to explain actions and outcomes, but to actually embody shared patterns in
how actors understand the world.

To summarise, theories consist of statements that reflect how the world works
(Meissner 2017), and causal mechanisms and understanding occupy a central role in
the practice of inquiry.

It is appropriate that linguists derived the word 'theory' from the Greek word
'spectator'. Just like 'spectators' at a rugby match, scientists are observers of what
is happening in the world. More than onlookers who are merely observing for
enjoyment, scientists attach meaning to the observed event or phenomenon.

If the observed occurrence, or fact, creates problems, we utilise explanations and
understanding to improve it. Like paradigms, theories are constructions of the mind
(Lynham 2002; Meissner 2017). It is here where theories start making practical
sense, since practitioners use them subliminally, or explicitly, in applied situations
(Du Plessis 2000; Meissner 2017). Theories, along with paradigms, help researchers
to propose solutions to difficulties, or to create opportunities for further exploitation.

One more thing needs to be considered, especially when linking paradigms and
theories to practice. Over the years, social scientists have disagreed over the definitions and theories of human behaviour (Geldenhuys 2004; Meissner 2017). This is not
a disadvantage, but rather, it is the foundation for integrating paradigms and theories,
in order to produce prospects and to address problems. In this regard, various scientific
paradigms and theories are not incommensurable (lacking a basis of comparison).
The conceptual and theoretical divisions could strengthen our understanding of how
practices develop and manifest themselves in society (Meissner 2017).

2.2.3   Productive Utility

From the above discussion, we see that paradigms and theories can be used produc-
tively. Paradigms, which consist of meta-theoretical building blocks, describe the
nature of the 'world', the individual's place therein and the range of possible relationships to the world (Pearse 1983; Guba and Lincoln 1994; Meissner 2017), as well
as the practical knowledge that we produce through science (Reus-Smit 2013). Since
paradigms provide the conceptual and operational approach for conducting research
(Pearse 1983; Meissner 2017), they have a knowledge-derived practical usefulness.

Since we are all theorists, the implication is that being theoretical is unavoidable
(Rosenau 2003). Ferguson (2003: 1) confirmed this when he claimed that:

However much the casual observer of world affairs, harried practitioner, or naïve scholar may
ignore or perhaps go so far as to mock 'theory' or 'theoreticians', the truth is that everyone
is deeply enmeshed in theory whether they like it or not. Theory in the sense of mind-set is
arguably implicit in every opinion we express and every action we take.

An example of 'everyday' theory is the advice that we receive from colleagues, friends and family (Lynham 2002; Meissner 2017). Should we act on their advice, the theory is implicit in the specific action that we decide upon. Their guidance gives us a sense of what we could potentially do; it extends our cognitive abilities, because others have already done the reasoning and are imparting their knowledge to us as advice. This is one of the ways that experience assists us in creating opportunities, solving problems or deepening our understanding of reality (Meissner 2017). In this sense, Gay (1979: 178) notes that 'theory provides the practitioner with a means of describing, interpreting and guiding … activities'.

With productive usefulness in mind, we are constantly calculating the observed rules and patterns through the cognitive processes that govern the characteristics of the tangible world, as well as the complex relationships that we have with one another and the natural environment (McGann 2008; Meissner 2017). We do not merely determine the rules, patterns and relationships for pleasure. Rules and patterns are important elements of day-to-day tasks, and associations entail the production of theories, as well as their use, in various situations and contexts (Meissner 2017). Paradigms and theories provide us with practical knowledge, and with practical knowledge we ask, 'How should I, we, or they act?' (Reus-Smit 2013: 589). Paradigms and theories play a central, albeit invisible, role in how we generate our knowledge into an action.

To answer the question of how we should perform, we need an appreciation of the world in which we would like to act (the empirical dimension) and have some sense of the goals that we want to achieve (the normative element) (Reus-Smit and Snidal 2008). Paradigms and theories are both empirical and normative, and there is an inherent interconnectedness between the two dimensions. What is more, scientific disciplines and practical occupations (for instance, consulting, engineering, environmental management, and public administration) contain a milieu of ideas. Paradigms and theories, infused with empirical and normative elements, inform our deeply interconnected ideas (Reus-Smit and Snidal 2008), which suggests that we cannot easily separate the empirical from the normative.

To explain this further, every paradigm and theory is, simultaneously, about what the world *is* like and what it *ought* to resemble. Because theories hold these features concurrently, paradigm and theory utilisation is a practical discourse that is concerned about 'How should we act?' (Reus-Smit and Snidal 2008). For instance, Hans Morgenthau, who is considered to be the pioneer of the IR theory, concerned himself with the right and duty of theorists to,

> exercise academic freedom as a critic of government power… For Morgenthau, the responsibility to hold governments to account by reference to the "higher laws" that underpin and legitimize democracy in its truest form, was a key function of the theorist in society (Molloy 2019: 1).

This means that the normative aspect of theory, or using theory to critique a government's actions, should be applied to the empirical elements of upholding a democracy. That said, theory and certain paradigms, like interpretivism and critical theory, have a central function as a critical practice in academia and society—armed with a theory, a theorist can assume the role of a dissident (Molloy 2019).

Every paradigm and theory emphasises a different issue that demands action (Reus-Smit and Snidal 2008), with some theories leaning more to the 'What is?' question and others to the 'What ought?' subject. Within any occupation, theorists formulate diverse conclusions about the different actions (Reus-Smit and Snidal 2008) that are required to address the issues. Whatever actions scholars demand, they are all concerned with the practical question of how we should act, in order to improve the human condition (Reus-Smit and Snidal 2008). That said, 'How we should act?' is the common ontological, epistemological and methodological denominator of paradigms and theories.

It is within this logic that a separation between 'empirical' and 'normative' theory is artificial, and this has a debilitating influence on the practical application of these constitutive and causal instruments. After all, paradigms and theories influence how we perceive and handle phenomena. The unwillingness of scientists to address ethical considerations, such as normative aspects, when utilising positivism, demonstrates that they frequently have little to say about significant present-day problems (Reus-Smit and Snidal 2008). These cognitive instruments often help us to question the accepted views about aspirations and needs (Seale 1998) and how to act on them. Problem analysis rests on both our ability to ask what can possibly be done, from an empirical perspective, and on our judgement as to who can act, and with what rights and obligations. If we want fields of inquiry to be practical, we need to ground our analyses on both an empirical and a normative foundation. 'Answering the question of "How we should act?" in terms of the values that we seek to realise, demands an appreciation of "How we can act?", in terms of the context of the action that we face' (Reus-Smit and Snidal 2008: 20). Reus-Smit and Snidal (2008: 20–21) continued, by arguing that:

> Answering the question how we should act requires empirical knowledge production that examines the causal, constitutive, and discursive structures and processes that frame political action. But it also requires normative knowledge production that examines who the "we" is that seeks to act, what principles we seek to realize, and what resources we are prepared to sacrifice to achieve our ends… With practical discourses, therefore, knowledge production cannot be confined to the "scientific" realm of empirical theory [only]; it is necessarily more expansive, encompassing normative inquiry just as centrally.

As mentioned earlier, the function of a theory is to explain and understand, through simplification (Grover and Glazier 1986; Kerlinger 1986; Koh 2013; Meissner 2017). From a scholarly perspective, paradigms and theories are important in every scientific discipline and research activity (Alderson 1998; Oliver 1998; Owusu 2016). The point is that, when using theory, it is not only necessary to focus on the empirical side, at the expense of the normative, and vice versa. For the purposes of this book, it is not so much how we describe the concepts 'paradigm' and 'theory' that is significant. What is essential is to realise that there are multiple paradigms and theories to explain phenomena (Seale 1998; Walt 1998, 2005), for practical purposes.

To reaffirm an earlier argument, where paradigms play a structural role in the sciences, the general purpose of theory is to create new empirical and normative knowledge by explaining the meaning, nature and challenges of the phenomena that we experience within the paradigm systems. We use such knowledge to act practically in a perceived, effective and informed manner (Lewin 1951; Whetten 1989; Strauss and Corbin 1990, Gioia and Pitre 1990; Lynham 2002; Meissner 2017). However, what is at stake is to what extent our practical knowledge is empirical or normative. Combining paradigms and theories, as multiple inquiry systems, is a way of affecting empiricism and normativity simultaneously.

2.3 Paradigms and Theories as Multiple Inquiry Systems

2.3.1 Epistemological and Practical Value

Since paradigms and theories provide a deeper comprehension, through explanation and understanding, they open the possibility of developing multiple inquiry systems. With the role of paradigms and theories in mind, I argue that scientists should first confront social issues, at a worldview level, before they go into the specific theorising of an issue, since paradigms contain several theories. This is a subliminal act, with the paradigms that we have acquired through our training and experience acting as our background knowledge. Christians (2018) argue that, in this sense, the issue of method plays a secondary part in the debate surrounding social matters, with the questions around paradigms coming first. Closer to the topic of this book, and emphasising the link between paradigms and practice, Katzenstein (1976: 13) declared that paradigms '... should be viewed not simply as a constant but as a variable which is closely interrelated with [government] policy'. The relationship between theory and day-to-day practices is also explained at an individual level by Chernoff (2007: 37), when he pronounced that: 'Decision-makers may choose a policy when they have a set of factual beliefs about conditions; a set of cause-and-effect beliefs about how [actors] interact...; and a set of objectives, goals, or values, which may be part of the theory'. As already alluded to by Katzenstein (1976), the epistemological value of paradigms and theories, as well as their influence on policy, are dynamic.

2.3.2 Actions and Situations

How we generate knowledge has a bearing on the way we solve problems and create opportunities. This means that paradigms and theories are the 'bricks and mortar' of a policy and its associated practice. It is not always possible to conduct an experiment to solve a problem. Without having a formal scientific method, cognition can also accomplish problem resolution and create opportunity creation. As researchers, we do not only have one 'dependable' way of knowing (Eisner 1990) and informing practice, but we have multiple techniques that are contained in paradigms. Consequently, it is possible to act based on multiple paradigms (Pretty 1994) and theories, but I would argue that it is a necessity. As Pretty (1994: 38) noted:

> But no scientific method will ever be able to ask all the right questions about how we should manage resources for sustainable development, let alone find the answers. The results are always open to interpretation. All actors, and particularly those stakeholders with a direct social or economic involvement and interest, have a uniquely different perspective on what is a problem and what constitutes improvement in [a]…system.

Although Pretty (1994) applied his argument to sustainable development and its associated problems, it is safe to say that his reasoning covers a variety of issues or areas. Since paradigms and theories influence practice, only one paradigm is legitimate (Lake 2011; Meissner 2017), nor is not only one theory the foundation for a complete explanation and understanding of social reality (Meissner 2017). Regarding comprehension, Cilliers (2005) argued, 'that different intellectual traditions have different understandings of what the nature and status of meaningful knowledge is'. Cox (1981) says that theory is 'always for someone and for some purpose'. In the context of legitimacy and comprehension, Selby (2018: 333) interpreted what Robert Cox said almost forty years ago, by mentioning the following about paradigms, theories and impactful research:

> They are always oriented to certain audiences or interlocutors ("users" and "beneficiaries" in the contemporary jargon); they always serve or at least complement specific functions, interests and values; and they are also, it may be added, always situated—that is, practised by individuals and their collaborators in ways which are inevitably affected by social locations, statuses, networks and biases.

This suggests that seemingly epistemologically objective research has certain aims in mind that are not always geared towards uncovering wholesome 'truth'.

To explain Selby's (2018) argument, let us consider emotions and the situatedness that is linked to positivism. Our emotions and situatedness within certain contexts have a bearing on how we conduct research. From a political science perspective, research can have a political function for an individual or group. Eisner (1990) takes this further when he eloquently wrote about understanding reality and truth, based on positivism: 'the truth is ultimately a mirage that cannot be attained because the [political] worlds we know are made by us'. Selby (2014: 949) reasoned that positivists believe they are producing an objective truth through empiricism, which is constituted by their function, interest and values. The latter two aspects stand central in the analyses of political and international relations (p. 106, 138). However, this belief is somewhat skewed due to 'the positivist fallacy that social relations are just as mechanistic as those that are the focus of the natural sciences'. Objective truth appears to be a contradiction in terms, not only scientifically, but also politically.

2.3.3   Configurations

Paradigms and theories function together in various configurations (Hayes and James 2014). Since not only one paradigm is legitimate, and because there are multiple paradigm and theory arrangements in practice, we can shift the way we think about knowledge, the mind, intelligence and cognition (Eisner 1990, 2017; Pretty 1995; Meissner 2017). Should we have multiple paradigms and theories in hand, we could uncover what was previously hidden in the conceptual elements. This notion relates to the impact of research that Selby (2018: 340) described as investigations that identify '...flaws within, or at least limitations to, existing socio-political knowledge and/or practice'. According to Schickore (2014), it is through the possession and visibility of multiple paradigms and theories, along with practices, that scientific discovery becomes possible. Nevertheless, I am not only concerned about scientific knowledge and its production.

If we recognise that all paradigms and theories are important for crafting knowledge, there could be consequences for our strategies and visions, as well as the way we evaluate practices and evidence-based policy. This also applies to the norms, or standards, of appropriate behaviour (Eisner 1990; Klotz 1995; Meissner 2004), since individuals experience the world in unique ways. These, in turn, constitute new forms of knowledge that influence their experiences and produce novel knowledge by means of an endless loop of knowledge production (Eisner 1990, 2017; Meissner 2016b, 2017).

If one brings policies into the equation, their content and intent express messages about how role players are supposed to act, how they view issues, as well as the nature of their beliefs and action expectations. Policy content also conveys an application of epistemology. As such, policies are not only 'a set of ideas reflecting certain values and beliefs that are created to guide decision-making' (Eisner 1990: 95), but they can also act as communication media that express certain attributes about an issue, as well as the thinking behind the construction of a problem, and the way in which practitioners go about creating change. Our policies are shaped by the 'beliefs about the kind of knowledge one can trust, and the kind of methods one can use to get such knowledge' (Eisner 1990: 96). That said, paradigm and theory configurations constitute policies through a complex array of observations, conceptualisations and interpretations. We can also add to this the function, interests and values of policy developers and implementers. This means that, should policy developers not trust a certain type of knowledge generation, it is likely that they will resist using it and view it as unreliable or illegitimate. The level of trust or suspicion is, equally, a construction of the human mind that is influenced by external factors, such as a practitioner's education, which concretises his/her biases over time. The psychological aspect of practitioners is potentially one of the most significant variables that influence the recognition of alternative paradigms and theories and how to integrate them with their dominant paradigms and theories.

2.3.4 Insightful Consequences

Recognising other paradigms and theories, and integrating them with those that are familiar, could have insightful consequences for policy processes. Considering and employing alternative paradigms could be a foundation for investigating the source of a practitioner's actions. Through positivism, for instance, we view action in a structured, linear and regulatory 'cause-and-effect' manner (Kurki 2008), while interpretivism brings to light a deeper understanding of the context and it's bearing on policy actions. With positivism as a frame of reference, action comes over as being well thought through, goal directed and systemic driven, and as such, it is often described as rational (Satz and Ferejohn 1994). Should a person want to act rationally, he or she must have specific goals. The nature of these objectives will determine the likely methods. In turn, researchers evaluate the method by using its effects to determine the link between the pre-specified goals and human behaviour. When one takes other paradigms and theories into consideration, the understanding of the source of an action becomes less neat and tidy, but more comprehensible. Messiness creeps in when situations are more ambiguous and complex than they are without these aspects (Eisner 1990; Meissner 2016b). Complexity is not only a theory (Anderson 1999; Cilliers 2000; Byrne 2002), but we also construct complexity when we add other paradigms and thinking to our thought processes. Complexity is defined according to how much it surrounds us (Meissner 2017). However, we should not view complexity as a parsimonious theory with the least assumptions and variables, but as a perspective that has the greatest explanatory power (more on this later).

Whatever the case may be, looking at issues through multiple paradigms and theories changes our conceptualisation and underpins our actions. This is because we are unable to constantly specify goals, which are at times difficult to articulate, due to the inherent dynamism of our multiple physical and social environments. Intuition, for instance, can play a part in our actions. In such contexts, paradigms other than, and together with, positivism can paint a more realistic and finely nuanced picture of a policy environment and of how a practitioner operates (Eisner 1990). What one paradigm ignores, another paradigm highlights.

In this respect, Rueschemeyer (2009: 116) argued that: 'To the positivist view, the effect of emotion on beliefs is mysterious'. He also states that emotions do interfere with cognition and, because of this capability, we cannot dismiss them as a mere nuisance when conducting research; we must take them into consideration during our investigations. Coicaud (2014) also reached this conclusion in his study on the role of emotions and passions in international politics. It is in these situations that paradigms, such as interpretivism and critical theory, bear fruit, by placing emotions central to what is, and what ought to be, questioned. In this sense, Rueschemeyer (2009: 114) contextualised the role of emotions in practice and science when he said that: 'Emotions engender wishful thinking and, when passionate, blind us to many features of the situation and to the consequences of rush actions... Emotions may define and "protect" the unthinkable; but a passionate search for the truth is driven by emotions as well'. This implies that emotions can play a significant role in research, as well as in the cognition of the actors, and that they can influence research endeavours (Rueschemeyer 2009) and practical situations.

Therefore, emotions play an important role in the recognition of, or resistance to, alternative paradigms. Either way, the participatory paradigm (Lincoln et al. 2011, 2018; Meissner 2017), could help to realise that subjective knowledge also has merit in the policy process (p. 40).

By including subjectivity in the explanation of policy processes, researchers could dilute the mystery of the influence of emotions on their beliefs and give credence to explaining aspects of policy progressions (Meissner 2017), which positivist- and postpositivist-type theories are unable to explain. Such a theory is rational choice, which assumes that human actors choose alternatives by evaluating their likely outcomes. Prior preferences and well-structured self-interests inform the instrumental rationality of the actors. Such actors '…are capable of comparing and calculating policy alternatives and choose the one that maximises their interests' (Qin and Nordin 2019: 5). Bringing intuition and emotion into the loop complicates matters, which means that a rational choice kicks these variables out and, by doing so, sketches an oversimplified representation of reality.

2.3.5 Science Speaking Truth to Power

The use of positivism and postpositivism puts these paradigms in a hegemonic position, with regard to scholars and their research products, and they act as a medium between the paradigm and the policy process. As a result, 'science speaking truth to power' (Wildavsky 1979; Bäckstrand 2004; Braun and Kropp 2010) becomes a central adage that guides research quests and influences government policy processes. It is inevitable that researchers try to influence government policy through the gathering of empirical evidence or by discovering the empiricism of the truths about social reality (Ruane 2018). Even so, as Ruane (2018) rightly observes, we need to recognise that there are many 'truths', and not just one truth (i.e. objectivity). We can learn valuable lessons from a concrete example of 'truth construction'. I will take an example from the history of foreign affairs in South Africa, when an expert in this domain apparently fell victim to science by 'speaking truth to power', when other scholars called for 'truths' in both his analyses and the South African foreign policies during the 1980s. Although this example is far removed from eThekwini's green and ecological infrastructure policy landscape, it highlights the serious academic repercussions when scholars link analyses to policy outcomes by 'speaking truth to power' through the mechanism of science.

Professor Deon Geldenhuys, who has now retired from the University of Johannesburg's Department of Politics and International Relations, is one of South Africa's foremost experts on the country's foreign policy (Chan 1989). Over the years, he has authored and published copious peer-reviewed articles and books on the subject, such as *The Diplomacy of Isolation*. Chan (1989: 374) notes the following about this book: 'In this work he [Geldenhuys] unveils the policy processes which have been secret and secretive. It has been, however, the very secretive nature of foreign policy formulation that has made Geldenhuys's published work so visible by contrast'. Because of the State's aggressive military intervention in southern African countries in the early 1980s, Geldenhuys became the target of international criticism, with some academics attributing a cause-and-effect relationship between his published works and South Africa's foreign policy exploits. According to Chan (1989: 374), some academics argued that Geldenhuys's 'work had inspired the policies of the securocrats [military beaurocrats] and had served as advisory documents to them'. We find one of the most famous, or dare I say, infamous, connections in Joseph Hanlon's 1986 publication *Beggar Your Neighbours: Apartheid Power in Southern Africa*. Hanlon (1986: 29) is entirely forthright about the cause-and-effect linkage between Geldenhuys's published work and the disreputable policy, by calling Geldenhuys 'a major theoretician of destabilisation, both explaining the concept and showing how it could be done'. In response to Hanlon's (1986) claims, Geldenhuys (1989) himself asked how Hanlon had come upon his insights and whether his work was a 'blueprint for destabilisation'?

Geldenhuys (1989:87) 'simply' did not know the answer to this question and he could not find any empirical evidence of Hanlon's claim, even after he asked him twice to supply such evidence. One of the unintended consequences of Hanlon's (1986) publication is that other authors took Hanlon's word as 'gospel' (Desmond 1987; Brittain 1988; Urdang 1989). Of these authors, only Brittain apologised to Geldenhuys, in a formal letter, after she realised that references to him as being 'one of President [P.W.] Botha's leading academic consultants on foreign policy' were inaccurate. This is an indication that several other researchers accepted Hanlon's (1986) observations, which were damaging and offensive to Geldenhuys, as being factually correct (Geldenhuys 1989).

With respect to 'science speaking truth to power', Geldenhuys (1989: 88) noted that there are 'risks involved in this type of research' in that '[t]he academic…has no control over what decision makers will do with his analysis—assuming that they do take note of it. Second, the…analyst should be aware of the danger of guilt by perceived association, of being branded an academic accomplice of an illegitimate regime …'. Geldenhuys (1898: 88) goes on to say that these risks should not deter academics from 'applying their minds to the formidable challenges facing South Africa, both domestically and externally'.

The case of Deon Geldenhuys is also a lesson on the consequences of the often-positivistic causal link that scholars create between theory and policy. By critically investigating matters, as Brittain (1988) did, and by employing a multiple paradigm and theory lens, we can avoid simplistic cause-and-effect relations like those of Hanlon. We could, furthermore, produce a deeper understanding of the role that research plays in policy processes and avoid ascribing too much to the notion that science speaks truth to power. In this sense, science speaking truth to power means that our scientific practices must inform our interaction in the public policy domain (Ntuli and Smith 1999), with the adage exhibiting the use of scientific knowledge in policy processes. It is possible that many scholars think that this is their central role in society, and that everything they analyse and report on must speak truth to power, but this is not always the case, as Molloy (2019) argues in his account of the role of researchers (p. 28).

In summary, the integration of paradigms has consequences for research products, for their uptake into policy processes and, by default, for the consequences of actions that are based on policies and the views of the wider epistemic community. As already mentioned, positivism and postpositivism are supposed to produce dependable prescriptions for action. Bringing other paradigms into the fold could increase the quality of a practitioner's deliberations, as analysts use multiple-informing paradigms and theories. For instance, the purpose of research, from an interpretivist/constructivist perspective with its inherent understanding function, moves from being prescriptive, to helping practitioners widen their thinking (Cronbach 1975; Eisner 1990) and to deepening their understanding of reality.

2.4 Pulse^3

In this section, I present and justify the analytical framework of PULSE^3 as an alternative way of analysing and conducting research, by using a paradigmatic and theoretical integrative mechanism. The framework is a tool that identifies paradigms and theories that influence policies and strategies, as well as their remedial action (Meissner 2014, 2016a, 2017, 2018). In this regard, PULSE^3 clarifies in visual form, the nature and extent of paradigmatic and theoretical puzzles that are at play within a policy arena. Beyer (2015: 489) noted that puzzles relate to one's collective and individual consciousness, and she argued that if one examines the interaction between global and individual consciousness, one could '[c]ompare this to pieces of a puzzle. If we think of puzzle pieces as individual pieces, they do not amount to much. But if they are put in an integrated order, they produce something that is larger than the sum of them, without an order'.

This is the case with the paradigms and theories that can be uncovered by using PULSE^3. The paradigms and theories contained in the analysed information resemble puzzle pieces that, together, help to explain how the information forms a whole. If one knows the nature of the puzzle pieces (e.g. the paradigms and theories), one will be able to deepen the understanding of the decision-makers on the issues that are expressed in the analysis. PULSE^3 enables the systematic reading of concepts and framings, and it offers a broad and robust review for assessing and aggregating a body of literature (Meissner 2017). The ensemble of knowledge provides researchers and practitioners with a spectrum of comprehension (Malacalza 2020), ranging from explanations to understanding.

In this section, I will outline the rationale of PULSE^3, which I will base on the dominant bias that researchers have towards positivism, as has already been indicated above. I will then describe the characteristics and capabilities of the framework by presenting its components, namely, the research paradigm assessment, the ethos of analytic eclecticism, the repertoire of theories, the causal mechanisms as well as the problem-solving and critical theory typology.

2.4.1 Rationale

As has already been argued, a single paradigm and/or theory has difficulty in dealing with fundamental social processes, ambiguities, uncertainties, paradoxes, contradictions and causal relationships. It is necessary to have a healthier appreciation of these, in order to address the challenges that are being faced by the various policy sectors. With complexity in mind, Cilliers (2000: 27) warns us 'that a theory of complexity cannot help us to take in specific positions, or to make accurate predictions'. In addition, the natural environment shapes and affects the changes that influence human society and the way we live in our environment (Berger and Luckmann 1966; Giddens 1984; Kooiman and Bavinck 2005; Gillings 2010). This dichotomous assertion may seem straightforward, but there is more than a mere relationship between humans and the environment. Certainly, relationships are one characteristic of a system, but another distinguishing factor is that society and the environment consist of multiple components, which, together with the relationships, in effect, make them complex systems (Urry 2005). If we see it in this way, we can conceptualise complexity as the interaction (Vasileiadou and Safarzyńska 2010) that takes place between the components, or agents, of a system, indicating that there is variation, without being random (Byrne 1998). It is within the ambit of explaining and understanding complexity that the inclusion of postpositivism, interpretivism/constructivism, critical theories, as well as the participatory paradigm, becomes relevant, due to the large number of components and the relationships between them. The addition of other paradigms is likely to assist in improving our understanding of events and the practicalities associated with situations (Meissner 2014, 2016a, 2017, 2018), which is the main foundation of PULSE³.

I developed PULSE³ in order to add value to research endeavours, to gain a deeper understanding of issues and phenomena and, ultimately, to assist in the creation of opportunities. This is where PULSE³ fills a paradigm and theory niche, by assisting practitioners to deepen their understanding of real-world challenges. Its purpose is not to make point predictions or to impose control over policy processes, but to assist practitioners in understanding phenomena, situations, issues and relationships. It also helps to explain the impact of the environment on human actions, and vice versa, in the face of changes, ambiguities, uncertainties, paradoxes and contradictions (Meissner 2017). With this rationale in mind, PULSE³ consists of various components.

2.4.2   Components

PULSE³ has two characteristics. The first, and most obvious, is that it is not devoid of theory, nor does it deny the existence of theory. As previously mentioned, these ideational instruments have a role to play in practice, and by acknowledging, and emphasising, the role of the relationship between various paradigms and theories, PULSE³ enriches theory and practice. Secondly, PULSE³ does not take a positivist stance, although the paradigm permeates the methodology of the framework.

PULSE³ can analyse practices, plans, projects and programmes on a paradigmatic and theoretical level, in order to highlight their inherent meta-theoretical and theoretical elements and, by implication, the gaps. When considering this, Grant (2018) asserts that the world feels a little different with the passing of time, as new issues and actors emerge. He goes on to say that, as times change '...some drivers of change

have been either overlooked—and so they appear to be "new" to many observers—
or unanticipated'. As I have already explained, theory influences how practitioners
see the world, their place in the world, as well as the possibilities for changing the
world. Nevertheless, not all theories are good at everything. Realism, an IR theory,
emphasises states and their leaders. It can be employed in explaining a problem, like
recurring violent inter-state conflict. However, when it needs to present solutions for
the problems, it might not be up to the task (Hayes and James 2014), because one
of its limitations is that it ignores, to a large extent, the non-state actors in world
politics, like humanitarian organisations.

Practitioners are influenced by generalised empirical observations, as well as
tentative or firm deductions and explanations that bring about a theoretical perspective. It is easy to see theory at work in the policy process (Morgan 2003); one just
needs to look for those familiar definitions or conceptualisations that describe the
structures of rule and the political behaviour of the actors (Meissner 2017). By way
of example, in the past, researchers in water governance and politics have used
concepts like a 'pivotal state' (Ashton and Turton 2009; Sebastian and Warner 2014)
and 'hydro-hegemony' (Turton 2005; Zeitoun and Warner 2006; Zeitoun 2007) to
describe inter-state relations in transboundary river basins. The researchers derived
these concepts from the realist and neo-realist concept of 'hegemon'. Using the
'hegemon' concept to model inter-state relations within a transboundary river basin
could paint a picture that such relationships are only composed of, and driven by,
state actors (Meissner and Jacobs 2016). This would also imply that practitioners
subliminally ignore other relevant actors as constituting actors in transboundary water
politics (Meissner 2016a). The explanation of the role of theory in policy processes
indicates what we might be 'missing' vital actors and, by default, relationships, by
taking a particular (realist) theoretical stance (Meissner 2004, 2005).

Practitioners will not always follow a theory to the letter in their interpretation of
situations and relations; however, a certain theoretical way of thinking could play an
overbearing role in governing systems (Morgan 2003). In this sense, PULSE³ is not
only a philosophical exercise; it highlights which substantive issues are absent when
researchers formulate recommended practices that influence policy and political
decisions.

Take, for instance, a scenario where two countries want to construct a large dam on
a transboundary river that is shared by both. From a liberal institutionalist perspective, what would be necessary for them both would be a cooperative agreement,
the necessary finances, expertise that could be sourced internally or internationally,
policy mechanisms, such as environmental impact assessments, and a process of relo-
cating people from the reservoir basin, with the necessary compensation. By looking
at the situation from this perspective, one could miss the nuances that might derail
the planned dam. By combining this theoretical perspective with an interest group
pluralism outlook, researchers could be forced to ask which non-state actors might
become involved in the construction of the dam? Asking such a question from the
latter theory could open an alternative view of the interest group's practices relating
to the dam. This could raise further questions that are more ethical, such as, 'How
might these non-state actors influence the transboundary politics of the large dam'?

and 'What do they feel is just compensation for local communities'? (Meissner 2004, 2005, 2016c). If interest groups oppose the dam, what necessary steps should the government take, in order to decide on the future of the dam and its influence on humans and the natural environment? These are ethical or normative questions that have practical relevance and implications, not only for the future of the dam, but also for the wider society that it is supposed to serve.

This means that PULSE³ recognises entities, such as individuals, interest groups, scientists and private companies, as being powerful actors alongside a state, its governing apparatus and its leaders. It values all of these entities, their role in society, as well as the fundamental social processes that they produce. PULSE³ also recognises psychological elements, such as emotions, and their importance in driving an action. Acknowledging psychological elements might assist researchers and practitioners to move away from one-sided paradigmatic and theoretical explanations, and towards a more inclusive research agenda that acknowledges a foundation of critical and interpretivist knowledge production. Such an appreciation could play a role in the development of scenarios that are not dominated by state entities and their leaders and that, simultaneously, bring to the fore normative elements to guide the practices. If we focus on the explanatory value of theories, we might better understand the drivers behind phenomena and situations, and how they shape society and various issues. This might assist in developing more emphatic policies, programmes and projects, while predicting the outcome of these actions be of lesser importance (Meissner 2017).

To recap, paradigms are not permanent features of the scientific landscape (Kuhn 1962). They get torn down when scholars replace them with alternative paradigms (Eisner 1990; Weber 2004; Lake 2011; Meissner 2014, 2017, 2018). For Eisner (1990: 89), alternative paradigms are '...views of mind and knowledge that reject the idea that there is only one single epistemology and that there is an epistemological supreme court that can be appealed to settle all issues concerning Truth'. However, a specific paradigm can persist for some time, which could lead to the paradigm becoming the dominant worldview that influences policy-related actions through active substantiation. Active substantiation, or confirmation bias, is a tendency to look for, and interpret, information in a way that is consistent with existing beliefs or expectations (Marks and Fraley 2006) in line with social stereotypes, attitudes and self-serving conclusions (Frey 1981; Holton and Pyszczynski 1989; Johnston 1996; Lundgren and Prislin 1998; Jonas et al. 2001). Active substantiation applies across a range of social settings, issues, situations, processes and relations in investigations ranging from scientific research to aircraft accident inquiries (ASC 2002; Robertson 2012). PULSE³ aims to improve active substantiation.

2.4.2.1 Paradigm Assessment

The foundation of PULSE³ rests on the difference between positivism, postpositivism, interpretivism/constructivism, critical theories and the participatory paradigm, as outlined by Lincoln et al. (2011, 2018) (Table 2.1). With this differen-

Table 2.1 Paradigm assessment index

Meta-theoretical Assumptions About	Paradigm				
	Positivism Realists or the so-called 'hard science' researchers	**Postpositivism** An altered form of positivism	**Interpretivism/ Constructivism** This type of researcher attains an understanding of subjects by interpreting the subject's perception	**Critical Theories** These researchers create change to benefit those oppressed by power, like feminists	**Participatory** These researchers try to achieve transformation, based on democratic participation between the research and the subject
Knowledge generation					
Ontology Defined as '[t]he worldviews and assumptions in which researchers operate in their search for new knowledge' (Schwandt 2007: 190 cited in Lincoln et al. 2018) Theories are about the nature of being or what we can say exists, how to categorise things, and how things relate to one another (Reus-Smit 2013) An ontology asks what the nature of reality is (Creswell 2007; Lincoln et al. 2011). Simply put, an ontology is our way of understanding reality	Raw realism. A 'real' reality exists, but it is one that is apprehensible (Lincoln et al. 2018) The researcher and reality are separate (Wendt 1999; Weber 2004) There is only one identifiable reality. Research is there to control and predict nature (Guba and Lincoln 2005; Lincoln et al. 2011)	Critical realism. A 'real' reality exists, but it is only imperfectly and probabilistically apprehensible (Lincoln et al. 2018) A single reality (Weber 2004), which we will never fully understand. We will never fully understand the nature of the reality or how to attain a full understanding. This is because there are hidden variables and a shortage of absolutes (Guba and Lincoln 2005; Lincoln et al. 2011)	Relativism. Local and specific constructed and co-constructed realities define relativism (Lincoln et al. 2018) The researcher and reality are inseparable (lifeworld) (Weber 2004). Realities are mental constructs. They are social and experienced based, local, specific, constructed and co-constructed. The realities also depend on the form and content of the person holding such realities (Guba and Lincoln 2005; Lincoln et al. 2011) There are multiple realities which are dependent on the individual (Guba 1996; Lincoln et al. 2011)	Historical realism claims that social, political, cultural, ethnic and gender values that has crystallised over time shape a virtual reality (Lincoln et al. 2018) There is a constant power struggle as human nature operates in this world. Interactions between individuals and groups are in the form of privilege and oppression. These interactions are based on ethnicity, race, socio-economic class, gender, mental or physical abilities, as well as sexual preference (Bernal 2002; Giroux 1982; Kilgore 2001; Lincoln et al. 2011)	A participative reality endures, that is characterised by the human mind and a given cosmos or surrounding landscape co-creating a subjective-objective reality (Guba and Lincoln 2005; Lincoln et al. 2011, 2018) There is also a freedom from objectivity emanating from a new understanding of the relationship between the researcher and others (Heshusius 1994; Lincoln et al. 2011) A subjective-objective reality exists. 'Knowers can only be knowers when known by other knowers'. Participation and participative realities form the foundations of the paradigm (Heron and Reason 1997; Lincoln et al. 2011)

(continued)

Table 2.1 (continued)

Epistemology Defined as the truths we seek and believe, as researchers The thinking process and the relationship between what we know and what we see (Lincoln et al. 2018) Theories of knowledge or how one gains a knowledge of issues, phenomena, and reality (Reus-Smit 2013) An epistemology asks what the relationship between the researcher and the researched is (Creswell 2007; Lincoln et al. 2011)	Dualist/objectivist with findings that are true (Lincoln et al. 2018) Objective reality exists beyond the human mind (Weber 2004)	Modified dualist/objectivist, with a critical tradition and community with findings probably true (Lincoln et al. 2018) We can only approximate nature. Research and statistics give incomplete data (Guba and Lincoln 2005, Lincoln et al. 2011)	Transactional/subjectivist with value-mediated findings (Lincoln et al. 2018) An actor's reference frame to the setting in which it is located construct social reality (Guba and Lincoln 1985) A person's lived experience constitutes a knowledge of the world (Weber 2004). Meaning is based on our interactions with our environment (Guba and Lincoln 1985; Lincoln et al. 2011) Research findings are the creation of the process of interaction between the inquirer and the inquired into (Guba 1990; Lincoln et al. 2011)	Transactional/subjectivist with value-mediated findings (Lincoln et al. 2018) The drivers of research are the study of social structures, freedom and oppression, as well as control and power. The knowledge we produce can change existing structures and get rid of oppression through emancipation (Merriam 1991; Cox and Sinclair 1996; Lincoln et al. 2011)	Characterised by a critical subjectivity in participatory transactions with the cosmos. An extended epistemology exists, categorised by experimental, propositional, and practical knowing with co-created findings (Lincoln et al. 2018). Critical subjectivity exists and is dependent on the transaction with reality. There is an extended epistemology consisting of experiential, propositional, as well as practical means of knowing. These facets co-constitute research findings (Heron and Reason 1997; Guba and Lincoln 2005; Lincoln et al. 2011)
Research objects Described as the relationship between the researcher and the objects	The research object has inherent qualities that exist independently of the researcher (Weber 2004)	Minimum interaction with the research objects. Distance between researcher and object to get objectivity (Guba and Lincoln 2005, Lincoln et al. 2011)	The researcher's lived experience interprets the research object (Weber 2004) Through its research methods there is adequate dialogue between the research subjects and objects to collaboratively construct a meaningful reality (Angen 2000)	Research objects such as structures, freedom and oppression and control and power inform the researcher's research agenda (Merriam 1991; Cox 1996; Lincoln et al. 2011). There is a distance between the researcher and the object, but only in terms of a physical separation. Other than that, the researcher could be imbedded in the research object (Meissner 2017)	There is a participatory transaction between the researcher and the object (Guba and Lincoln 2005; Lincoln et al. 2011)

(continued)

Table 2.1 (continued)

Methodology Described as the research process (Creswell 2007; Lincoln et al. 2018) or the means of how we discover new knowledge or the principle of our investigation and how such investigations should continue (Lincoln et al. 2018)	Experimental/manipulative, as well as the verification of hypotheses and mainly quantitative methods (Lincoln et al. 2018). Statistics, content analysis, laboratory experiments, field experiments and surveys (empirical data gathering and analysed through statistical analyses) (Weber 2004) The scientific method is king and a belief in the falsification principle (results and findings are true until disproved) (Merriam 1991; Lincoln et al. 2011). Researchers use maps, photographs and sketches to indicate one identifiable and objective reality, an independent reality and a correspondence theory of truth (Meissner 2017)	Modified experimental/manipulative with critical multiplism, the falsification of hypotheses that may include qualitative methods (Lincoln et al. 2018) Statistics, content analysis, laboratory experiments, field experiments and surveys (empirical data gathering and analysed through statistical analyses) (Weber 2004) Statistics are important to visually interpret findings Scientists use the hypothetical deductive method (hypothesis, deduce and generalise). Scientific method is king. Because of the unknown variables, postpositivists ask more questions than positivists Merriam 1991; Guba and Lincoln 2005; Lincoln et al. 2011). Researchers use maps, photographs and sketches to indicate one identifiable and objective reality, an independent reality and a correspondence theory of truth (Meissner 2017)	Hermeneutical/dialectical. Researchers use qualitative methods (Lincoln et al. 2018) Hermeneutics (interpretation of, for instance, the recognition and explanation of metaphors), phenomenographic studies, case studies, ethnographic studies and ethnomethodological studies (Weber 2004) Interpretive approaches rely on interviews, observations, and analyses of existing texts (Angen 2000) Meanings also emerge from the research process (Lincoln et al. 2018) A hermeneutic cycle in research exists whereby actions lead to the collection of data, which then results in the interpretation of data, driving action based on the data (Lincoln et al. 2018) Researchers use maps, photographs and sketches to indicate an interpreted and constructed reality. Researchers also use these to interpret reality according to the researcher's lived experience with maps, photographs and sketches, as representations of the interaction that the researcher has had with his/her environment (Meissner 2017)	Dialogic/dialectical (Guba and Lincoln 2005; Lincoln et al. 2018) Researchers actively search for participatory research to empower the oppressed and support social transformation and revolution (Merriam 1991; Lincoln et al. 2011). Researchers use maps, photographs[a] and sketches as representations of inequality and injustice (Meissner 2017)	Characterised by political participation in collaborative action inquiry, with the practical being of prime importance. Participants ground language use in a shared experiential context (Lincoln et al. 2018). Deconstructing, face-to-face learning, and democratic dialogue (Heron and Reason 1997) Collaborative action inquiry emphasises the practical, using language grounded in a shared experiential context (Guba and Lincoln 2005). Research participants share maps, photographs and sketches as experiences (Meissner 2017)

(continued)

Table 2.1 (continued)

Theory of truth	Correspondence theory of truth: one-to-one mapping between research statements and reality (Weber 2004). Only one truth or reality (Lincoln et al. 2011)	One truth exists but we will never fully understand it. In addition, the researcher is in control of the process of inquiry that produces the truth (Lincoln et al. 2011)	Truth as intentional fulfilment: interpretations of research object match lived experience of object (Weber 2004). There are no permanent standards to know truth in a universal manner (Lincoln et al. 2011)	Researchers find truth in the struggle for equality and social justice. Social science indicates the oppression of people (Lincoln et al. 2011)	Knowledge is based on the transformation and experience gained through shared research inquiry between researchers and subject (Lincoln et al. 2011)
Time	Time is linear, neutral and unitary (clock-time) with history repeating itself exactly. Theories have 'time-less' qualities for the researcher (McIntosh 2015; Meissner 2019)	Time is linear, neutral and unitary (clock-time), (McIntosh 2015; Meissner 2019) and although history repeats itself, it is not an exact repetition. Theories have 'time-less' qualities for the researcher (McIntosh 2015; Meissner 2019) but context and fluidity also play a central role (Meissner 2017)	Time is linear, neutral, unitary (clock-time) and a construction based on context and temporal fluidity (McIntosh 2015; Meissner 2019). Historical events do not repeat themselves exactly because contexts change together with temporal fluidity. Theories have a temporary quality because of changing contexts and temporal fluidity (Meissner 2017)	Time is linear, neutral and unitary (clock-time) (McIntosh 2015; Meissner 2019) and history is replete with oppressive moments. Theories have 'time-less' qualities for the researcher (McIntosh 2015; Meissner 2017 2019)	Time is linear, neutral, unitary (clock-time) (McIntosh 2015; Meissner 2019), and an objective-subjective construction. History repeats itself not exactly and we can learn from history by constructing practical ways of knowing more about the world. Theories have a temporary quality because of changing contexts and temporal fluidity, based on objective-subjective intersections (Meissner 2017)
Scale	The researcher views scale as a natural/geographic scale, level, or size having an embedded spatial hierarchy, with global forces having more agency than what local actors have (Bulkeley 2005; Neumann 2009; Meissner 2019; Warner et al. 2014)	The researcher views scale as a natural/geographic scale, level or size having an embedded spatial hierarchy, with global forces having more agency than what local actors have (Bulkeley 2005; Neumann 2009; Meissner 2019; Warner et al. 2014)	The researcher problematises scale and hierarchy as traditional concepts (Meissner 2017 2019). Scale is a social construction (Norma et al. 2012) with networks being central features (Bulkeley 2005)	Scale is 'contingent on political struggle' (Norma et al. 2012)	Scale is an objective-subjective construction arrived at through discussions between the researchers and the research participants (Meissner 2017)

(continued)

Table 2.1 (continued)

Validity					
The research truly measures what the researcher intends on measuring or, stated differently, how truthful the research results are (Golafshani 2003). Validity is not a universal, but a contingent construct based on specific research methodologies and projects, as well as processes (Winter 2000)	Certainty: data truly measures reality (Weber 2004) Researchers can proof their collected data (Lincoln et al. 2011)	The validity of research comes from peers and not the subjects that we study (Guba and Lincoln 2005; Lincoln et al. 2011)	Defensible knowledge claims (Weber 2004), depending on the methods used (Lincoln et al. 2011) These are extended constructions of validity through consensus, which is based on participants and the inquirer (Guba and Lincoln 2005; Lincoln et al. 2011)	Validity comes about when research creates action (action research) or participatory research that creates positive social change (Guba and Lincoln 2005; Merriam 1991; Lincoln et al. 2011)	Validity is in the ability of the knowledge to become transformative, according to the findings of the experiences of the subject. Extended validity of constructions exists (Guba and Lincoln 2005; Lincoln et al. 2011)
Reliability					
The extent to which results are consistent over time, the accurate representations of the total population one is studying and, if it is possible, to reproduce the same results using a similar methodology. In other words, reliability entails the repeatability or replicability of observations or results (Golafshani 2003)	Defined as replicability, which means that researchers can reproduce their results (Weber 2004)	There is a statistical confidence level and the researcher produces objectivity in data through inquiry (Lincoln et al. 2011)	Defined as interpretive awareness, which means that researchers recognise and address implications of their subjectivity (Weber 2004)	The value of the inquiry is in the subversion of privileges and its ability to impart action for the creation of a just and fair society (Giroux 1982; Guba and Lincoln 2005; Lincoln et al. 2011)	There is a correspondence between experiential, presentational and practical knowing. This leads to action for the transformation of the world in the service of human well-being (Guba and Lincoln 2005; Lincoln et al. 2011)

(continued)

Table 2.1 (continued)

Training Defined as how researchers are prepared to conduct research	Training is technical and quantitative, with substantive theories that prescribe the scientific method (Guba and Lincoln 2005; Lincoln et al. 2011, 2018). In the social sciences, the training has a strong bias towards the scientific method	Training is technical, quantitative, and in some instances qualitative, with substantive theories. Researchers can conduct mixed-method research (Guba and Lincoln 2005; Lincoln et al. 2011, 2018)	Researchers are resocialised, can do quantitative and qualitative research, study history, values of altruism, liberation and empowerment (Guba and Lincoln 2005; Lincoln et al. 2011, 2018)	Training is in qualitative and quantitative methods. Researchers study history and social science to understand the nature of liberation and empowerment (Guba and Lincoln 2005; Lincoln et al. 2011, 2018)	Researchers initiate the other researchers into the research/inquiry process by the researcher or facilitator. The co-researchers learn through active participation. To do this, the researcher or facilitator must have emotional competence, a democratic personality, as well as appropriate skills (Guba and Lincoln 2005; Lincoln et al. 2011, 2018)
Recommended actions					
Organising question	Who governs and who benefits (Hobson and Seabrooke 2007)?	Who governs and who benefits (Hobson and Seabrooke 2007)?	Who acts and what are the consequences of their actions (i.e. how are their actions enabling change)? (Hobson and Seabrooke 2007)	Who acts to bring about a just, fair and equal society (Meissner 2017)?	Who is in dialogue to bring about change to better the human condition (Meissner 2017)
Unit of analysis	Hegemons/great powers, international regimes, ideational entrepreneurs, capitalist world economy, structures of rule (Hobson and Seabrooke 2007)	Hegemons/great powers, international regimes, ideational entrepreneurs, capitalist world economy, structures of rule (Hobson and Seabrooke 2007)	Everyday actors interacting with elites and structures (Hobson and Seabrooke 2007)	Everyday actors and the epistemic community interacting with one another and subverting structures of injustice and discrimination (Meissner 2017)	Researchers, the elite, practitioners and everyday actors from different economic and social spheres that are in dialogue to better the human condition (Meissner 2017)
Prime empirical focus	The supply of order and welfare maximisation by elites, as well as the maintenance of the powerful and the unequal distribution of benefits (Hobson and Seabrooke 2007)	The supply of order and welfare maximisation by elites, as well as the maintenance of the powerful and the unequal distribution of benefits (Hobson and Seabrooke 2007)	The social transformative and regulatory processes enacted, or informed, by everyday actions of individuals (Hobson and Seabrooke 2007)	Social transformations that will bring about justice, equality, non-discrimination and revolution (Meissner 2017)	The supply of order and welfare maximisation by elites (Hobson and Seabrooke 2007), researchers and practitioners from different spheres of the economy and society (Meissner 2017)

(continued)

Table 2.1 (continued)

	Top-down (Hobson and Seabrooke 2007)	Top-down (Meissner 2017)	Bottom-up (Hobson and Seabrooke 2007)	Bottom-up (Meissner 2017)	Horizontal (Meissner 2017) and vertical
Locus of agency Described as the hierarchical position from which an actor produces or projects power					
Level of analysis	Systemic (Hobson and Seabrooke 2007)	Systemic (Meissner 2017)	Complex, holistic, (Hobson and Seabrooke 2007) and detailed (Meissner 2017)	Complex, holistic, and detailed (Meissner 2017)	Complex, holistic, detailed and systemic (Meissner 2017)
Ontology	Structuralist (Hobson and Seabrooke 2007)	Structuralist (Meissner 2017)	Agential or structuralist (Hobson and Seabrooke 2007)	Agential and/or structuralist (Meissner 2017)	Agential, structurationist and structuralist (Meissner 2017)
Recommendations based on specific theoretical assumptions	Positivist, postpositivist **or** interpretivist (Hobson and Seabrooke 2007)	Positivist, postpositivist **and** interpretivist	Interpretivist, postpositivist, subjectivist and/or positivist (Hobson and Seabrooke 2007)	Subjectivist and interpretivist (Meissner 2017)	Interpretivist, subjectivist, postpositivist and/or positivist (Meissner 2017)
Voice Described as the researcher who recites or narrates the produced research. The language of the researcher and the ability to present the researcher's research material, together with the research subject's story (Lincoln et al. 2018)	The data speak for themselves A 'disinterested scientist' informs decision-makers, policy practitioners and change agents. This means that the researcher does not show an interest in the influence of his/her research findings (Lincoln et al. 2011, 2018). Use of the third person passive voice (Meissner 2017)	Researchers inform populations by using their data produced by their research (Lincoln et al. 2011, 2018). Use of the third person passive voice (Meissner 2017)	The researcher is a passionate participant and acts as a facilitator of multivocal reconstruction, including that of culture. The advocate and activist are a 'transformative intellectual' (Lincoln et al. 2011, 2018). 'This means that while critical theorists attempt to get involved in their research to change the power structure, researchers in this paradigm attempt to gain increased knowledge regarding their study and subject by interpreting how the subjects perceive and interact within a social context' (Lincoln et al. 2018: 124). Use of first-person active voice (Meissner 2017)	Researchers create the data with the intention of producing social change and conveying a social justice that produces equal rights for all. The advocate and activist are 'transformative intellectuals'. This means that critical theorists attempt to get involved in their research to change the power structure (Lincoln et al. 2011, 2018), Use of first-person active voice (Meissner 2017)	The researcher *and* practitioners create a voice of 'reason' that informs policies, programmes and plans. Use of first-person active voice (Meissner 2017)

(continued)

Table 2.1 (continued)

Ethics Defined as the relationship or interaction between the subject and the researcher, together with the influence of research on populations (Lincoln et al. 2018)	Regarding ethics, there is a belief that the data drive the side effects of any research endeavour. The research studies nature and not to influence how nature affects populations (Guba and Lincoln 2005; Lincoln et al. 2011, 2018)	The research process must be as statistically accurate as possible during interpretations of reality. Researchers do not take the consequences on others into consideration because they conduct their work to gain accuracy, not to influence populations (Lincoln et al. 2011, 2018)	Research develops the research process to reveal special problems (Guba and Lincoln 2005; Lincoln et al. 2011)	Researchers link their research to specific interests towards a just society (Giroux 1982; Lincoln et al. 2011)	Research develops the research process to solve practical problems, among others (Guba and Lincoln 2005; Lincoln et al. 2011)
Hegemony Described as the influence of researchers on others or who has the power during research and who decides what to research (Lincoln et al. 2011, 2018)	Positivists believe that research should be influential and not the researcher or scientists that conduct the inquiry. The aim is to produce the truth and not to provide actions for the reality to affect others in society (Lincoln et al. 2011)	Researchers produce data that decision-makers use for the foundation of their decisions. Researchers produce the data through the statistical analysis of reality. The researcher is in control of the process of inquiry (Guba and Lincoln 2005; Lincoln et al. 2011)	Researchers seek input into practices and recognition. They also offer to change existing paradigms (Guba and Lincoln 2005; Lincoln et al. 2011)	The endeavours of researchers' indicate how interactions of privilege and oppression stand in relation to race, ethnicity, gender, class, sexual orientation, physical or mental ability and age (Kilgore 2001)	Power is a variable of what and how we know (Kilgore 2001; Lincoln et al. 2011)
Axiology[b] Defined as the theory of values, and more specifically, the way in which researchers act based on their research and the criteria of values and value judgments particularly in ethics (Lincoln et al. 2011, 2018)	There should be a distance between the researcher and the research subject so that researchers' actions do not influence populations. The laws that researchers produce should influence populations (Guba and Lincoln 2005; Lincoln et al. 2011, 2018)	Researchers should, as far as possible, get a better understanding of reality and get as close as possible to the truth. Researchers perform this task utilising statistics that explain and describe what is known as reality (Guba and Lincoln 2005; Lincoln et al. 2011)	Propositional and transactional knowledge is instrumentally valuable to achieve social emancipation (Guba and Lincoln 2005; Lincoln et al. 2011)	Researchers seek to change issues and social institutions' practices and policies (Bernal 2002; Lincoln et al. 2011)	Researchers use their practical knowledge to flourish with a balance of autonomy, cooperation and hierarchy as an end. The purpose of creating reality when we participate in research is to change the world and impart responsibility (Heron and Reason 1997; Lincoln et al. 2011, 2018)
Action Defined as the way researchers produce beyond the data or how society uses the generated knowledge (Lincoln et al. 2011, 2018)	Researchers must remain strictly objective. They are not concerned with the actions that come about because of their research (Guba and Lincoln 2005; Lincoln et al. 2011)	Researchers must remain strictly objective. They are not concerned with the actions that come about because of their research (Guba and Lincoln 2005; Lincoln et al. 2011)	The research can mandate training in political action, should the participants not understand the political processes and systems. If the research does not have an education purpose, then it can compel people to act politically (Guba and Lincoln 2005; Lincoln et al. 2011)	The research must produce social change, change in human thinking and be an examiner of human existence (Creswell 2007; Lincoln et al. 2011)	The research can mandate training in political action should the participants not understand the political processes and systems. The research process is an enabler of action (Guba and Lincoln 2005; Lincoln et al. 2011)

(continued)

Table 2.1 (continued)

| Control
Described as who dictates the research methodology and how users employ the research (Lincoln et al. 2011) | The researchers conduct control without any inputs coming from participants or society (Guba and Lincoln 2005; Lincoln et al. 2011) | The researchers conduct control control without any inputs coming from participants or society (Guba and Lincoln 2005; Lincoln et al. 2011) | The researcher and participants share the research. Without equal or co-equal control, the research would be impossible (Guba and Lincoln 2005; Lincoln et al. 2011) | Control can be shared by the researcher and the subjects. The subject can have a say in how to conduct the research (Bernal 2002; Lincoln et al. 2011) | The researcher and participants share the research. Without equal or co-equal control, the research would be impossible. Knowledge expresses power embodied in research participation (Kilgore 2001; Lincoln et al. 2011) |

[a] According to Bosch (2019) 'Photography is a versatile medium that is able to freeze a single moment in time as well as provide insight into [the] zeitgeist of a longer period. Therein lies part of the value of the medium of photography, as well as a political application.' That said, the photographer is able to present a piece of truth and examine 'visual normativity', memory as well as everyday practices (Bosch 2019). Taking this latter point into consideration, photographs do not only supply empirical evidence but can also, when interpreted within a broader context, guide us in developing normative explanations of what the image represents

[b] Linguists developed the word 'axiology' from the Greek words 'axios' or worth and 'logos' or theory, reason. Axiology investigates problems and issues that involve our thinking about conditions of life, the structure of reality, the order of nature and human beings' place therein. That said, our quest for various values or the things and conditions beneficial to our survival and improvement of life was probably the wellspring of our quest for knowledge or reality (Hart 1971)

tiation in hand, we can profile knowledge construction that is intrinsic to policies, programmes and practices, or any written document, for that matter (Meissner 2017).

The paradigm assessment tool that is described in Table 2.1 considers the way by which researchers or practitioners generate knowledge and express agency, using the five paradigms. In other words, the main objective of PULSE³ is to study the knowledge content of a situation, as it is expressed in written form. I define 'agency' as any action that involves human activity and that sets into motion general or specific ideas, operations or recommendations. It will suffice to include a note on the 'positivist' way of determining the influence of paradigms and of identifying them in analysed information. So far, I have argued for a move away from positivism to a more inclusive ontological, epistemological and methodological view of studying reality. Because of this argument, when some commentators look at the paradigm assessment component of PULSE³, they might accuse me of double standards. I believe that methodological pluralism is the way to go if we want to understand the ontologically complex social and biophysical environments. Quantitative methods can point us to interesting patterns of observable data (Meissner 2017). However, in order for us to explain the patterns of data, a qualitative approach is necessary because it highlights the causal processes that take place in a more distinct manner (Kurki 2006; Meissner 2017). Having said that, PULSE³ utilises a quantitative formulation to identify paradigms, theories, causal mechanisms, as well as problem-solving and critical theories, in a text, after which the researcher then uses a qualitative methodology to interpret the results (Meissner 2017).

In this way, the framework forces a mixed-method approach, which Johnson and Onwuegbuzie (2004) argue, is likely to enable more effective research. They based their reasoning on a dynamic research environment that is more interdisciplinary and complex and that compels researchers to compliment one method with another. What is more, with the outbreak of 'paradigm peace', epistemological and ontological issues are less prominent than they were during the 'paradigm wars', which was characterised by the debate between qualitative and quantitative research. The dispute saw qualitative and quantitative researchers pitted against each other who often used competing methods, which were based on fundamentally distinct research principles (Bryman 2006). Beach and Kaas (2019) founded the popularity of multimethod research on the complex nature of the disciplines. This means that multifaceted issues are ripe for a mixed-method approach because of their multifarious epistemological and ontological nature.

To determine the extent of a research paradigm (Table 2.1) for underpinning policy-related knowledge, I use a simple scoring method that I have described in previous publications (Meissner 2014, 2017, 2019). I received funding from the National Research Foundation (NRF) to conduct research on eThekwini's green and ecological infrastructure policy landscape. This enabled me to contract a CSIR software developer, Dr. Marc Pienaar, to develop a digital version of PULSE³, based on the R software package, which utilises Java as its programming language. The software enables statistical computing and graphics generation, which are presented in Chap. 3. In short, PULSE³ employs a scoring system that counts the absence, or presence, of a paradigm's metatheatrical assumptions within a text (Meissner

2017). After counting the meta-theoretical assumptions, the software programme produces a bar graph, which ranks the paradigms according to their scores (Fig. 3.3). I employ the same scoring method to determine the presence or absence of theoretical perspectives, causal mechanisms and problem-solving and critical theories.

Why undertake a paradigm assessment? Empirical research is based on a conceptual inquiry (Wight 2006; Kurki 2008; Kurki and Wight 2013). This means that it is necessary to have an adequate understanding of the concepts, as theoretical building blocks that we use in evidence-based policy formulation, in order to know why we use certain notions over others and to recognise the strengths and weaknesses of the definitions. In the absence of such an understanding, peers will not adequately justify the research and will perceive it as mere 'fact-finding'. It also means that we will not appreciate how and why researchers engage in research and use the same paradigms and theories, which could have an undesirable influence on a constructive debate with those who have different perspectives (Kurki 2008: 9). Be that as it may, herein lies the crux of the paradigm assessment, according to Kurki (2008: 9):

> '…whenever we make factual, explanatory or normative judgements about…' [the world around us] '…important meta-theoretical filters [or meta-theoretical assumptions] are at work in directing the ways in which we talk about the world around us, and these filters are theoretically, linguistically, methodologically, and also potentially politically consequential'.

Arguing from a certain paradigm '…creates intellectual blinders and institutional barriers' (Sil and Katzenstein 2011: 481), which is one of the reasons why academics are side-lined when it comes to practical problems (Nye 2009; Sil and Katzenstein 2011; Lake 2011). According to Kurki (2008), this could have a spill-over effect on policy formulation and implementation. The paradigm assessment assists in identifying potential red flags and, furthermore, it highlights where current research or policy formulation is focusing its attention and where the potential pitfalls for practicalities lie. Kurki (2006: 213) went on to say that:

> …metatheoretical framings of explanatory frameworks have direct effects on the kinds of explanations we advance for concrete…political processes: indeed, theoretical and conceptual lenses "constrain and enable"…the kinds of explanations we can construct.

Therefore, investigating which paradigm is dominant could enable us to generate more holistic and open explanations (Kurki 2006), with regard to the challenges that practitioners experience. That being said, the meta-theoretical assumptions that are utilised to assess the paradigmatic stance of knowledge are important, because they show us not only how practitioners think about issues, but also how they construct a way of dealing with problems or creating opportunities (Meissner 2017). In summary, the paradigm assessment counts the meta-theoretical filters that are at work in how we describe a situation (Kurki 2008), as well as the actions contained in such explanations.

The Ethos of Analytic Eclecticism

If positivism, or any other paradigm or methodology for that matter, has blind-spots, what is the proposed alternative? The answer lies in analytic eclecticism, that is, a need to go beyond the paradigms (Sil and Katzenstein 2010; Bennett 2010; Meissner 2017, 2019) and to see the world through multiple worldviews (Haas 2010a). According to Sil (2009), Friedrichs (2009), Friedrichs and Kratochwil (2009) and Cornut (2015), analytic eclecticism is a pragmatist alternative to scholarship that is grounded in current research traditions. In other words, it is a pragmatic method-ology to the traditional scholarly knowledge, the stuff that existing and dominant paradigms, such as positivism, postpositivism and interpretivism/constructivism, are based on (Sil 2009; Franke and Weber 2011; Cornut 2015). Therefore, analytic eclec-ticism is 'an intellectual viewpoint that combines various theoretical frameworks or that 'finds the best assumptions for capturing the nature of humans' (Zyla 2019: 3). Put in another way, '...it is a general approach and perspective in viewing a particular research problem, or a given set of inquiries in...scholarship' (Regilme 2018: 16). This means that the approach to researching issues holds the potential to 'deal with problems characterised by uncertainty' (Haas 2010b). Analytic eclecticism, there-fore, is problem-driven (Regilme 2018), since it includes the extraction, adaptation and integration (not synthesis) of hidden concepts, mechanisms, logical beliefs, as well as interpretive actions, that are embedded in research traditions. Traditionally, these paradigms identify separate styles of research and reflect dissimilar combina-tions of ontological and epistemological principles. Other characteristics include the extraction and integration of factors, causal narratives, assumptions and 'interpretive moves' (Sil 2009). This is also the case with the numerous social theories that I have already explained in the above discussion on the distinction between the use of paradigms and theories (Meissner 2017).

Purpose

The purpose of analytic eclecticism is to avoid any paradigmatic limitations. Such confinement can lead to a disconnection between researchers and their practical contributions. Sil and Katzenstein (2010) argued that a paradigm can become an obstacle of understanding, even if it gives powerful insights through active substan-tiation (Meissner 2017). Positivist-dominated research, for instance, introduces and informs the foundational aspects of research (Kurki 2008), with the result that researchers develop research questions and set boundaries for their investigations, based on a priori assumptions, and they often conduct investigations to reflect these prior assumptions. When they are done, they claim that, as arguments progress within a specific bounded research tradition, there is paradigmatic progression characterised by ever-increasing sophistication. However, explaining the complexities of problems are victim (Sil and Katzenstein 2010).

   Analytic eclecticism does not discard established paradigms or traditions; instead, it discovers an applicable relationship between them. It then reveals the invisible connections between the perceived mismatched and paradigm-bound theoretical elements. The commitment is to produce novel insights that will influence policy

debates and practical problems. This requires alternative thinking about the relationships among assumptions, concepts, theories, research activities and problems (Sil
and Katzenstein 2010). To better understand the complexities of real-world problems and to assist practitioners, it is necessary to step outside of the theoretical and
paradigmatic boundaries and to engage in such problems, from multiple perspectives.
This is the underlying ethos and function of PULSE[3] (Meissner 2017).

Complexity

One way of discounting analytic eclecticism is to say that the complexity theory is
all we need to address problems. The argument states that the complexity theory is
all-encompassing and explains all, or at least several, aspects found in nature and
society. Such an argument is bound to run into trouble. Academics often consider
complexity thinking to be superior to alternative theories for explaining and solving
problems. However, by doing so, it runs the risk of a high degree of error (Sil and
Katzenstein 2010), since it is also a theory that has epistemological and ontological
limitations.

Complexity, as a theory, has limited scope, for it is only one perspective. A
researcher who argues from a complexity stance, could 'miss' other viewpoints
because of the misperception that complexity is all-encompassing in explaining
and understanding what is happening. Complexity might be useful in illuminating
numerous elements that are at work. However, it has little to offer decision-makers
who are interested in bringing about change in society, since it does not rest on a
critical theory or participatory ontology. In other words, the complexity theory is not
a critical theory (Cox 1981; Cox and Sinclair 1996) with an emancipating agenda
(Meissner 2017), but a problem-solving perspective that explains the world as it is
(Cox 1981; Cox and Sinclair 1996). The theory is not a reliable blueprint and can
desensitise us to the possibility that a specific theory could be wrong (Hirschman
1970; Tetlock 2005, in Sil and Katzenstein 2010).

I believe that the placating ability of complexity thinking lies in its apparent epistemological and ontological legitimacy, which are informed by what it is supposed to
explain, namely, complexity. Grenstad (2007: 121) hypothesised that, 'If empirical
complexity exists, several sets, or paths, of causal conditions will be identified. If
empirical simplicity exists, a smaller set of causal conditions will prevail'. We can
turn Grenstad's (2007) first hypothesis on its head: for example, if we explain that
the world is complex, we might think that it is always, and under all circumstances,
so. What limits complexity thinkers is the epistemological and ontological simplicity
of the theory, namely, that almost everything is complex, and we should explain it
as such. The irony is that we oversimplify reality with complexity, by using active
substantiation as a biasing device. Our perception of a complex reality forces us to
use complexity parsimoniously. According to Richardson (2008: 19),

> Despite all the rhetoric about reshaping our worldview, taking us out of the age of mechanistic
> (linear) science into a brave new (complex) world, many complexity theorists...have actually
> inherited many of the assumptions of their more traditional scientific predecessors [such as
> positivism] by simply changing the focus from one sort of model to another, in very much
> the same way as some [business] managers jump from one fad to another in the hope that
> the next one will be the ONE' (emphasis in the original).

Jolly and Chang (2019) speak of a flatland fallacy, where the parsimony of our low-dimensional theories reflects the reality of many higher dimensional problems. The flatland fallacy forces us to seek simplified explanations for complex phenomena. As a theory, complexity thinking is epistemologically reductionist (Richardson 2008) and ontologically low-dimensional, since it represents the 'complexity' of reality, but it ignores the high importance of political and psychological occurrences.

Transdisciplinarity

What is more, analytic eclecticism and transdisciplinarity are different. Transdisciplinarity deals with the breaking down of disciplinary silos in and between the natural and social sciences. It is recognised that transdisciplinarity should guide research practices at several levels and deal with issues found in both sciences (Jacobs and Nienaber 2011). We also need transdisciplinarity at a functional level, where the relationship between ecosystems and society needs deeper understanding.

A link exists between transdisciplinarity and complexity. Transdisciplinarity 'facilitates a deeper understanding of complexity and complex problems by examining different facets of reality through the lens of multiple perceptions' (Jacobs and Nienaber 2011: 670). It operates at the intersection between agents and structures, and, as such, it is situated at the nexus between problem-solving and values, ethics, norms, cultures and beliefs (Sil 2000b; Lawrence and Depres 2004; Max-Neef 2005; Jacobs and Nienaber 2011). To some extent, transdisciplinarity also deals with the epistemological question of how to generate knowledge. Transdisciplinarity and complexity operate on a level where the methodological meets the theoretical (Max-Neef 2005), but not the meta-theoretical elements of the paradigms. To explain this further, strong transdisciplinarity asks that researchers grapple with the binary, linear and rationalist limitations that dominate science. Analysts need to recognise that there are multiple subjective realities that they can access by means of various perception levels. In doing so, they need '...to be constantly reflexive about their [paradigmatic] positionality, the subjective perceptions they hold, their worldviews and their belief systems' (Jacobs and Nienaber 2011: 9).

Therefore, transdisciplinarity is a foundation for analytic eclecticism. Analytic eclecticism operates on the meta-theoretical level, where scholars and practitioners address questions of knowledge-generation and agency. Practical claims about phenomena do not separate paradigms. The fault line is on their meta-theoretical assumptions on 'how such claims should be developed and supported' (Sil and Katzenstein 2010: 4). With meta-theoretical divisions in mind, complexity and transdisciplinarity are silent on the meta-theoretical aspects of their assumptions. Analytic eclecticism makes it possible for teams to work together and it also affords the researcher the opportunity to work solo in multiple research traditions or paradigms, even if they are incompatible (Laudan 1977, in Sil and Katzenstein 2010). Jacobs and Nienaber (2011) highlight the importance of teams throughout their research; however, they are silent about the role of the individual researchers that can transcend paradigm boundaries on their own. Linking analytic eclecticism with the use of transdisciplinarity by teams and solo researchers, investigators cannot be transdisciplinary without eclecticism, but they can be eclectic without transdisciplinarity. This means that individual researchers and teams with eclecticism can integrate paradigms simultaneously across different disciplines, yet they cannot practice eclecticism, with only transdisciplinarity in hand (Meissner 2017). It also entails that it is possible to integrate analytic eclecticism and transdisciplinarity at both a paradigm's meta-theoretical micro-level and the overall science macro-vision that are contained in its overarching belief systems.

Theoretical synthesis

Furthermore, analytic eclecticism is not theoretical synthesis (Sil 2009; Friedrichs and Kratochwil 2009). As Sil and Katzenstein (2010: 17) explained:

> …analytic eclecticism [is] a flexible approach that needs to be tailored to a given problem and to existing debates over aspects of this problem. As such, it categorically rejects the idea of a unified synthesis that can provide a common theoretical foundation for various sorts of problems.

In other words, there is not just one theoretical approach for tackling all sorts of problems, but it is more a case of there being a diverse convergence of theoretical elements for varied problems (Sil 2009; Sil and Katzenstein 2010). In this regard, analytic eclecticism resists an elegant theory with a small number of assumptions that explain a wide range of phenomena and develop panaceas for a variety of problems (Meissner 2017).

After all is said and done, complexity asks for a better appraisal of societal and ecological challenges, while transdisciplinarity propagates the breakdown of scientific silos. Scientists promote synthesis from a positivist viewpoint, to assist with the challenges that occur in the natural environment. The problem is that they do not transgress the divide between positivism, postpositivism, interpretivism/constructivism, critical theories and the participatory paradigm.

Three pillars

One of the characteristics of analytic eclecticism is that it rests on three pillars. The first is open-ended problem formulation, which considers the difficulty of phenomena and issues. This type of research preparation has no intention of advancing or filling gaps in paradigm-bound research. Secondly, it is a middle-range causal account that integrates multifaceted interactions among multiple causal mechanisms and logics that are drawn from more than one paradigm. This causal account relates to the middle position of analytic eclecticism along the agent-structure axis, on the one hand, and the material-ideational axis, on the other. Thirdly, the findings and arguments connect the scholarly debates and real-world dilemmas that decision-makers face. Analytic eclecticism engages both academic and practical concerns (Sil and Katzenstein 2010; Hayes and James 2014; Pohl and van Willigen 2015). Let us unpack this further by outlining what analytic eclecticism promises, in order to counteract the argument that it is the same as, or similar to, the complexity theory, transdisciplinarity and synthesis.

Regarding the second pillar, analytic eclecticism does not slice up complex social phenomena just to make them simple and easy to analyse. In other words, reductionism is not an underlying premise, and analytic eclecticism can help us to move away from causal and theoretical oversimplification, to a more holistic understanding of processes (Kurki 2006). Important substantive questions with relevant real-world applications are possible by integrating empirical observations and causal stories; in other words, by incorporating the five paradigms. This brings about the 'promise of richer explanations' (Sil and Katzenstein 2010: 3) and deeper understanding. For instance, material resources matter and are an important ingredient in social relations. Even so, material resources come about through social processes, which involve actors and principles that socialise these actors. Formal causes have a different influence in different causal contexts (Kurki 2006) (more on this later).

To put things differently, analytic eclecticism facilitates the quantum leap from singular explanations of real-world problems to fuller clarifications, alternatives and solutions. Whereas paradigms have blind-spots, at the same time, they also have useful insights into issues, challenges and opportunities (p. 28). With analytic eclecticism as the foundation for research practice, there are connections and complementarities between the paradigms that can be exploited. This could lead to a situation where researchers generate more useful theoretical and empirical insights (Sil and Katzenstein 2010), which can service the practitioner in a more meaningful manner (Meissner 2017).

Through analytic eclecticism, it is possible to produce narratives or theories that have a practical impact on social and biophysical conditions and that rest on prevailing ideas. This is analytica eclecticism's third pillar. Analytic eclecticism brings forth useful insights. Such understanding needs to go beyond scholarly work and add value to policy debates and normative discussions. The purpose of analytic eclecticism is not to create new arguments, for the sake of argumentation, but to show that it is possible to confront problems with a variety of paradigms and theories (Cornut 2015).

The determination of analytic eclecticism is not to produce a new line of analysis that goes beyond classification within an existing paradigm. Rather, it investigates how paradigm-bound research generates insights for developing causal stories that capture the complexity, contingency and messiness of the environment in which the actors work. Furthermore, analytic eclecticism should also produce implicit recommendations, and it must have 'some clear implications for some set of policy debates or salient normative concerns that enmesh leaders, public intellectuals, and other actors in a given political setting' (Sil and Katzenstein 2010: 22).

To reiterate, analytic eclecticism is about the integration of different paradigms and theories that are relevant to the issue at hand, which, in this case, is eThekwini's green and ecological infrastructure policy landscape. To do this, analytic eclecticism should produce a more thorough investigation of phenomena or issues (Teddlie and Tashakkori 2011).

Because of this, we need to assess the prevailing paradigm configuration within a problem area (Meissner 2017), as the first step of a more thorough investigation, by starting with an open-ended analysis and not the formulation of a hypothesis. Based on such an appraisal, one can move forwards in a meaningful way and apply analytic eclecticism where it matters. It will not only show what researchers and practitioners are missing, but it will also apply eclecticism as a foundation for various theoretical elements, in order to illuminate issues and perspectives (Meissner 2017), for practical reasons. In other words, analytic eclecticism provides an approach and perspective that will help to gain a deeper explanation and understanding of the issues that plague society.

2.4.2.2 Causation

The central element of PULSE[3] is the role of causation in paradigms and theories. Here, I will discuss the philosophical views of causes in the environmental and social sciences, based on the thinking of David Hume and Aristotle borrowed from the scholarly work of Milja Kurki and, to a certain extent, Richard Ned Lebow, and I will link these notions of causation to the agent-structure and the material-ideational axes mentioned above.

For Kurki (2008), positivism views causation in a strict and narrow manner. This understanding is based on the philosophical works of David Hume (1711–1776), who based the conception of cause on several assumptions. The first is that scholars link causal relations to regular patterns of occurrences and the study of the patterns of regularities. Secondly, causal relations are regularity relations of patterns of observables that rely on constant conjunctions. Thirdly, causal relations are determined by regularity. This means that, based on certain observed regularities, when one type of event occurs, then we often assume another type of episode will follow (in a probabilistic way, at least). Regular determinism is the fountainhead of prediction or logical deduction. Fourthly, causes refer to moving causes in that they are efficient causes that push and pull (Kurki 2006, 2008; Lebow 2014).

According to Kurki (2008: 6), 'These assumptions about the concept of cause are deeply embedded in the modern philosophy of science and social science…'. This has led to an empirical and positivist interpretation of causation. In addition, many natural and social scientists see this explanation as the only acceptable way of conducting a causal analysis. Because of the dominance of Hume's conceptualisation of cause in positivism, there have been repercussions for theory and theory development. The practical sense that is contained in theoretical assumptions has evaporated, so to speak, and a narrower approach to cause has taken root (Kurki 2008). The focus emphasises the instrumentality of causation and not so much its role and involvement in practice. Here we see how assumptions from a specific paradigm about causation can shift the attention onto the cause, at the expense of practice.

Enter Aristotle (384–322 B.C.E.), whose account of cause and causation was much broader and deeper than the later account of Hume. The Greek philosopher developed a typology of causes that does not focus on Hume's instrumentality of causes. Aristotle's classification includes the material, formal, agential or efficient and final causes. An example of a material cause is the marble, from which a sculptor shapes a statue. Matter is quite fundamental in any explanation and Aristotle saw it in the light of its 'indeterminate potentiality'. Matter is the cause of something through the provision of the material used to create objects. Without marble, an artist cannot sculpt a statue. Also of importance are the properties, or the substance, of the material, since these characteristics can enable or constrain how we shape it. The material is insignificant, when it is considered on its own: it has no intelligence and needs an action to become a statue. This brings us to formal causes, namely, the form, idea or essence of things. The formal cause of the statue would be the idea, image or shape thereof. These reside in the mind of the artist, with the formal cause being the pattern or form of something (Kurki 2008). According to Kurki (2008: 27) '...formal causes define and "actualise" material potentiality into things or substances'.

Agential or efficient causes are the primary sources of change. These foundations could be any 'agential mover', or an 'act of doing something' (Kurki 2008: 27), namely, a practice. That said, the efficient or agential cause of the statue is the sculptor or the act of sculpting, while the final cause is the purpose that guides the change. For instance, we walk and do other exercises to be healthy and by stating this, we assign a cause to the action (Kurki 2008).

Lebow (2014) has added a fifth type of cause, namely, inefficient causation, which rests on the notion of singular causation. According to Lebow (2014: 5–6, 36), when events take place:

> [w]e can construct causal narratives about [their] outcomes, but they cannot be explained or predicted by reference to prior generalisations or narratives. Nor do they allow us to predict future events. Singular cause refers to events that are causal but non-repetitive.

According to Lebow (2014), David Hume denied the existence of singular causes and, in the process, this restricts our thinking about non-material types of cause, such as norms, rules and emotions.

The typology of causes, as described above, is 'flexible and sensitive to pragmatic concerns of explanation' (Kurki 2008: 28). Unlike the Humean restrictive notion of cause, the different types of causes, as outlined by Aristotle (Kurki 2008) and Lebow (2014), bring into focus the intangible forces behind the cause, namely, the ideas, norms, principles and beliefs, as well as the paradigms and theories. These can be formal causes because they define the structure of social relations when constituting practices, in that they relate agents and their social roles to each other, and the meaning that is inherent in them. These ideational elements 'describe the rules and relations that define social positions and relationships, and hence', they are '"that according to which" social reality works' (Kurki 2006: 207). In a sense, researchers can see these non-material sources of cause 'constraining and enabling' other triggers (Kurki 2006).

Paradigms, Theories and Causation

Why are research paradigms and theories important when considering cause and causal relations? Since paradigms and theories are formal causes in society and they have 'constraining and enabling' effects on social relations, we are confronted with a deeper sense and level of causation in the social world than what we often recognise to be the case (e.g. observable patterns of behaviour) (Kurki 2006). By focusing on them as such, we broaden our notion of cause. In other words, by studying paradigms and theories, our ontological horizons are widened with respect to what constitutes cause, and we recognise the agential power of paradigms and theories.

Paradigms and theories also contain and explain the intentions and reasons for actors behaving in a certain way. Because paradigms and theories are causal, intentions and reasons are, by default, also causal '…in the sense that they signify a contributory cause that "for the sake of which" something is done' (Kurki 2006: 209). By focusing on the role and intention of paradigms and theories, it is possible that intentions and reasons are the wellspring of practices (i.e. daily actions) as well as policy actions, be it in the public or private sector (Meissner 2017).

Broadening the types of causes could help us to understand the various assumptions contained in theories in a more nuanced manner. For example, where a theoretical assumption gives a Humean explanation of something that causes another thing, another theoretical assumption could explain, in an Aristotelian way, the reason for the primary cause taking place in the first place.

For instance, in the South African water sector, water managers and researchers place much faith in Catchment Management Agencies (CMAs) to bring about a more sustainable way of managing the country's water (Stuart-Hill 2015; Meissner et al. 2017; Stuart-Hill and Meissner 2017). The argument behind this is that CMAs, as structures of rule, are less restrictive because they are a decentralised way of managing water resources, as opposed to the 'command and control' way of doing things under the old (Apartheid) dispensation. Catchment management agencies are closer to the public, are established in a more participatory manner, and therefore, they rely on a more inclusive governing model than the previous system (Rogers et al. 2000; Meissner et al. 2014, 2017; Meissner and Funke 2014).

Based on this description, it therefore follows that CMAs will be more effective, as structures of rule (Hobson and Seabrooke 2007), in governing water in their respective water management areas (Rogers et al. 2000; Stuart-Hill 2015). The restrictive Humean notion of cause is evident in that a move from the old command and control system to a more inclusive and participatory way of governing will be good, in terms of supplying more, and a better quality of, water for the people, the environment and future generations. If one brings in Lebow's (2014) notion of inefficient cause, then the narrative changes. The South African government plans to establish nine CMAs (Stuart-Hill 2015). Each CMA will be a singular cause event, in that it will differ in its demographic, economic, financial (material), management (structural), ideology and values (ideational), geographic, hydrological, climatological (material) and societal actor (agential) profile. Their establishment and functioning could therefore differ and there will be no generalisations or predictions on how efficient they will be in bringing about the possible sustainable management of water resources. We just do not know how successful they will be at managing water in a sustainable way, under certain circumstances.

AIMS

As we are unable to say with certainty that a specific structure of rule, for instance, catchment management agencies, will bring about positive changes in water resources management, how can we harness the eclectic notion of cause and causal relations as a PULSE[3] component? The Agential, Ideational, Material and Structural typology (AIMS) is a starting point (Sil and Katzenstein 2010). Although scholars devised this causal mechanism typology in IR to analyse world politics, it is suitable for other levels on which political activity occurs (Meissner et al. 2018). Causal mechanisms are not a specific level of analysis, but they are components of the philosophy of the natural and social sciences (Hedström and Ylikoski 2010; Meissner et al. 2018) and, as such, they are linked to scientific disciplines and not to the levels of analysis.

At any level, causal mechanisms are processes that are characterised by the interaction and relation of a system's parts, structure and environment (Hedström and Ylikoski 2010; Meissner et al. 2018) and that '...uncover the underlying social processes that connect inputs and outcomes' (Falleti and Lynch 2009: 19). In this sense, one can talk of 'complex combinations of causal conditions' (Jager 2016: 282). Kurki (2008: 233-234) spoke of:

> 'complexes of causes' after arguing that we should not '...seek...to define mechanisms in a fixed way [but] rather as a vague metaphor... In the light of the Aristotelian plural conception of ontology and causal powers, it could be argued that mechanisms are usefully thought of as the particular kinds of, often relatively stable, interactions that take place between certain types of causal forces. Mechanism explanations, then, can be seen as accounts of the processes of interaction between different elements that bring about given events or processes. On such a definition we can refer to various causal interactions or processes as mechanisms: from market mechanisms (not seen as a logically necessary system but made up of various socially embedded and positioned agents and structures coming together in certain ways) to mechanisms of discursive reproduction (for example, variously socially positioned and shaped strategies of media representations).'

Kurki (2008: 234) views this description as '...an open definition for an already vague term, but it also allows us to separate causes from mechanisms'. This implies that mechanisms do not define causes...' but give credence to the '...accounts of the interaction of causes...' through the notion of 'causal processes'.

Whatever the case may be, when returning to the AIMS typology and speaking of agential causes, I refer to the mechanisms or processes that are brought on by the actions of the actors (Kurki 2008; Meissner 2017, 2018), such as building a water infrastructure and purifying the water. Ideational causal mechanisms include perceptions, anticipation and ideologies, or the cognitive aspects that result in processes and action. Material mechanisms include the resources needed to implement activities, such as money, technology and human resources, while structural causal processes include policies, strategies and the rule of law (Kurki 2008; Sil and Katzenstein 2010; Meissner 2017, 2018). We should not assume that structural causal mechanisms are the opposite of agential causal processes, because 'structures are often revealed through agency' (Selby 2018: 339).

Through the AIMS technique, researchers and practitioners can get a more nuanced and deeper understanding of issues and their causal substance. This is because '[a] causal mechanism provides an explanatory account of observed results by describing the mediating process by which the target factor could have produced the effect' (Koslowski et al. 1989: 1317). In this respect, knowledge aims to understand phenomena and issues, while causal mechanisms provide an understanding of the 'What if?' questions (Hedström and Ylikoski 2010).

Causal mechanisms are important for understanding causation (p. 55), and vice versa (Falleti and Lynch 2009), in any field of study and at any level. Elster (1998: 45, cited in Guzzini 2011; Elster 2007: 36) conceptualised causal mechanisms as 'frequently occurring and easily recognisable causal patterns that are triggered under generally unknown conditions or with indeterminate conditions. They allow us to explain, but not to predict'. Guzzini (2011) argued that not all causal mechanisms are observable and observed. We can assign observable mechanisms to true causal statements (Elster 2007), but causal mechanisms are often 'hidden' (Guzzini 2011). Elster continued, by asserting that:

> To cite a cause is not enough: the causal mechanism must also be provided, or at least suggested. In everyday language, in good novels, in good historical writings, and in many social scientific analyses, the mechanism is not explicitly cited. Instead, it is suggested by the way in which the cause is described.

As Goertz and Mahoney (2010) observed, when people see data relating to the association between two variables, they often request additional information on the mechanisms, before concluding that the association is causal in nature.

Causal mechanisms describe the relationships or actions among the units of analysis, or the cases under investigation, within a study (Falleti and Lynch 2009, cited in Guzzini 2011). Causal mechanisms show why something has happened (Guzzini 2011) by providing 'an explanatory account of observed results by describing the mediating process by which the target factor could have produced the effect' (Koslowski et al. 1989: 1317). Should researchers reduce causal mechanisms to variables, they would be denying the 'possibility of a combinational or configurational explanation', with the interpretation being part of the methodology of causal mechanisms (Guzzini 2011: 333). For Falleti and Lynch (2009: 1143), causal mechanisms are transferable concepts that explain how and why a contextualised cause primes certain outcomes. Their study 'defines context as the relevant aspect of a setting in which an array of initial conditions leads to an outcome of a defined scope and meaning, via causal mechanisms' (Falleti and Lynch 2009: 1143).

Researchers therefore need to be attentive to the interplay between context and causal mechanisms, irrespective of the method that they employ, be it small-sample, formal, statistical or interpretive (Falleti and Lynch 2009). Hedström and Ylikoski (2010) affirmed this when they said that causal mechanisms are the processes that are characterised by the relations and interactions of a system's parts, its structure and environment, or context. In his definition of a causal mechanism, Rueschemeyer (2009: 21) noted that it is '...a condition, a relation, or a process that brings about certain events and states'. These occasions and conditions play out in specific contexts. In this regard, Rueschemeyer (2009), just like Falleti and Lynch (2009), highlighted two other characteristics of causal mechanisms, namely, condition and relation.

When reflecting on these definitions, we see that causal mechanisms institute both the processes of causation and the 'things' that constitute change, and because of this, causal mechanisms show not only why something is happening (Guzzini 2011), but also that something is happening and how it is taking place (Koslowski et al. 1989; Meissner et al. 2018). In a critique of the concept of causal mechanism, Gerring (2010: 1500–1501) asserted that:

> In prior work, I demonstrate that "causal mechanism" may refer to (a) the pathway or process by which an effect is produced, (b) a micro-level (microfoundational) explanation for a causal phenomenon, (c) a difficult-to-observe causal factor, (d) an easy-to-observe causal factor, (e) a context dependent (tightly bounded or middle-range) explanation, (f) a universal (i.e. highly general explanation), (g) an explanation that presumes probabilistic, and perhaps highly contingent, causal relations, (h) an explanation built on phenomena that exhibit law-like regularities, (i) a technique of analysis based on qualitative or case study evidence, and/or (j) a theory couched in formal mathematical models.

Indeed, Goertz and Mahoney (2010) stated that the literature around causal mechanisms contains heterogeneous definitions, which, in some way, affirm Gerring's (2010) conclusion. Considering what Gerring (2010) had to say about the various meanings or typology of 'causal mechanism', my argument in this regard is that causal mechanisms are all of these elements, and more, depending on the paradigmatic plain from which a researcher views the world. As was rightfully pointed out by Gerring (2010), the reason for this is that several of these definitions are contradictory, because causal mechanisms cannot simultaneously be difficult to see and easy to observe (Guzzini 2011). Taking contradiction into consideration, Schaffer (2000) contended that one can 'wire' causal mechanisms in many ways. The various 'wirings' depend on the researcher's training, which constitutes his or her paradigmatic stance, with respect to research objects and the context in which causal mechanisms are enacted. For a person wearing a positivist lens, it is easy to see the direct connection between a clogged sewer line and sewerage pollution spilling into a river or stream. Nevertheless, to uncover some of the *hidden* causal mechanisms, one might want to look into the other 'causes', which may constitute a blocked sewer. What is the source of the blockage; household litter or disused diapers? How did the source of the blockage end up in the sewer line in the first place? Is it because the consumer of the diaper did not know that disposing of it through the sewer system might cause a blockage? These are questions that not only highlight the 'difficult-to-observe causal factors' but that also need to be asked to ascertain how causal mechanisms arise (Imai et al. 2011). To answer these questions, there needs to be an interplay between ontology, epistemology and methodology, from various paradigmatic perspectives. On the meta-theoretical functions of causal mechanisms, Turner (2013: 536) went on to say that:

The epistemic function of [causal] mechanisms is this: they are free-floating intelligibility-producing devices that fill in between inputs and outputs in a way that is more satisfying—more understanding-producing—than "predictors which cannot be excluded a priori as "no affect" relations.

When we identify a causal mechanism, the integration of an isolated piece of causal knowledge to a much larger body of knowledge takes place. This larger body of knowledge assists in answering numerous follow-up questions about the conditions that are necessary for the causal dependence to occur, which I have already referred to above. Hedström and Ylikoski (2010: 10) formulate this question in the following way: '…what are the necessary background conditions and what are the possible intervening factors that have to be absent for the effect to be present?' By asking and answering this question, we expand our understanding of the causal mechanisms operating within a specific context (Hedström and Ylikoski 2010).

These notions of causal mechanisms and their functions shed light on the nature of a causal mechanism as an explanatory account, condition, relation and process. In a previous publication (Meissner et al. 2018), these descriptions were utilised to identify the nature of specific causal mechanisms that were contained in the face-to-face interviews that we conducted in two municipalities on water security, namely, the eThekwini Metropolitan and Sekhukhune District Municipalities. This enabled the research team to isolate causal variables that influence people's perceptions of what constitutes water security, at a local government level, and to see whether their perceptions are positive or negative and characterised by either a short- or long-term view of water (in) security (Meissner et al. 2018, 2019). Such an analysis could shed further light on what practitioners could do to change negative situations into more positive ones. I will be using the same procedure to identity the causal mechanisms expressed by the respondents that were interviewed during the green and ecological policy landscape research project.

2.4.2.3 Theories for Practice

For analytic eclecticism to progress in a meaningful manner, one needs a repertoire of theories (see Table 2.2), because eclectic studies utilise various theories to

Table 2.2 Repertoire of theories

Agential power	Interactive governance theory (Governability)	Normative commensalism
Ambiguity theory of leadership	Interest group corporatism	Political ecology or Green politics
Complexity theory	Interest group pluralism	Risk society
Cultural theory of International Relations	Marxism	Social constructivism
Everyday international political economy	Modernity	Strategic adaptive management or adaptive management
Feminism	Neo-liberalism (Liberal pluralism)	Theory of social learning and policy paradigms
Hydrosocial contract theory	Neo-realism (Realism)	

analyse problems (Cornut 2015). As already stated, not one single theory can explain everything or even an event. We need alternative approaches and traditions in order to construct a collective understanding (Hayes and James 2014). The late Elinor Ostrom (2007: 15181) propagated the '...serious study of complex, multivariable, nonlinear, cross-scale and changing systems,' instead of relying on '...simple, predictive models of social-ecological systems...'

In order to take up Ostrom's (2007) call, we need a plurality of theories and not just a few or, at best, one, like complexity thinking, to constitute a collective understanding. Table 2.2 contains a list of the interpretivist, critical, postpositivist and positivist theories that are applicable to the practicalities of the governance and politics of society and the natural environment. This list is by no means exhaustive, since it identifies only the theories that I have come across during my research career. Ideally, one should be able to expand on it and refine it. What's more, practitioners do not adhere to one theory when executing practices or policies. In this light, theories 'describe modes of thinking and not so much objective patterns of behaviour' (Hayes and James 2014: 408). We also need a plurality of theories, because practitioners are likely to move from one approach to another and, in so doing, they rely on a variety of perspectives. This will inevitably generate contradictions when issues are investigated (Hayes and James 2014).

I would now like to discuss the pluralist philosophy that lies behind my list of theories. My opinion, which is in agreement with that of Lebow (2011: 1225–1226, cited in Rengger 2015), is that: 'Pluralism must be valued not only as an end in its own right, but also as an effective means of encouraging dialogue across different approaches, something from which we have something to learn'. This means that the pluralistic use of theories brings about a variety of explanatory forms of what is happening in the world (Rengger 2015). This claim is also in line with the ethos of analytic eclecticism (Meissner 2017). If we look closely at Lebow's (2011) claim of pluralism in scholarship and theory, what is this 'something' that we might learn? I am of the view that this 'something' goes beyond the fact that there is a plurality of theories for explaining and understanding social phenomena. For me, this 'something' is the inherent value that is contained in each theory, which widens our knowledge horizons and does not restrict us to mono-paradigmatic and mono-theoretical explanations. In other words, as far as I am concerned, we will gain more knowledge with which we can explain and deepen our understanding of issues, and with which we can open the dialogue between disciplines and fields of study, which is something we can then utilise in the service of assisting practitioners and the public. For me, this service of scientists is the ultimate 'something' that we will learn, because it will take us further away from our mono-theoretical convictions and prevent us from blindly following the so-called sages, or gurus, who put forward theories for others to follow blindly, without questioning their basic assumptions (Meissner 2017).

The dialogue that Lebow (2011) refers to can take several forms, and it is not just the 'genuine and reasoned communication between equals' (Rengger 2015: 3). Therefore, if the dialogue can take many forms, I would like to start my own by continuing my support of paradigmatic and, particularly, theoretical pluralism, as opposed to empiricism and the blind following of the sages, both of which are easily followed in the academia and in practice. I need to qualify this by saying that empiricism does have a role to play in research, especially when investigating the biophysical quality of water resources, as it is expressed in their pH, conductivity and turbidity, as well as the impact of low quality water on human and ecosystem health.

However, when we deal with the social aspects of water resources, empiricism has severe limitations with respect to being able to adequately explain what is happening. Mearsheimer and Walt (2013: 427) went so far as to say that simplistic hypothesis testing is bad, because it emphasises the discovery of empirical regularities which, in their view, emphasises that simplistic hypothesis testing over theory creation, testing and utilisation 'is a mistake'. Their reasoning is that '…insufficient attention to theory leads to mis specified models or misleading measures of concepts'. Consequently, simplistic hypothesis testing, over theory creation, testing and utilisation, is widening the chasm between the ivory tower of scholarship and the real world (Mearsheimer and Walt 2013). A situation like this does not bode well for the dialogue that was mentioned earlier and the service that scientists are supposed to render to practitioners and the public.

In support of my repertoire of theories, it is necessary that scholars must have a good and solid grasp of theory and, just as importantly, they must use theories to guide their research (Mearsheimer and Walt 2013) and influence debates. Regarding my selection of theories, Mearsheimer and Walt (2013: 430) argued that: '…many kinds of theory…can be useful for helping us understand how [society] works. In our view, a diverse theoretical ecosystem is preferable to an intellectual monoculture'. They went further by saying that '…we believe progress in the field depends primarily on developing and using theory in sophisticated ways' (Mearsheimer and Walt 2013: 430).

With that being said, researchers need to consider that it is not only about the presentation of a collection, but also about the utilisation of theories. In my opinion, using analytic eclecticism will be a good starting point in the sophisticated use of the theory repertoire. The purpose of the collection is not merely to club together a few theories, but to identify as many concepts and assumptions as possible, which one theory may omit and another theory may highlight (Meissner 2017).

There is another reason for presenting this list, namely, to prevent PULSE[3] from becoming a panacea. Employing these theories to policies, programmes and practices produces 'messy analyses'. What I mean by this is that by utilising different theories all at once in one case study, or in a set of cases, different interpretations of the context will come to light. Such an analysis will come across as unstructured, because jumping from one theoretical explanation to another can place high demands on the reader. What is more, in order to provide a better explanation of events or a situation, we must abandon single ontological factors (ideas, material concerns, agents and structures) that may have an influence on the processes. Causal factors are not independent. By incorporating more than one theoretical explanation and understanding, we can ask more open and multi-causal questions, which requires that we move away from theoretical reductionist explanations (Meissner 2017).

In order to identify reductionist tendencies in a text, we also need to consider a wider classification of theories, namely, the problem-solving and critical perspectives.

2.4.2.4 Problem-Solving and Critical Theories

According to Cornut (2015: 12), the analytical eclecticism outlined by Sil and Katzenstein is 'rather fuzzy, and it is not easy to understand what epistemological or methodological criteria are used to adjudicate by a jury of peers in an eclectic "court"'. The steps that I outline below are an important way forward in making the ethos of analytic eclecticism and the repertoire of theories more robust as methodological considerations. A simple classification, along the lines of problem-solving and critical theories, will suffice.

Problem-solving theories explain reality as it is and then suggest ways and means of solving the problems that are encountered by scholars. These theories have a positivist inclination (Cox 1981; Cox and Sinclair 1996). Critical theories describe the world and its structures, and, in return, they suggest how they can be modified. They are not positivist or postpositivist, but they fit the critical theory paradigm (Cox 1981; Cox and Sinclair 1996).

Problem-solving and critical theories can further be categorised into grand, middle-range and mathematically oriented theories (Mearsheimer and Walt 2013). The five theory types will be sufficient for the purposes of this framework, since the field of study that is covered here involves both the natural and social sciences. We find the five forms in both sciences, with mathematically oriented theories being more common in the natural sciences than in the social sciences. The classification will initially give an indication of the factors, actors and variables that are highlighted by a theory to explain an issue or phenomenon.

Grand theories give explanations of broad patterns of behaviour. In the discipline of International Relations, these are theories, such as realism or liberalism, which explain state and non-state behaviour in its broadest sense in world affairs. Middle-range theories, on the other hand, spotlight more narrowly defined issues and phenomena, such as management, collective and individual psychology, coercion and so on (Mearsheimer and Walt 2013).

Programme theory explains how practitioners understand the contribution of an intervention, such as a project, programme, policy, strategy or practice, to a number or chain, of results that produce intended or real impacts (Funnel and Rogers 2011). The theory focuses on the logical building blocks of a programme by highlighting its processes (Kadiyala et al. 2009). This perspective has two components, namely, a theory of change and a theory of action:

> The theory of change is about the central processes or drivers by which change comes about for individuals, groups, or communities [such as] social processes, physical processes, and economic processes. The theory of change could derive from a formal, research-based theory or an unstated, tacit understanding about how things work. The theory of action explains

how programmes or other interventions are constructed to activate these theories of change. (Funnel and Rogers 2011)

This theory type can include both positive (beneficial) and negative (detrimental) impacts, and it can indicate other factors that contribute to the cause of impacts, such as the context and other policies, projects and programmes. Programme theory, as an epistemological perspective, highlights causal mechanisms, such as norms, that lead to behavioural changes and that are instrumental in delivering the intended results of programmes. Programme theory can assist practitioners as a tool for evaluating the impact of a programme (Funnel and Rogers 2011), and I will use it as an epistemological device throughout.

Mathematically oriented theories use the language of mathematics, as opposed to the other two sub-types, which utilise a written language (Mearsheimer and Walt 2013). Examples of mathematically oriented theories include algorithms in computers science and software programming, equations in physics and quantum physics (Gutman et al. 2016), equations of motion in super space in M theory (Howe et al. 1997) and M theory as a matrix model (Banks et al. 1997). According to Ye (2009), M theory is characterised by a super-symmetry string theory, which pays attention to mathematical forms. He goes on to describe it as 'the mainstream of contemporary theoretical physics, M theory is an overall theory embracing not only quantum gravity but also matter and force' (Ye 2009). In other words, it is a perspective of theoretical physics (Nakayama 2010).

To highlight the operation of these theory types, I will briefly consider the case of climate change. Liberal institutionalism is a grand, problem-solving theory that explains inter-state behaviour and that proposes solutions to ameliorate various problems at numerous levels. It also puts forwards the construction of structures, such as the United Nations Framework Convention on Climate Change (UNFCCC) and the Intergovernmental Panel on Climate Change (IPCC), for supporting the global response to climate change through a variety of activities and practices (UNFCCC 2020). A middle-range theory would explain certain scientific measures for dealing with climate change and its associated problems, such as adaptation and mitigation, while a critical theory would, for instance, explain the issue in terms of wealth inequality and how this perpetuates the situation and its associated solutions. This theory then suggests how various actors could change to bring about more equality between states and, by so doing, it could influence the debate between developed and developing countries regarding the matter. Mathematically oriented theories play a critical role in developing climate change models, when scientists develop algorithms to run their models.

2.5 Conclusion

Paradigms and theories that are linked with an action help to guide our thinking of the unexplained and to bring us closer to an explanation and understanding of the mysterious; they are a means of uncovering that which is invisible. With multiple paradigms and theories, scholars can extend their investigations to include other theoretical phenomena relating to the issues under investigation. Therefore, paradigms and theories go hand-in-hand in explaining and understanding natural and social scientific concepts. Furthermore, as cognitive instruments, paradigms and theories assist academics in their work of defining and conceptualising a variety of phenomena. Paradigms provide the background knowledge or frames that allow scholars to test and measure their theories. Because of this quality, a paradigm could contain several theories within its structure. The purpose of theories is to explain phenomena, based on certain criteria. That said, paradigms and theories operate concurrently, although their differences provide the mainstay of science. With limited information in hand, theories extend the ability of our cognitive processes to explain that which is often unexplained. Paradigms and theories inform human practices, from conducting research, to developing and implementing green and ecological infrastructure projects. Just as a textile loom produces multi-coloured fabrics and a variety of designs, our minds also bring forth multiple explanations of how occurrences are connected in the tangible world (Myers 1887; Sherrington 1942; Meissner 2017). To reiterate, we produce multiple theories to not only solve mysteries and explain or understand our multiple natural and social environments, but also to assist us in our daily lives. Theories are inherent to the never-ending cognitive processes by which we recognise, experience and think about things, as well as how we understand them and act (Meissner 2017).

PULSE^3 consists of several components, namely, the paradigm assessment, the repertoire of theories, causal mechanisms, as well as the problem-solving and critical theory typology. It gives a pragmatic perspective on the utilisation and role of paradigms and theories in conducting analyses. The main objective of the analytical approach is to study the knowledge content of situations and issues that are expressed in written form. The framework utilises an open-ended problem analysis, which means that it does not employ the traditional hypothesis formulation and analysis methods. The hypothesis testing of complex social matters is frequently the victim of active substantiation, when specialists identify and use these variables to validate the hypothesis and discard the evidence that contradicts it (Meissner 2017). They then have the tendency to predict outcomes that are based on pre-identified variables that might have only a small influence, or no bearing at all, on the problem at hand. This means that no scientist can claim to offer an objective and unbiased situational analysis.

For PULSE^3, the purpose of science is not the discovery of absolute truths, but a commitment to constant critique by means of open-ended problem formulation. An open-ended analysis starts with collecting information on a phenomenon and does not attempt to prove or disprove a hypothesis that is linked to the occurrence. Rather than predicting, in order to assume an unobserved event, an open-ended inquiry embraces the complexity of the event and the complex interactions of its different aspects. The findings need to span both the practical problems and the scientific and research debates. In order to do this, an open-ended investigation shows how the causal forces from different perspectives co-exist as part of a complex framework of understanding. At the same time, open-endedness helps to make sense of social situations and phenomena, and to help practitioners gain a deeper insight into their inherent complexities. In other words, an open-ended analysis explains 'what makes things tick'. In the following chapter, I will apply PULSE[3] to the matter of eThekwini's green and ecological infrastructure policy landscape.

References

Alderson P (1998) The importance of theories in health care. BMJ 317:1007–1010

Anderson P (1999) Perspective: complexity theory and organization science. Organ Sci 10(3):216–232

Andreotti V (2006) 'Theory without practice is idle, practice without theory is blind': the potential contributions of post-colonial theory to development education. Dev Educ J 12(3):7–10

Angen MJ (2000) Evaluating interpretive inquiry: reviewing the validity debate and opening the dialogue. Qual Health Res 10(3):378–395

Ashton P, Turton A (2009) Water and security in sub-Saharan Africa: emerging concepts and their implications for effective water resource management in the Southern African Region. In: Brauch HG, Spring UO, Grin J, Mesjasz C, Kameri-Mbote P, Behera NC, Chourou B, Krummenacher H (eds) Facing global environmental change: environmental, human, energy, food, health and water security concepts. Springer, Berlin

Aviation Safety Council (ASC) (2002) Aircraft accident report: crashed on a partially closed runway during takeoff, Singapore Airlines Flight 006, Boeing 747-400, 9 V-SPK, CKS Airport, Toayuan, Taiwan, October 31, 2000. Aviation Safety Council, Taipei, Taiwan

Bäckstrand K (2004) Science, uncertainty and participation in global environmental governance. Environ Polit 13(3):650–656

Banks T, Fischler W, Shenker SH, Susskind L (1997) M theory as a matrix model: a conjecture. Phys Rev D 5(8):5112

Beach D, Kaas JG (2019) The great divides: Incommensurability, the impossibility of mixed-methodology, and what to do about it. Int Stud Rev viaa016

Béland D, Cox RH (2013) The politics of policy paradigms. Gov Int J Policy Adm Inst 26(2):193–195

Bennett A (2010) From analytic eclecticism to structured pluralism. Qual Multi-Method Res 8(2):6–9

Berger PL, Luckmann T (1966) The social construction of reality: a treatise in the sociology of knowledge. Anchor Books, Garden City

Bernal DD (2002) Critical race theory, latino critical theory, and critical race-gendered epistemologies: recognising students of color as holders and creators of knowledge. Qual Inq 9(1):105–126

Beyer AC (2015) Insights from para-psychology for international relations. Peace Rev 27(4):484–491

Blyth M (2013) Paradigms and paradox: the politics of economic ideas in two moments of crisis. Gov Int J Policy, Adm Inst 26(2):197–215

Bosch AWM (2019) Regarding visuality: understanding apartheid photography as a tool for resistance through the visual turn in IR. Unpublished Master of Arts Thesis, Department of International Relations, Leiden University, Leiden, The Netherlands

Braun K, Kropp C (2010) Beyond speaking truth? Institutional responses to uncertainty in scientific governance. Sci Technol Human Values 35(6):771–782

Brittain V (1988) Hidden lives, hidden deaths: South Africa's crippling of a continent. Faber and Faber, London

Bryman A (2006) Of methods and methodology. Qual Res Organ Manag Int J 3(2):159–168

Bulkeley H (2005) Reconfiguring environmental governance: towards a politics of scales and networks. Polit Geogr 24(8):875–902

Burrell G, Morgan G (1979) Sociological paradigm and organisational analysis: Elements of the sociology of corporate life. Ashgate, Aldershot

Byrne D (1998) Complexity theory and the social sciences. Routledge, London

Byrne D (2002) Complexity theory and the social sciences: an introduction. Routledge, London

Carstensen MB, Matthijs M (2018) Of paradigms and power: British economic policy making since Thatcher. Governance 31:431–447

Chan S (1989) The strategist in isolation: the case of Deon Geldenhuys and the South African military. Def Anal 5(4):374–376

Chernoff F (2007) Theory and metatheory in international relations: concepts and contending accounts. Palgrave Macmillan, New York

Christians CG (2018) Ethics and politics in qualitative research. In: Denzin NK, Lincoln YS (eds) The sage handbook of qualitative research, 5th edn. SAGE Publications, Thousand Oaks, CA

Cilliers P (2000) What can we learn from the theory of complexity? Emergence 2(1):23–33

Cilliers P (2005) Complexity, deconstruction and relativism. Theory, Cult Soc 22(5):255–267

Cloete F, de Coning C (2018) Models, theories and paradigms for analysing public policy. In: Cloete F, de Coning C, Wissink H, Rabie B (eds) Improving public policy for good governance, 4th edn. Van Schaik Publishers, Pretoria

Coicaud J-M (2014) Emotions and passions in the discipline if international relations. Jpn J Polit Sci 15(3):485–513

Cornut J (2015) Analytic Eclecticism in Practice: a method for combining international relations theories. Int Stud Perspect 16(1):50–66

Cox RW (1981) Social forces, states and world orders: beyond international relations theory. Millen J Int Stud 10(2):126–155

Cox RW, Sinclair TJ (1996) Approaches to world order. Cambridge University Press, Cambridge

Creswell JW (2007) Qualitative inquiry and research design: choosing among five approaches. Sage, Thousand Oaks, CA

Crewe K, Forsyth A (2003) LandSCAPES: a typology of approaches to landscape architecture. Landscape J 22(1):37–53

Cronbach L (1975) Beyond the two disciplines of scientific psychology. Am Psychol 30:116–127

Desmond C (1987) 'South Africa's own sanctions', and Reginald Herbold Green 'Sanctions and the SADCC economies' in third world affairs 1987. Third World Foundation, London

Du Plessis A (2000) Charting the course of the water discourse through the fog of international relations theory. In: Solomon H, Turton A (eds) Water wars: enduring myth or impending reality. Durban, The African Centre for the Constructive Resolution of Disputes

Eisner EW (1990) The meaning of alternative paradigms for practice. In: Guba EG (ed) The paradigm dialog. Sage publications, Newbury Park, CA

Eisner EW (2017) The enlightened eye: qualitative inquiry and the enhancement of educational practice. Teachers College Press, New York

Falleti TG, Lynch JF (2009) Context and causal mechanisms in political analysis. Comp Polit Stud 42(9):1143–1166

Ferguson YH (2003) Illusions of superpower. Asian J Polit Sci 11(2):7–20

Franke U, Weber R (2011) At the Papini Hotel–on pragmatism in the study of international relations. Eur J Int Relat 18(4):669–691

Frey D (1981) The effect of negative feedback about oneself and cost of information on preferences for information about the source of this feedback. J Exp Soc Psychol 17:42–50

Friedrichs J (2009) From positivist pretence to pragmatic practice varieties of pragmatic methodology in IR scholarship. Int Stud Rev 11(3):645–648

Friedrichs J, Kratochwil F (2009) On acting and knowing: How pragmatism can advance international relations research and methodology. Int Org 63(4):701–731

Funnell SC, Rogers PJ (2011) Purposeful program theory: effective use of theories of change and logic models. Wiley, San Francisco

Gay G (1979) The continuing quest for rationality in curriculum practice. Educ Leadersh 37(2):178–183

Geldenhuys D (1989) Notes on the risks of policy analysis. Politikon 16(2):86–88

Geldenhuys D (2004) Deviant conduct in world politics. Palgrave Macmillan, Houndmills

Giddens A (1984) The constitution of society. University of California Press, Berkeley

Gillings A (2010) How earth made use: water. BBC, London

Gioia DA, Pitre E (1990) Multiparadigm perspective on theory building. Acad Manag Rev 15(4):584–602

Giroux HA (1982) Theory and resistance in education: a pedagogy for the opposition. Bergin and Garvey, Boston

Golafshani N (2003) Understanding reliability and validity in qualitative research. Qual Rep 8(4):597–607

Grant JA (2018) Agential construction and change in world politics. Int Stud Rev 20:255–263

Grenstad G (2007) Causal complexity and party preference. Eur J Polit Res 46:121–149

Grover R, Glazier J (1986) A conceptual framework for theory building in library and information science. Libr Inf Sci Res 8:227–242

Guba E (2002) Comments on some basic Qigong concepts. Taijiquan Qigong J 8(2)

Guba EG (1990) The alternative paradigm dialog. In: Guba EG (ed) The alternative paradigm dialog. Sage publications, Newbury Park, CA

Guba EG (1996) What happened to me on the road to Damascus. In: Heshusius L, Ballard (eds) From positivism to interpretivism and beyond: tales of transformation K. in educational and social research. Teachers College Press, New York

Guba EG, Lincoln YS (1985) Naturalistic inquiry. Sage, Newbury, CA

Guba EG, Lincoln YS (1994) Competing paradigms in qualitative research. In: Denzin NK, Lincoln YS (eds) Handbook of qualitative research, 1st edn. Sage Publications, Thousand Oaks, CA

Guba EG, Lincoln YS (2005) Paradigmatic controversies, contradictions, and emerging confluences. In: Denzin NK, Lincoln YS (eds) The sage handbook of qualitative research, 3rd edn. Sage, Thousand Oaks, CA

Gutman I, Gültekin I, Şahin B (2016) On Merrifield-Simmons index of molecular graphs. Kragujevac J Sci 38:83–95

Haas PM (2010a) Introduction. Qual Multi-Method Res 8(2):5–6

Haas PM (2010b) Practicing analytic eclecticism. Qual Multi-method Res 8(2):9–14

Hall PA (1993) Policy paradigms, social learning, and the state: the case of economic policymaking in Britain. Comp Polit 25(3):275–296

Hanlon J (1986) Beggar your neighbours: apartheid power in Southern Africa. Indiana University Press, Bloomington, Indiana

Hart SL (1971) Axiology-theory of values. Research 32(1):29–41

Hayes J, James P (2014) Theory as thought: Britain and German unification. Secur Stud 23:399–429

Hedström P, Ylikoski P (2010) Causal mechanisms in the social sciences. Ann Rev Sociol 36:49–67

Heron J, Reason P (1997) A participatory inquiry paradigm. Qual Inq 3:274–294

Heshusius L (1994) Freeing ourselves from objectivity: managing subjectivity or turning toward a participatory mode of consciousness? Educ Res 23(3):15–22

Hirschman AO (1970) The search for paradigms as a hindrance to understanding. World Polit 22(3):329–343

Hobson JM, Seabrooke L (2007) Everyday IPE: revealing everyday forms of change in the world economy. In: Hobson JM, Seabrooke L (eds) Everyday politics of the world economy. Cambridge University Press, Cambridge

Holton B, Pyszczynski T (1989) Biased information search in the interpersonal domain. Pers Soc Psychol Bull 15(1):42–51

Howe PS, Sezgin E, West PC (1997) Covariant field equations of the M-theory five-brane. Phys Lett B 399(1–2):49–59

Jacobs IM, Nienaber S (2011) Waters without borders: Transboundary water governance and the role of the 'transdisciplinary individual' in Southern Africa. Water SA 37(5):665–678

Jager NW (2016) Transboundary cooperation in European water governance—a set-theoretic analysis of international rivers basins. Eur Policy Gov 26:278–291

James NR (2003) The theoretical imperative: unavoidable explication. Asian J Polit Sci 11(2):7–20

Johnson RB, Onwuegbuzie AJ (2004) Mixed methods research: a research paradigm whose time has come. Educ Res 33(7):14–26

Johnston L (1996) Resisting change: information-seeking and the stereotype change. Eur J Soc Psychol 26:799–825

Jolly E, Chang LJ (2019) The flatland fallacy: moving beyond low-dimensional thinking. Top Cogn Sci 11:433–454

Jonas E, Schulz-Hardt S, Frey D, Thelen N (2001) Confirmation bias in sequential information search after preliminary decisions: an expansion of dissonance theoretical research on selective exposure to information. J Pers Soc Psychol 80(4):557–571

Kadiyala S, Rawat R, Roopnaraine T, Babirye F, Ochai R (2009) Applying a programme theory framework to improve livelihood interventions integrated with HIV care and treatment programmes. J Dev Effectiveness 1(4):470–491

Katzenstein PJ (1976) International relations and domestic structures: foreign economic policies of advanced industrial states. Int Org 30(1):1–45

Kerlinger FN (1986) Foundation of behavioural research, 3rd edn. Holt, Rienhart and Winston, New York

Kilgore DW (2001) Critical and postmodern perspectives in learning. In: Merriam S (ed) The new update of education theory: new directions in adult and continuing education. Jossey-Bass, San Francisco

Klotz A (1995) Norms reconstituting interests: global racial equality and U.S. sanction against South Africa. Int Organ 49(3):451–478

Koh K (2013) Theory-to-research-to-theory strategy: a research-based expansion of radical change theory. Lib Inf Sci Res 35:33–40

Komisarczuk P, Welch I (2006) A board game for teaching internet engineering. Aust Comput Educ (ACE) 52:117–123

Kooiman J, Bavinck M (2005) The governance perspective. In: Kooiman J, Bavinck M, Jentoft S, Pullin R (eds) Fish for life: interactive governance for fisheries. Amsterdam University Press, Amsterdam

Koslowski B, Okagaki L, Lorenz C, Umbach D (1989) When covariation is not enough: the role of causal mechanism, sampling method, and sample size in causal reasoning. Child Dev 60(6):1316–1327

Kuhn TS (1962) The structure of scientific revolutions. University of Chicago Press, Chicago

Kurki M (2006) Cause of a divided discipline: rethinking the concept of cause in international relations. Rev Int Stud 32(2):189–216

Kurki M (2008) Causation in international relations: reclaiming causal analysis. Cambridge University Press, Cambridge

Kurki M, Wight C (2013) International relations and social science. In: Dunne T, Kurki M, Smith S (eds) International relations theories: discipline and diversity, 3rd edn. Oxford University Press, Oxford

Lake DA (2011) Why "isms" are evil: theory, epistemology, and academic sects as impediments to understanding and progress. Int Stud Quart 55:465–480

Laudan L (1977) Progress and its problems: towards a theory of scientific growth. University of California Press, Berkley, CA

Lawrence RJ, Depres C (2004) Introduction: futures of transdisciplinarity. Futures 36:397–405

Lebow RN (2014) Constructing cause in international relations. Cambridge University Press, Cambridge

Lebow RN (2011) Review article: philosophy of science. Int Aff 87(5):1219–1228

Lewin K (1951) Field theory in social sciences. Harper Row, New York

Lhabitant F-S (2004) Hedge funds: quantitative insights. Wiley, Chichester, UK

Lincoln YS, Lynham SA, Guba EG (2011) Paradigmatic controversies, contradictions, and emerging confluences, revisited. In: Denzin NK, Lincoln YS (eds) The SAGE handbook of qualitative research, 4th edn. Sage, Thousand Oaks, CA

Lincoln YS, Lynham SA, Guba EG (2018) Paradigmatic controversies, contradictions, and emerging confluences, revisited. In: Denzin NK, Lincoln YS (eds) The SAGE handbook of qualitative research, 5th edn. Sage, Thousand Oaks, CA

Lloyd HA (2017) Theory without practice is empty; practice without theory is blind: the inherent inseparability of doctrine and skills. In: Edwards LH (Comp.). The Doctrine Skills Divide: Legal Education's Self-Inflicted Wound. Carolina Academic Press, Durham

Lundgren SR, Prislin R (1998) Motivated cognitive processing and attitude change. Pers Soc Psychol Bull 24:715–726

Lynham SA (2002) The general method of theory-building research in applied disciplines. Adv Dev Hum Res 4(3):221–241

Malacalza B (2020) The politics of aid from the perspective if international relations theories. In: Olivié I, Pérez A (eds) Aid power and politics. Routledge, London

Marks MJ, Fraley RC (2006) Confirmation bias and the sexual double standard. Sex Roles 54(1/2):19–26

Max-Neef MA (2005) Foundations of transdisciplinarity. Ecol Econ 52:5–16

McGann K (2008) The story of maths: the language of the universe. BBC Four Documentary, London

McIntosh C (2015) Theory across time: the privileging of time-less theory in international relations. Int Theory 7(S3):464–500

Mearsheimer JJ, Walt SM (2013) Leaving theory behind: why simplistic hypothesis testing is bad of international relations. Eur J Int Relat 19(3):427–457

Meissner R (2004) The transnational role and involvement of interest groups in water politics: a comparative analysis of selected Southern Africa case studies. University of Pretoria, Faculty of Humanities, D.Phil. Dissertation

Meissner R (2005) Interest groups and the proposed Epupa Dam: towards a theory of water politics. Politeia 24(3):354–369

Meissner R (2014) A critical analysis of research paradigms in a subset of marine and maritime scholarly thought. In: Funke N, Claassen M, Meissner R, Nortje K (eds) Reflections on the state of research and technology in South Africa's marine and maritime sectors. Council for Scientific and Industrial Research, Pretoria

Meissner R (2016a) Paradigms and theories in water governance: the case of south Africa's national water resource strategy. Water SA 42(1):1–10

Meissner R (2016b) The relevance of social theory in the practice of environmental management. Sci Eng Ethics 22:1345–1360

Meissner R (2016c) Hydropolitics, interest groups and governance: the case of the proposed Epupa Dam. Springer, Dordrecht

Meissner R (2017) Paradigms and theories influencing water policies in the South African and international water sectors: PULSE[3], a framework for policy analysis. Springer International Publishing, Cham, Switzerland

Meissner R (2018) Ocean governance for human health and the role of the social sciences. Lancet Planet Health 2(7):e275–e276

Meissner R (2019) Towards an individual-centred water security theory. In: Meissner R, Funke N, Nortje K, Steyn M (eds) Understanding water security at local government level in South Africa. Palgrave Macmillan, London

Meissner R, Funke N (2014) The politics of establishing catchment management agencies in South Africa: the case of the Breede-overberg catchment management agency. In: Huitema D, Meijerink S (eds) The politics of river basin organisations: coalitions, institutional design choices and consequences. Edward Elgar Publishing, Cheltenham

Meissner R, Jacobs I (2016) Theorising complex water governance in Africa: the case of the proposed Epupa Dam on the Kunene River. Int Environ Agreements: Polit Law Econ 16:21–48

Meissner R, Funke N, Nienaber S, Ntombela C (2014) The status quo of research on South Africa's water resources management institutions. Water SA 39(5):721–731

Meissner R, Stuart-Hill S, Nakhooda Z (2017) The establishment of catchment management agencies in South Africa with reference to the *Flussgebietsgemeinschaft Elbe*: Some practical considerations. In: Karar E (ed) Freshwater governance for the 21st Century. Springer, Dordrecht

Meissner R, Steyn M, Moyo E, Shadung J, Masangane W, Nohayi N, Jacobs-Mata I (2018) South African local government perceptions of the state of water security. Environ Sci Policy 87:112–127

Meissner R, Steyn M, Jacobs-Mata I, Moyo E, Shadung J, Nohayi N, Mngadi T (2019) The perceived state of water security in the Sekhukhune District Municipality and the eThekwini Metropolitan Municipality. In: Meissner R, Funke N, Nortje K, Steyn M (eds) Understanding water security at local government level in South Africa. Palgrave Macmillan, London

Menahem G (1998) Policy paradigms, policy networks and water policy in Israel. J Pub Policy 18(3):283–310

Menahem G (2001) Water policy in Israel 1948-2000: policy paradigms, policy networks and public policy. Israel Aff 7(4):21–44

Merriam SB (1991) How research produces knowledge. In: Peters JM, Jarvis P (eds) Adult education. Jossey-Bass, San-Francisco

Molloy S (2019) Realism and reflexivity: morgenthau, academic freedom and dissent. Eur J Int Relat 26(2):1–23

Morgan PM (2003) National and international security: theory then, theory now. Asian J Polit Sci 11(2):58–74

Myers FWH (1887) Multiplex personality. Nineteenth Century Mon Rev 20(117):648–666

Nakayama Y (2010) No forbidden landscape in string/M-theory. J High Energy Phys 2010(1):30

Neumann RP (2009) Political ecology: theorizing scale. Prog Hum Geogr 33(3):398–406

Norma ES, Bakker K, Cook C (2012) Introduction to the themed section: water governance and the politics of scale. Water Altern 5(1):52–61

Ntuli P, Smith JA (1999) Speaking truth to power: a challenge to South African intellectuals. Alternation 6(1):1–20

Nye JS (2009) Scholars on the sidelines. Washington Post, April 13: A15

Oliver M (1998) Theories of disability in health practice and research. BMJ 317:1446–1449

Ostrom E (2007) A diagnostic approach for going beyond panaceas. Proc Natl Acad Sci 104(39):15181–15187

Owusu G (2016) On philosophical foundations of ecohydrology: the challenge of bridging natural and social science. Ecohydrology 9:882–893

Oxford Advanced Learners Dictionary (2013) Oxford dictionaries: language matters. Accessed at: http://www.oxforddictionaries.com/definition/english/. Accessed on: 2 June 2014

Pearse H (1983) Brother, can you spare a paradigm? The theory beneath the practice. Stud Art Educ 24(3):158–163

Pohl B, van Willigen (2015) Analytic eclecticism and EU foreign policy (in)action. Glob Soc 29(2):175–198

Pretty JN (1994) Alternative systems of inquiry for sustainable agriculture. IDS Bull 25(2):37–48

Pretty JN (1995) Participatory learning for sustainable agriculture. World Dev 23(8):1247–1263

Qin Y, Nordin AHM (2019) Relationality and rationality in Confucian and Western traditions of thought. Cambridge Rev Int Aff 32(5):601–614

Regilme SSF (2018) Beyond paradigms: understanding the South China Sea dispute using analytic eclecticism. Int Stud 55(3):213–237

Rengger N (2015) Pluralism in international relations theory: three questions. Int Stud Perspect 16(1):32–39

Reus-Smit C (2013) Beyond metatheory? Eur J Int Relat 19(3):589–608

Reus-Smit C, Snidal D (2008) Between utopia and reality: the practical discourse of international relations. In: Reus-Smit C, Snidal D (eds) The oxford handbook of international relations. Oxford University Press, Oxford

Richardson KA (2008) Managing complex organisations: complexity thinking and the science and art of management. E:CO 10(2):13–26

Ridley MJ (2003) Database systems or database theory—or why don't you teach Oracle. Paper presented at the LTSN-ICS Workshop on Teaching Learning and Assessment in Database, Coventry, United Kingdom, July 2003

Robertson I (2012) Singapore airlines flight 006—caution to the wind. Cineflix, Montreal, Canada

Rogers K, Roux D, Biggs H (2000) Challenges for catchment management agencies: lessons from bureaucracies, business and resource management. Water SA 26(4):505–511

Ruane JM (2018) Should sociologists stand up for science? Absolutely! Sociol Forum 33(1):239–241

Rueschemeyer D (2009) Usable theory: analytic tools for social and political research. Princeton University Press, Princeton and Oxford

Satz D, Ferejohn J (1994) Rational choice and social theory. J Philos 91(2):71–87

Scheiner SM, Willig MR (2008) A general theory of ecology. Theor Ecol 1(1):21–28

Schickore J (2014) Scientific discovery. In: Zalta En (ed) The stanford encyclopaedia of philosophy (Spring 2014 Edition). Accessed at: http://plato.stanford.edu/archives/spr2014/entries/scientificdiscovery/. Accessed on: 17 July 2014

Schultz M, Hatch MJ (1996) Living in multiple paradigms: the case of paradigm interplay in organizational culture studies. Acad Manag Rev 21(2):529–557

Schwandt TA (2007) The SAGE dictionary of qualitative inquiry, 3rd edn. Sage, Thousand Oaks, CA

Seale C (1998) Theories and studying the care of dying people. BMJ 317:1518–1520

Sebastian AG, Warner JF (2014) Geopolitical drivers of foreign investment in African land and water resources. Afr Identities 12(1):8–25

Selby J (2014) Positivist climate conflict research: a critique. Geopolitics 19(4):829–856

Selby J (2018) Critical international relations and the impact agenda. Br Polit 13:332–347

Sherrington CS (1942) Man on his nature. Cambridge University Press, Cambridge

Sil R (2000a) The questionable status of boundaries: the need for integration. In: Sil R, Doherty EM (eds) Beyond boundaries? Disciplines, paradigms, and theoretical integration in international studies. University of New York Press, Albany, NY

Sil R (2000b) The foundations of eclecticism: the epistemological status of agency, culture, and structure in social theory. J Theor Polit 12(3):353–387

Sil R (2009) Simplifying pragmatism: from social theory to problem-driven eclecticism. Int Stud Rev 11(3):648–652

Sil R, Katzenstein PJ (2010) Beyond paradigms: analytic eclecticism in the study of world politics. Palgrave Macmillan, Houndmills, Basingstoke

Sil R, Katzenstein PJ (2011) De-centering, not discarding, the "Isms": some friendly amendments. Int Stud Quart 55(2):481–485

Strauss A, Corbin J (1990) Basics of qualitative research: grounded theory procedures and techniques, 2nd edn. Sage, Thousand Oaks, CA

Stuart-Hill SI (2015) Mainstreaming adaptation to climate change into decision making in the water sector: concepts and case studies from South Africa. Unpublished Ph.D. thesis, School of Agricultural, Earth and Environmental Sciences, University of KwaZulu-Natal, Pietermaritzburg

Teddlie C, Tashakkori A (2011) Mixed method research. In: Denzin NK, Lincoln YS (eds) The SAGE handbook of qualitative research, 4th edn. Sage Publications, Thousand Oaks, CA

Tetlock P (2005) Expert political judgement: how good is it? How can we know?. Princeton University Press, Princeton, N.J.

Turton A (2005) Hydro hegemony in the context of the orange river basin. Paper presented at the Workshop on Hydro Hegemony, School of Oriental and African Studies, University of London, 20–21 May 2005, London

United Nations Framework Convention on Climate Change (UNFCC) (2020) About the Secretariat: what is the purpose of the Secretariat? Rio de Janeiro, Brazil and New York, US: United Nations

Framework Convention on Climate Change. Accessed at: https://unfccc.int/about-us/about-the-secretariat. Accessed on: 6 Apr 2020

Urdang S (1989) And still they dance: women, war, and the struggle for change in Mozambique. Monthly Review Press, New York

Urry J (2005) The complexity turn. Theory, Cult Soc 22(5):1–14

Vasileiadou E, Safarzyńska K (2010) Transitions: taking complexity seriously. Futures 42:1176–1186

Walt SM (1998) International relations: one world many theories. Foreign Policy 110(29–32):34–46

Walt SM (2005) The relationship between theory and policy in international relations. Annu Rev Polit Sci 8:23–48

Warner JF, Wester P, Hoogesteger J (2014) Struggling with scales: revisiting the boundaries of river basin management. WIREs Water 1:469–481

Warner J, Waalewijn P, Hilhorst DH (2002) Public participation in disaster-prone watersheds: time for multi-stakeholder platforms? Wageningen University. Irrigation and Water Management Group. Disaster Studies

Weber R (2004) The rhetoric of positivism versus interpretivism: a personal view. MIS Q 28(1):iii–xii

Wendt A (1999) Social theory of international politics. Cambridge University Press, Cambridge

Wight C (2006) Agents, structures and international relations: politics as ontology. Cambridge University Press, Cambridge

Wildavsky A (1979) Speaking truth to power: the art and craft of policy analysis. Little, Brown and Company, Boston

Winter G (2000) A comparative discussion of the notion of 'validity' in qualitative and quantitative research. Qual Rep 4(3):1–14

Ye FY (2009) From chaos to unification: U theory vs. M theory. Chaos, Solitons Fractals 42(1):89–93

Zais R (1976) Curriculum: principles and foundations. Thomas Y. Crowell, New York

Zeitoun M (2007) The conflict vs. cooperation paradox: fighting over or sharing of Palestinian-Israeli groundwater? Water Int 32(1):105–120

Zeitoun M, Warner J (2006) Hydro-hegemony—a framework for analysis of trans-boundary water conflicts. Water Policy 8:435–460

Zyla B (2019) Eclecticism and the future of the burden-sharing research programme: why trump is wrong. Int Policy Sci Rev

Chapter 3
eThekwini's Green and Ecological Infrastructure Policy Landscape

Abstract In this chapter, I will present the results of an investigation that I conducted in August and September 2018 with experts inside and outside the eThekwini municipality who were involved in it's green and ecological infrastructure policy landscape. Although a green infrastructure plays an important role in the municipality's urban and peri-urban environment, an ecological infrastructure has a more prominent place beyond its political boundaries. The uMngeni River, which supplies the bulk of Durban's water, originates in the rural areas of eThekwini's municipal neighbours, particularly the uMgungundlovu District Municipality and, to a certain extent, the uThukela District Municipality. This river, as well as others, supply not only valuable goods and services, but they can also cause hardship. It is within this context, as well as that of long-term anthropogenic climate change, that eThekwini frames its green and ecological infrastructure policies. It is my conclusion that positivism dominates the thinking and practice of municipal officials and members of the epistemic community that collaborate with the municipality. A theory that stands out is liberal institutionalism. What is also noticeable is that the municipality's Environmental Planning and Climate Protection Department consists of numerous experts who focus on climate change adaptation as the foundation that underpins the theory. It would appear, therefore, that an epistocracy is operating within the municipality and driving its climate adaptation strategy, by using green and ecological infrastructure initiatives.

Keywords eThekwini metropolitan municipality · Climate change adaptation · Green infrastructure · Ecological infrastructure · Positivism · Liberal institutionalism

3.1 Introduction

In this penultimate chapter, I will report on the results of the PULSE[3] analysis. The respondents that I interviewed during 2018 were from a host of institutions; for example, the Council for Scientific and Industrial Research (CSIR), the eThekwini Metropolitan Municipality, the Institute of Natural Resources (INR), the Palmiet

R. Meissner, *eThekwini's Green and Ecological Infrastructure Policy Landscape*, https://doi.org/10.1007/978-3-030-53051-8_3

River Watch (PRW), the Wildlife and Environment Society of Southern Africa (WESSA) and the University of KwaZulu-Natal (UKZN). The municipal officials who were interviewed are all directly involved in green and ecological infrastructure enterprises, and the interviewees from the other institutions are, to a certain extent, mainly involved in collaboration with the municipality's green and ecological infrastructure projects. The individuals who were questioned have a first hand knowledge of the natural environment and municipal governing systems that are linked to the infrastructure types.

The chapter consists of several sections, the first of which gives a socio-economic and bio-physical description of the municipality. In Sect. 3.3, I describe the methodology that was used to gather the information, as well as how the data were analysed, using the PULSE[3] analytical framework. In Sect. 3.4, I will present the analysis of the paradigm and theory assessment, as well as the perspectives, causal mechanisms and critical and problem-solving theories, before ending with a conclusion.

3.2   The eThekwini Metropolitan Municipality

The eThekwini Metropolitan Municipality is the local authority that governs the city of Durban, which is situated on the east coast of South Africa in the KwaZulu-Natal Province . Regarding its geographical reach and population size, eThekwini covers an area of 2297 km^2 and in 2016 it was home to 3,555,868 people. The municipality has pockets of rural communities on the western, southern and northern outskirts of the densely populated urban centre (StatsSA 2016; Meissner et al. 2018a, 2019).

In the 2016 municipal elections, the African National Congress (ANC) won a majority of 56.01% of the votes and 126 seats on the municipal council. The second-largest party, the Democratic Alliance (DA), garnered 26.9%, giving it 61 council seats. The Inkatha Freedom Party (IFP) received 4.2% and 10 seats, followed by the Economic Freedom Fighters' (EFF) with 3.4% and 8 seats (IEC 2016). During this election, the ANC retained all 126 seats, while the DA gained 18 new councillors (Hogg 2016).

In terms of the South African economy, Durban is an important port city (see Figs. 1.4, 3.1 and 3.2) (Corbella and Stretch 2012), with a large shipping container terminal regarding container shipping into and out of the port of Durban. Mutombo (2014) indicates that it handled 2.6 million Twenty-foot Equivalent Units (TEUs) in 2012, compared to the 0.7 million TEUs in each of the Cape Town, Port Elizabeth and Ngqura ports. Long-term forecasts show that Durban's capacity could increase to 9.5 million TEUs in 2043, compared to 5.4 million TEUs for Port Elizabeth and Ngqura, and 2 million for Cape Town (Mutombo 2014). These figures confirm the port's status as South Africa's largest container terminal. The port is also closer to South Africa's economic hub in Gauteng than the other two major ports, i.e. it is 570 km away, as opposed to Port Elizabeth and Ngqura and Cape Town, which are 1070 km and 1400 km away, respectively (Mutombo 2014).

Fig. 3.1 The Port of Durban's oil storage facilities

Fig. 3.2 The Durban Marina

Apart from playing a key role in the South African economy, Durban is also a popular tourist destination and is, by far, the most-visited place in KwaZulu-Natal by domestic and foreign visitors (Schalkwijk et al. 2017). This was confirmed by the 2015 statistics of Tourism KwaZulu-Natal (TKZN). Of the 743,615 foreign visitors to the province, Durban attracted some 59%, followed by 13, 11and 10% for the North Coast, Pietermaritzburg and the Elephant Coast, respectively. On average, foreign visitors spent R737.00 per day, which resulted in a total market value of R4.7 billion. In the same year, domestic visitors made about 4.98 million trips to the province, of which 39% were to Durban, compared to 19 and 11% to Zululand and Pietermaritzburg/the Natal Midlands, respectively. On average, these visitors spent R1 108.00 per trip, which resulted in a total direct value of R5.5 billion (TKZN 2016).

From a socio-economic development perspective, 90% of the eThekwini population has access to a municipal source of water and, by default, good quality potable water (StatsSA 2016; Meissner et al. 2018a, 2019). According to statistics from the 2011 Census, eThekwini has a young population, with 66% of its people being below 35 years (StatsSA 2016; Meissner et al. 2019). The municipality's population consists mainly of persons from the African community (73.8%), followed by Indians (16.7%), Whites (8%) and Coloured (2%) (Roberts and O'Donoghue 2013). According to Roberts and O'Donoghue (2013: 301), 'these percentages are important, as challenges such as poverty and unemployment are associated predominantly with the African group'. A General Household Survey found that eThekwini was one of the metros in the country with the lowest percentage of households that had improved access to sanitation facilities (83.5%) (StatsSA 2016; Meissner et al. 2019). This is revealing, since Durban is South Africa's poorest metropolitan area, with 64% of its households earning less than US$440 in 2010–2011 (Roberts and O'Donoghue 2013).

From a green and ecological infrastructure perspective, it is not only important to describe the municipality by using demographic and socio-economic data and political party representation statistics, but also from a natural environment stance, since nature is the fountainhead of both green and ecological infrastructures. The municipality has a sub-tropical climate that is characterised by extreme humidity in summer. Winter temperatures, which are seldom below 10 °C (C), are warm compared to other parts of the country, particularly the Western Cape, which experiences intermittent snowfalls during winter, and South Africa's Highveld region, which is characterised by dry and cool-to-cold winters. eThekwini's summer temperatures can be in the low- to mid-thirties, while the humidity ranges between 55 and 75% (Dray et al. 2006; Meissner et al. 2019). In terms of precipitation, Durban usually receives more than 1000 mm per annum, mainly during the summer (eThekwini Municipality 2007; Meissner et al. 2019).

eThekwini has a diverse topography, which ranges from steep escarpments in the west of the city, to a relatively flat coastal plain in the east, where the Indian Ocean forms 98 km of its boundary. eThekwini contains 17 river catchments that flow into the sea via 16 estuaries and in-shore marine environments. The area in which the municipality is located supports a wide variety of terrestrial and aquatic ecosystems. The water that flows from the 17 catchments, such as the Isipingo, oHlanga, Palmiet, Umhlangane, Umhlatazana, Umlazi, uMngeni and Umkomazi Rivers, influences

Durban's water quality and quantity (eThekwini Municipality 2007; Meissner et al. 2019). The uMngeni is the largest of these rivers, and it supplies Durban's residents with much-needed water.

3.3 Methodology

I have already described eThekwini's demographic, economic, natural and political characteristics above, and in this section, I will discuss my research methodology. I will very briefly describe my data collection and coding strategies and will follow a predominantly qualitative methodology throughout, with the quantification being limited to electronic coding, using PULSE[3].

The qualitative methodology has several advantages, as it focuses on self-reflexivity, the context and elaborate description. From a practical perspective, using this methodology requires one to get out into the 'field' to study the contexts that generate personal curiosity. Qualitative data also provide an insight into numerous activities that a researcher might miss when using structured surveys or experiments (Tracy 2013), which is the downside of using a quantitative methodology. Furthermore, a qualitative methodology allows scholars to uncover salient issues that they can later study by using structured methods. As a practice, field research, with its related activities, also uncovers and gives glimpses into regularly guarded worlds and affords one the opportunity to tell a story about them. In this regard, the methodology is also well suited for accessing a tacit and intuitive understanding of cultures (Tracy 2013) or groups of people, which is often taken for granted.

Therefore, qualitative research is not merely about asking people about what they do, but it gives the scholar an opportunity to see and hear what they do. Using a qualitative methodology also exposes the researcher's understanding of the values of the participants and how they regularly live these values. Moreover, qualitative researchers construct explanations about the participants' experiences. In this way, they assist policy practitioners to understand the world, society and institutions, and by doing so, they serve humanity in various ways (Tracy 2013).

There is, however, a downside to being a qualitative researcher. In this section, I will focus on one of these issues, which happens on a normative level. Because this type of research consists of a set of interpretive activities, it resists the temptation to prefer one methodological practice over another. The concept 'qualitative research' is difficult to define, since it leans towards discourse and discussion and has no distinct paradigm or theory of its own. This means that qualitative research is open-ended in nature, and this resists the lure to impose a single overarching paradigm on the research (Denzin and Lincoln 2018). Because of these traits, it stands in contrast to traditional (natural) scientific research, which is based on empiricism and positivism (Kurki 2008).

As a result, the scientific community does not share qualitative research widely, and this resistance manifests itself in politics. Therefore, we can analytically distinguish the political role of a methodology from its procedural character. The procedural characteristics of qualitative research define how scholars use such methodologies to produce knowledge and these features often intersect. Politicians and so-called 'hard scientists' often refer to qualitative researchers as journalists or 'soft' scientists, and define their work as unscientific, exploratory only, or subjective. Epistemologically, critics also do not refer to qualitative research as theory, but as criticism, and often they interpret it politically, as Marxism or secular humanism. The reason for this is that the 'home' of qualitative research is the world of lived experience, since it is here where the beliefs and actions of individuals intersect with their culture (Denzin and Lincoln 2018) and politics. My aim is to uncover the lived experience of the interviewees, in order to gain a deeper comprehension of how their thinking influences the municipality's green and ecological infrastructure landscape.

3.3.1   Data Collection

In August and September of 2018, I conducted 16 interviews with eThekwini Municipal officials and other stakeholders, mainly researchers who are either directly or indirectly involved in the implementation of green and ecological infrastructure projects and policies. Several of these scholars are in the process of collaborating with the municipality on numerous green and/or ecological infrastructure enterprises.

I employed a participatory and iterative data collection strategy that incorporated a snowball technique for respondent identification (Meissner et al. 2018a), and I administered a structured open-ended questionnaire during face-to-face interviews. I was unable to meet two of the respondents face-to-face and therefore ran the questionnaire either telephonically or via Skype. I interviewed the respondents about their understanding of the concept of a 'green infrastructure'. A follow-up question from this was whether they viewed green and ecological infrastructures as two different elements of the same phenomenon. I also asked them about their knowledge of municipal policies, plans, bylaws or standards for implementing green and ecological infrastructures that are directly related to water security. Furthermore, I questioned them about specific green and ecological infrastructure initiatives that are being implemented by the municipality to improve water security. I then asked them to evaluate the content, implementation, and the financial and social costs and benefits of the operations. The final question was about which stakeholders the municipality collaborates with to implement such water improvement projects. I also conducted site visits to the Buffelsdraai Landfill, part of the Palmiet River catchment, the River Horse wetland, eThekwini's green roof initiative in Durban's inner city and the Durban Botanical Gardens. These visits afforded me the opportunity to get a first-hand view of some of the initiatives in and around the city and to photograph them. Photographs provide a visual discourse for understanding the subject matter (Ownby

2013). I utilised photographs that I had taken on previous field visits to Durban, as well as photographs from some of the stakeholders that depict certain incidents, such as the municipal waste pollution on the beaches after the 2019 rainstorm.

3.3.2 Data Coding

After conducting the interviews, I transcribed them and saved them in a Portable Document Format (PDF). I emphasised the sections that I wished to analyse by using the PDF highlighter function. Using the PULSE[3] digital version that was developed by Dr. Marc Pienaar, I uploaded each interview into the programme. The digital version recognises the tinted text, which enables the analyst to evaluate it by using the PULSE[3] analysis selection capability, which consists of a drop-down menu of its components, as outlined in Chap. 2. These options include paradigms (Table 2.1), perspectives (theories) (Table 2.2), causal mechanisms (p. 58) and critical and problem-solving theories (p. 63). With these available options, one can analyse the highlighted text in terms of the dominant paradigms, theories, causal mechanisms, as well as the critical and problem-solving theories, that are contained in the transcribed interviews.

3.4 Analysis

In this section, I will present the graphs and diagrams that were produced during the analysis. I will first discuss the dominant paradigm (research worldview) contained in the interviews, followed by the prevailing theories (perspectives) that are present in the analysed information, before considering the causal mechanisms. Following this discussion, I will report on the critical and problem-solving theory types of the analysed data.

3.4.1 Paradigm Profile

Figure 3.3 indicates the paradigm assessment score. With respect to the way in which the respondents generated the knowledge contained in their answers, the positivist research paradigm received the highest score, with 3632. The interpretivism/constructivism paradigm followed, with a score of 814, followed by postpositivism, with 444. Neither the critical theory or participatory paradigms received any points.

Most of the recommendations that the interviewees expressed showed positivist characteristics, with the positivist paradigm scoring 153, followed by the interpretivism/constructivism paradigm, with 32, and the postpositivism paradigm, with a

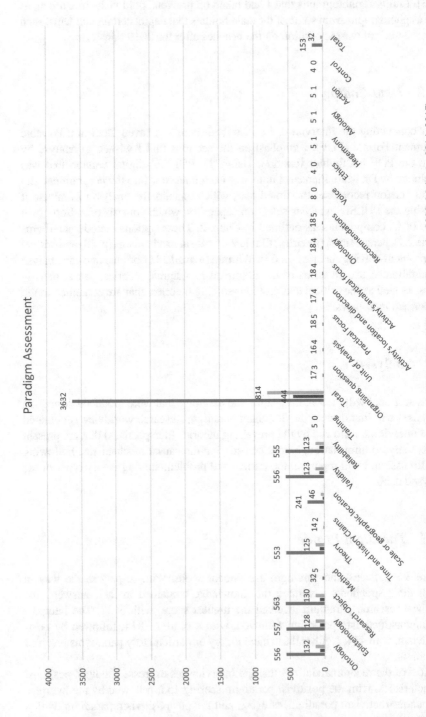

Fig. 3.3 Research worldview assessment (derived from Lincoln et al. 2011, 2018; Hobson and Seabrooke 2009; Meissner 2014, 2017)

score of 10. Again, neither the critical theory nor the participation paradigms scored any points.

3.4.1.1 Knowledge Generation

Positivism

Several explanations exist for the dominance of positivism, the first of which relates directly to the subject matter at hand. Green and ecological infrastructures fit neatly within the natural sciences, especially when it comes to the notion of biodiversity. For instance, Respondent 1, who is a municipal engineer, coined his understanding of the concept 'green infrastructure' by considering biodiversity and its function in retaining flood water. He remarked that: 'There [in the retention ponds] you have all your various grasses. This is what the environmental people like and there is lots of biodiversity, but it also functions as a storage place for the storm flow'.

Respondent 3, a researcher from the UKZN, who works closely with eThekwini on green infrastructure matters, said that she had been grappling with the green infrastructure concept quite a lot until she had a conversation with a colleague who was busy with research on food security. This colleague remarked to her that he had, 'in the last ten years been framing things around ecosystem services, biodiversity and the ecological infrastructure'. Furthermore, she said that:

> eThekwini is very good in understanding the difference between biodiversity, ecosystem services and ecological infrastructure. Biodiversity is part of that [ecological infrastructure], but it is also different from it. So, I am linking to those…terms and I add green infrastructure. Then it becomes…another kind of framing.

Respondent 9, a municipal official, was very explicit when I asked about her understanding of the 'green infrastructure' concept. She indicated that she worked within the 'biodiversity department', and that 'the department has the function to protect biodiversity and, for me, that biodiversity is our [ecological infrastructure]'.

When asked about the municipality's policies, plans and bylaws, the opinion of Respondent 10, a researcher from the INR, was that: 'I think the intention is around biodiversity conservation…in terms of their mandate under Section 24 of the Constitution, where people have the right to a clean environment'. Respondent 16, another municipal official who works closely with biodiversity matters, had the following to say: 'Then there is the biodiversity theme, which is also a very good example of policy [constituting green infrastructure initiatives]'.

Biodiversity

With respect to the concept of 'biodiversity', Slootweg (2005:37) noted that scientists treat it as an 'uncomplicated concept' and as 'a visible reality that surrounds and supports us'. Slootweg's (2005: 37) understanding is that:

> Biodiversity encompasses all living organisms on earth, their interactions, and the interaction with the physical environment (soil, water, air). Human beings can be considered to be part

of biodiversity since they are living creatures, in continuous interaction with the environment, and dependent on biodiversity for their survival. Biodiversity produces the oxygen we breathe, regulates water supplies, sequestrates greenhouse gases and thus is responsible for regulating climate, provides food, such as fish, shellfish, and (bush)meat, and also provides products, such as timber or raw materials for pharmaceutical industries, constitutes a source of genetic material of use to maintain the long-term productivity of our agricultural crops (themselves also being part of biodiversity), is a source of inspiration, recreation and greatly contributes to the quality of our day-to-day lives.

Several of these characteristics of the concept 'biodiversity' lend themselves to its measurable outcomes, such as the targets envisaged in the 17 Sustainable Development Goals (SDGs). The other normative aspects, such as inspiration, recreation and quality of life, are less measurable, although they are still part of the description of what constitutes biodiversity and how it functions and gives meaning to human existence. It is interesting to note that none of the respondents mentioned these intangible and non-measurable biodiversity functions.

Regarding the measurement of biodiversity, Bilderbeek et al. (2016) stated that worldwide biodiversity loss is not only threatening food security and human health, but it is also hostile towards the sustainable development agenda itself. According to biodiversity experts, because of these threats, it is important to mainstream biodiversity into sustainable development policies, the effective implementation of the 14 (life below water) and 15 (life on land) SDGs, as well as the Aichi targets of the Convention on Biological Diversity (CBD). Note that three of the structures of rule that Bilderbeek et al. (2016) mentioned contain measurable elements (SDGs 14 and 15 and the Aichi targets). Even biodiversity loss can be quantified, although it is declining rapidly (Possingham and Wilson 2005), which enables governments and non-state entities to set targets for reducing biodiversity losses (Scholes and Biggs 2005). In fact, Scholes and Biggs (2005) advocated that their proposed 'Biodiversity Intactness Index (BII)' helps to assess the progress of biodiversity forfeiture targets. According to them, the BII is 'simple and practical—but sensitive to important factors that influence biodiversity status—and which satisfies the criteria for policy relevance set by the Convention on Biological Diversity'.

Biodiversity is, therefore, a concept that can be described as a concrete reality that is apprehensible (Lincoln et al. 2018) and measurable, not only in terms of its 'real' characteristics, but also regarding its reduction and consequent improvement from a reduced biodiversity state and its negative impact on humankind. This is what Lincoln et al. (2018) call 'raw realism', or the pretended 'hard science', where the researcher and reality are separate, and the purpose of research is to control and predict (through logical deduction) nature and its influence on people. This type of science is steeped in empiricism and positivism, with objectivism being the central feature. It is within the nature of reality (ontology) that the respondents frame eThekwini's green or ecological infrastructure initiatives.

Interpretivism/Constructivism

When contemplating the results of the PULSE[3] paradigm assessment (Fig. 3.3), it is important to note the prominence of the interpretivism/constructivism paradigm.

This indicates that the respondents are not only defining the concept of green and ecological infrastructures within the frame of a 'raw realism', but they also view the concept in relativistic terms, as defined by the local and specifically constructed and co-constructed realities (Lincoln et al. 2018). That said, these framings fall within the structural and measurable epistemology and ontology of biodiversity.

When Respondent 1 was asked about his understanding of the concept 'green infrastructure', he gave a bit of his history, before answering in the following way:

> I started my career in catchment management… Then I went into rivers and storm-water systems in built-up areas and then into the coastal side… So, for me green infrastructure or ecological infrastructure would be, from a storm-water point of view, all those natural systems—everything from percolation areas, to wetlands, to reed beds.

When I asked Respondent 2, a researcher at the Oceanographic Research Institute (ORI), about her understanding of a green infrastructure, she immediately answered by saying:

> I think it depends on the context. I mean, an ecological infrastructure is fairly straightforward, in my mind, in that we are talking about natural resources or natural habitats that provide protection. However, a green infrastructure could be interpreted as the infrastructure that we put in place so that man-made infrastructures are designed in an environmentally friendly and sustainable manner to improve or support ecosystem functioning for the purpose of protection.

Although at first she mentioned the context, she was quite sure about her own conceptualisation of ecological infrastructures, and agreed with the meaning that other researchers have given to the concept. Since Respondent 2 works at the ORI, I also pulsed her about the green coastal infrastructure and said that I had not heard much talk of this being a type of green or ecological infrastructure. Her opinion was as follows: 'I think it is because the inland systems [mountains, wetlands, rivers, streams and flood reduction] are better understood'. Respondent 1 also referred to the city's estuaries, because for him, estuaries are part of a green infrastructure. He noted that:

> …[storm flows] come down to river systems and flood plains and one of the big issues for us is that guys just want to build on the flood plains. We try and keep them out because they are part of the green infrastructure. We need those flood plains, right down to the estuaries. We must teach people how to protect them and get them to work properly. Estuaries are heavily impacted—we've got sand mining in our rivers, which impacts the estuaries. We've got pollution, which is a big issue. We've got quite a lot of alien invasive species, which are in the estuaries because of nutrient levels, and the nutrient levels are there because of sewerage pollution in the main.

Respondent 2 indicated that she often collaborates with Respondent 1 on coastal protection matters. This informs her understanding of coastal systems like estuaries, and their importance as green and ecological infrastructures.

From this, we see a bias towards inland ecological infrastructure systems, which are linked with the structuralism and measurable benefits of biodiversity, at the expense of coastal ecological infrastructures, like estuaries. A likely reason for this predisposition is the emphasis that South African academia place on wetlands and mountains, for instance, the purification and supply of water (Nel et al. 2013) through natural processes (p. 91). This shows that scholars understand the concept of a 'green and ecological infrastructure' not only within the disciplinary structures, but also in terms of South Africa's emphasis on inland water systems. Since these are prioritised, they tend to be at the top of the water governance research agenda, which is not the case with coastal aquatic systems. This implies that there is a hierarachy among the research topics, with inland systems at the top of the list and coastal ecological infrastructures at the bottom.

For some of the respondents, the answer to the question 'What is your understanding of the concept "green infrastructure?"' was not straightforward, which is a further indication of the interpretivism/constructivism paradigm. Respondent 3, from UKZN, answered in the following way:

> When you say 'green infrastructure', it is always a bit difficult, because I think that people use these terms interchangeably. Sometimes, green infrastructure is used…more in the policy or applied fields. However, I tend to use ecological infrastructure more. I've learned about it from a science point of view, which brings it into that space. In the work that we're doing in eThekwini, I am working with Graham [Jewitt] on a WRC project, and we have framed it as an 'ecological infrastructure'. So, for me, when it comes to green infrastructures, in my mind I would think of when cities or planners are thinking of putting in green open spaces to try and provide a whole lot of ecosystem services. However, the problem is also that when you use the term 'green infrastructure', it immediately kind of 'separates' the environment, in a sense… You know, if you go down that road of people having to defend a [green infrastructure]…it sort of breaks that socio-ecological combination in some ways. I think from a socio-political point of view, I would always be a bit weary 'of saying [green infrastructure]' because people immediately jump onto that green bandwagon, and then it becomes a battle for the environment again.

It is interesting to note that she defines an ecological infrastructure from a 'scientific point of view', which indicates the element of consensus that is mentioned above. When it comes to a green infrastructure, she frames her answer from a political perspective, in that political formations are manifested in activism for green infrastructures. Here we see a clean break: an ecological infrastructure is a scientific concept, and a green infrastructure leans more towards a political framing that is embedded in a political context. In this sense, science frames an ecological infrastructure with empiricism in mind.

According to Respondent 3, it is evident that the concept of a green infrastructure can become a political tool that is informed by, and that informs, people's norms and interests. We see this in Respondent 4's opinion of the concept of an 'ecological infrastructure':

> I have strong opinions around the concept of a green infrastructure and what we are trying to sell. Let me offer an opinion. Well, for me, and this is about an ecological infrastructure, but we are trying to reinvent the term nature to appeal to different audiences. So, we started out life with nature. We then said we need to appeal to a whole bunch of economists. We got into all these concepts like ecosystem services, and all these economic terms, like natural capital, and bring all these economic terms into the nature debate. And then we said that we now needed to appeal to an whole bunch of engineers and, if we call it 'infrastructure', we can also bring them into the pot. Now I can understand that strategy. But what I am objecting to is that we are manufacturing terminology to suite other disciplines. It is, therefore, understandable that

businesses talk about ecosystems in a completely different context. For me, an ecological infrastructure is like a sex doll. It has all the passion, all the mystery, all the nuances and everything else stripped out of it. It just becomes a material entity. Nature is my lover. For me, when we talk about an ecological infrastructure, we become quite mechanistic about the way we think about nature. One of the reasons I am pushing this is that if we want to connect with society, we need to use terms that people will understand.

Respondent 4's answer highlights the way that the concept of an 'ecological infrastructure' is used. For him, the term functions not only as a scientific concept, but also as a reconstruction of nature to involve various disciplines and professions, in order to effectively implement ecological infrastructure initiatives. The fact that society wants to involve economists and engineers shows the pervasiveness of a positivist inclination towards solving problems, by utilising nature as an 'infrastructure' type.

Understanding a green infrastructure within a specific context also came to the fore during Respondent 5's answer. She is a municipal official, who specified that:

Well, by 'green infrastructure', I understand reducing, re-using and recycling. That is my initial understanding. I would relate an ecological infrastructure more to rivers and streams, securing our riparian zones and that we should have enough plants and matter to clean the water…

Respondent 7, a senior researcher at the CSIR said that, for him, a green infrastructure is '[e]ssentially…investment and development in projects in the green economy space'. When I asked him if there is a difference between green and ecological infrastructures, he answered by saying:

In theory…there should be a difference. An ecological infrastructure, to me, is more about things like floating wetlands, which are very specific projects, technologies or infrastructures that are developed to assist with ecological aspects, like water functioning in the water space and others. A green infrastructure is something like a green building, or it could be a roof top garden [Figure 3.4], or solar panels on the roof… It is at a higher level and talks to the whole idea of the sustainable development debate.

He defines a green infrastructure within the context of sustainable development, which, by implication, relates to biodiversity protection and ecosystem services, as mentioned above. For him, a green infrastructure is a component of sustainable development that lends a structural meaning to the way in which he frames the concept.

Respondent 8, a researcher from WESSA, answered in narrative form when I asked him about his understanding of the concept of a green infrastructure. He uses the concept 'ecological infrastructure', instead of 'green infrastructure', and noted that he is, just like Respondent 15, 'quite comfortable with the SANBI definition' (p. 5). In his account, he stated that:

For me, it's the natural infrastructure, a non-man-made and non-built infrastructure that supplies goods and services to people. So, for me this means fresh air, water and food. It is also linked to the energy nexus… A colleague of mine met with Cyril Ramaphosa just before he was elected President, and Ramaphosa asked him: "What is an ecological infrastructure?" And [my colleague]… took the whole forum to the window. It was there, in the Union buildings, that he said: "Do you see the hills?… [T]hose are the Magaliesberg [Figures 3.6, 3.7 and 3.8] and some people might say that they are underdeveloped." He also

Fig. 3.4 eThekwini's roof-top garden initiative

said, "You could say that it is a water factory that produces water for society" That is what an ecological infrastructure is! So, for me, that would then support the green economy.

His framing falls within the Water-Energy-Food (WEF) nexus and green economic thinking. We can describe both the nexus and green economy as positivist episte-mological and ontological structures. In a Water Alternatives blog critiqueing the WEF nexus, Gyawali (2020) argues that at the core of the concept 'lie much-ignored aspects of governance: moving away from technocratic fixes of "wicked" problems, recognizing complex trade-offs; replacing faith in full control with flexibility and adaptative management; and giving equal primacy to not-easily quantifiable values of ethics and justice arrived at by listening to marginalized grassroot voices'. In response to Gyawali's (2020) blog, I argued that empiricism and positivism entrenches WEF nexus thinking (Meissner 2020), since it favours a quantitative approach and the limited use of social science methods (Albrecht et al. 2018). The fact that Gyawali (2020) indicated that normative notions around governance and ethics and justice are not the mainstays of WEF nexus thinking, bears witness to a leaning towards structural empiricism and positivism. We can say the same for green economic thinking.

I also conducted an interview with a representative from a Durban-based interest group, who acts in the interests of the Palmiet River Catchment (Fig. 3.5) and the people in and around the area (Respondent 11). When I asked him about his understanding of the concept of a 'green infrastructure' he indicated that:

This is an important question about a "buzz-word", which, along with others, may be woven into proposals to access funds, without necessarily delivering on what should be realized: by reversing the accumulated environmental damage caused by past planning, design and maintenance regimes that have impaired, if not obliterated, the natural environment, so that it no longer provided goods and services... I understand "green infrastructure" to be

Fig. 3.5 The Palmiet River

Fig. 3.6 A thunderstorm over the Magaliesberg silhouetted by lightening at Hartbeespoort

the natural (and man-made systems) that enable and enhance nature's ability to provide goods and services and that allow nature to correct the imbalances. Vast natural forests (not plantations), large bodies of water, the earth's atmosphere and its crust all provide vital goods and services, which mankind has seriously impaired in the name of economic and social development, with little, if any, regard for the environment. The introduction of "green infrastructures", on a vast scale, would help to correct some of the imbalances.

Like most of the other respondents, he defines a green infrastructure in the context of the goods and services that emanate from nature and natural processes, which is in line with the SANBI definition. So far, the respondents' definition of the concepts 'green and ecological infrastructure' rests on the instrumental and mechanistic role of the environment in human society.

Respondent 12, another municipal official, linked his conceptualisation of a green infrastructure with a 'natural infrastructure', a concept used by the eThekwini officials during the launching of the UEIP in February 2013 (p. 6). He said that:

A green infrastructure…could mean a number of things. Those low carbon buildings can be considered as green infrastructure. But when you look at it more specifically, I think this links more to the work that our adaptation colleagues undertake. I think you might be referring to it in terms of your natural infrastructure.

Fig. 3.7 Snowfall over Hartbeespoort, with the Magaliesberg in the background, August 2012

It is safe to say that eThekwini's understanding of an ecological infrastructure as a 'natural infrastructure' functions as a consensus creator (p. 90) within the municipality, influencing members of the epistemic community who collaborate with the municipality (for example, Respondent 3). We see this concept of a 'natural infrastructure' from Respondent 13, a municipal official who works at the Durban Botanical Gardens (Fig. 3.9), when he says that:

> My understanding of a green infrastructure is everything that makes up green space, including your natural ecosystems, such as the nature reserves in the city, and cultivated areas, like parks, that act as a way of providing services and that work naturally to advance good quality air and water, and the social benefits of that. They stand in contrast to the typical engineered surfaces of the city, which are the structural benefits of the service, namely water retention and the cleaning of water as a resource. It is almost like a bio-mimicry of nature within the city. It speaks to a city's resilience.

When I asked him whether the Botanical Gardens is a green infrastructure system, or a component thereof, he said that: 'I think it is, to a large extent; it is very old green infrastructure and it is located within the city, so it is connected with the surrounding residential area'. For him, a green infrastructure is not something that occurs naturally, but something that is located within an urban environment. In this sense, the urban geography provides another structural mechanism that influences his construction of the concept 'green infrastructure'.

Fig. 3.8 The Dome Pools of the Magaliesberg near Mooinooi, September 2019

When she answered my question about her understanding of the concept, Respondent 14, a researcher from the INR, spoke about the operational context of a 'green infrastructure' in the following way:

> That is difficult, especially considering the context you are working in. My understanding is basically that they are your natural ecosystems, in all their states and forms. When I say

Fig. 3.9 Durban Botanical Gardens

natural, it could be your natural ecosystems, like your artificial wetlands and not your natural wetlands… In my head, I can [differentiate between a green infrastructure and an ecological infrastructure], because for me, the green infrastructure incorporates everything that it is; what would you call ecosystems that are not natural? I don't know what to call them. These are your artificial systems. So, in my head I don't think that the artificial systems are part of ecological infrastructure, but rather, they are part of a green infrastructure.

When asked about her understanding of a 'green infrastructure', Respondent 15, a lecturer at UKZN, answered in a relativistic way, by linking her understanding to the SANBI definition (p. 5):

Now I am thinking about the SANBI definition and about the built, non-built, and mimicked infrastructures. 'Mimicked' means when you build an artificial wetland. So, is it an ecological infrastructure or not? These are the questions I am asking myself so that I can answer the question. It is linked to the [uMngeni] Catchment.

She continued by saying that:

I am familiar with the term 'ecological infrastructure' and haven't given the term 'green infrastructure' any thought. I don't know if it would be any different; I would have to think about it. But if it is meant with the similar idea of environmental services from a [river] catchment, then it is about all the elements of a catchment, be they terrestrial [e.g. Figure 3.10] or atmospheric, and that they provide services to people with regard to water quantity and quality.

From the above, we see that the respondents define green or ecological infrastructures according to their actual experiences in which they work. In this sense, their reality of what constitutes a green or ecological infrastructure are mental constructs, together with relativism, that are defined by local and specifically constructed and co-constructed realities (Lincoln et al. 2018) and which play a central role in their

Fig. 3.10 The Drakensberg from the Monk's Cowl Nature Reserve, with a wetland in the foreground

conceptualisations. They link their understanding to positivistic assumptions through several structural concepts associated with nature and natural processes, like biodiversity, ecosystem services, green economy, sustainable development, SANBI's definition and the WEF nexus. Furthermore, the constructed and co-constructed realities of the respondents are social- and experienced-based, as well as local and specific. In other words, the certainties are also dependent on the form and content of the person holding to such a reality, although they may be influenced by the epistemic structural notions of the concepts. This means that there are multiple realities that are dependent on the individuals who embrace such realities (Guba and Lincoln 2005; Lincoln et al. 2011, 2018; Meissner 2017), and these are, in turn, constituted by the constructed and co-constructed elements. Having said that, their interpretivist/constructivist notions of the concepts is in line with systemic constructivism that is informed by empiricism and positivistic structuralist notions.

Five of the respondents (Respondents 3, 8, 10 [p. 96], 11 and 15) define an ecological infrastructure in terms of the SANBI (2019) definition. A few other researchers are familiar with this definition because they were either employed by SANBI, like Respondent 14, or have colleagues that work for the Institute. Many of the interviewed municipal officials are also familiar with the definition because they have collaborated with SANBI on ecological infrastructure projects in the past, particularly within the UEIP structure (p. 131). The definition, therefore, becomes part of the respondents' life-world, either through their direct or indirect affiliation to

SANBI, or their past employment. From this interpretation, we can deduce that there are several empirical and normative elements in their green and ecological infrastructure conceptualisations (Tables 3.1 and 3.2).

What is also noteworthy, is that many of the respondents defined green or ecological infrastructures in relation to their geographical features. Here, the uMngeni River (Figs. 3.11 and 3.12) is prominent, because it is a vital source of water for eThekwini and other upstream municipalities (Fig. 3.13). The uMngeni is also an important research subject for many of the researchers that were interviewed (Respondents 3, 4, 7, 8, 10 and 15). Furthermore, several respondents are part of the UEIP. In fact, Respondent 7 indicated that: 'I have the good fortune of sitting on the UEIP steering committee and am involved in the research project as well. So, my knowledge of green infrastructures has grown somewhat exponentially'.

Not only is the river a research subject for some of the respondents, it is also a source of major concern, especially with regard to the quality of its water (Fig. 3.14).

Table 3.1 The elements of green infrastructure conceptualisations

Empirical	Normative
Green open spaces—open spaces and the Botanical Gardens	'Buzz word' to access funding and not to reverse 'accumulated environmental damage'
Low carbon buildings	A word invented to accommodate engineers and economists in green infrastructure initiatives so that they can positively influence implementation
A concrete plan to realise climate adaptation	A manufactured word to suit other disciplines
Biodiversity that provides goods and services	A word used in the policy or applied field
Infrastructure we put in place so that an environmental and eco-friendly 'man-made infrastructure' functions as protection	A term that separates the environment and allows people to get on the 'bandwagon' in the battle for the environment
Reduce, re-use and recycle	
Natural ecosystems in all their states and forms, such as artificial wetlands	
The mimicking of nature	

Table 3.2 The elements of ecological infrastructure conceptualisations

Empirical	Normative
Biodiversity and wetlands, mountains, rivers and streams	Conservation
A scientific term	
Natural resources or natural habitats providing protection	
A concrete plan to realise climate adaptation	
Rivers and streams, and not septic tanks	
Floating wetlands—specific infrastructure to assist ecological aspects	
Natural infrastructure, like mountains, supplying goods and services	

Fig. 3.11 Howick and the Howick Falls from the air, with Midmar Dam on the uMngeni River in the background

Fig. 3.12 The headwaters of the uMngeni near Nottingham Road in the Natal Midlands

Hay (2017: 19), who is the Director of the INR, reports on the Dusi Canoe Marathon, an annual canoe race between Pietermaritzburg and Durban on the Msunduzi River, a tributary of the uMngeni, as well as on the uMngeni itself:

What the impact of reduced water quality on human health might be, is uncertain. A survey of Dusi Canoe Marathon paddlers immediately after the 2016 marathon indicated that 40%

Fig. 3.13 The Inanda Dam supplies water to the eThekwini Metropolitan Municipality. *Sources* Nortje et al. (2019), Meissner et al. (2018a, 2019)

of the field contracted mild to severe gastro-intestinal infections. This is not surprising, as paddlers were subjected to elevated pathogen levels on all three days of the race. Some experienced paddlers take antibiotic medication as a prophylaxis in anticipation of poor water quality.

Postpositivism

The paradigm that scored the third highest is postpositivism. This worldview is an altered form of positivism (Lincoln et al. 2011, 2018; Meissner 2017). The ontology of postpositivism rests on critical realism or a reality that exists, but one that is only imperfectly and probabilistically apprehensible (Lincoln et al. 2018). This means that there is a single reality, but we will never fully understand it or how to attain a full understanding of it, because of the hidden variables and the shortage of absolutes (Guba and Lincoln 2005; Lincoln et al. 2011, 2018; Meissner 2017).

When I asked Respondent 6, a municipal official, about eThekwini's policies, programmes and bylaws, he informed me about the following regarding Sustainable Drainage Systems (SuDS):

The storm water policies essentially give details about the type of materials and pipes. They will also deal with new developments in that…they will provide a storm water lens…We don't specify how, but we specify that we need to maintain previous development flow rates. Our present bylaws don't cover quality issues. Cape Town has incorporated quality issues

Fig. 3.14 The Msunduzi River, a tributary of the uMngeni River (Note the solid household waste pollution in the bottom picture)

in their bylaws, and we are hoping to learn from them. One of the challenges that they have is maintaining their standards, but is anybody going back to see if those standards are being met by the SuDS? This is part of what we and the [WRC], are trying to monitor. So, there are some assumed numbers; for example, if you have a wetland then you will be getting certain benefits, and if you have a roof garden, then you will benefit in another way. But no-one has sat down to qualify the benefits. Therefore, we are learning from the WRC and they are learning from us.

Considering the qualification of the benefits, eThekwini is in the process of attaining a fuller understanding by learning from Cape Town and the WRC about storm-water quality and how these will influence future bylaws. This means that eThekwini is aspiring to gain a fuller understanding through the WRC's empirical research and best practices in Cape Town. The notion of an ecological infrastructure also plays a role here, and specifically how wetlands will play a future role in the SuDS. We see that eThekwini has an 'incomplete understanding' of the role of wetlands and of SuDS, when Respondent 6 speaks about the River Horse (Fig. 3.15) initiative and the benefits thereof:

The idea is to carry out a long-term maintenance plan of the wetland. We know what the [River Horse] initiative will cost, but we cannot evaluate it at the moment because it only ended in 2017. We now need to work out what the benefits and costs will be. It is really about getting an understanding of what is happening in the catchment because it was ultimately a catchment management issue.

The answers provided by Respondent 6 indicate that they do not yet have a full understanding, therefore they plan to conduct studies to know the full extent of the benefits of the River Horse initiative. More of the 'hidden variables' will most probably be uncovered when the municipality has completed its assessment of the

Fig. 3.15 The River Horse wetland

benefits. Therefore, the future benefitial outcomes of the implemented project are uncertain.

The way in which Respondent 10 frames her understanding of the concept 'green infrastructure' falls within the postpositivist paradigm, as can be seen by her answer:

> They [green and ecological infrastructures] don't necessarily mean the same thing. A green infrastructure generally refers to an artificially created infrastructure, for example, artificial wetlands that are not naturally occurring features. My understanding is that their definition includes largely natural systems. However, I think you can view an ecological infrastructure as having a green infrastructure component, but not the other way around. An ecological infrastructure could potentially incorporate a green infrastructure, but it is not explicit in the SANBI definition of an ecological infrastructure.

The statement by Respondent 10 that '[a] green infrastructure generally refers to…', that 'an ecological infrastructure could potentially incorporate a green infrastructure' and that the SANBI definition 'is not explicit' about this incorporation, suggests a probabilistic and/or imperfectly apprehensible reality about the separation of green and ecological infrastructures. It also suggests that we will never fully understand the differences between green and ecological infrastructures, and that where the one ends, the next one begins. This is also evident from the discussion on the various elements contained in the two definitions. In the next section, I will elaborate further on the recommended actions that are expressed by the interviewees and how they are framed, from a paradigmatic point of view.

3.4.1.2 Recommended Actions

When the respondents spoke about recommended actions, they also framed these in positivist terms. The positivist paradigm received the highest score of 202, followed by interpretivism/constructivism, with a score of 32. Since the score for postpositivism was too low, I will only discuss positivism and interpretivism/constructivism.

From a positivist perspective, Respondent 5 spoke about the education of communities with regard to the importance and rationale of having green infrastructure projects that eradicate alien vegetation and rainwater harvesting. She said that the municipality does not 'educate people on the importance of why we need to clear alien invasives'. Furthermore, she indicated that it is important to educate the community about '[t]heir impact, as well as water security and how they use JoJo tanks. People also need to become aware of the strategies and practices that are used to save water, which is, in my opinion, a bit lacking'.

Respondent 7 also alluded to more research to highlight the economic costs and social benefits that green infrastructure initiatives could have for communities and the wider public:

> For me, it is an under-researched area because the impact of these projects is poorly understood. So, whilst you and other institutions might have a good understanding of the initiatives, I don't think that there is enough communication or a sufficient amount of feedback to the affected people that have benefited from the projects. Therefore, if you clean up the Msunduzi River, for example, there are many communities that will benefit from this. I am not just

talking about a local community living in a suburb; I am also talking about the multi-national companies that are situated on that river. For me, right now, the impact of these projects is extremely poorly understood… I think that for the Msunduzi and eThekwini [Municipalities] to engage in these projects, more studies need to be done on their social impact, and it needs to be made more tangible and readable for the lay person… Therefore, if we say we have saved 50,000 litres of water, and we write it in a newspaper article for the benefit of the community, there will be a readable benefit for the community. The experts understand the consequences and benefits of, for example, clearing three kilometres of alien vegetation. In other words, the faculty of people in UEIP understand the benefits, or see the benefit of an M.Sc. student studying the benefit of water quality in a particular river, but is the lack of water quality, or high *E. coli* count, being translated into terms that will have an impact on the people who are directly affected by, or who use, that resource? There is a gap! I think there is a space for us to improve that. My scientific mind tells me that we need to not only communicate it better, but also to understand it better. It is good to write scientific papers, but sometimes there needs to be more.

Respondent 9 related the municipality's plan to set up a water fund, in collaboration with The Nature Conservancy (TNC), which is a transnational interest group. According to the TNC, it uses science for the advancement of policies that address the problems faced by the environment and people (TNC 2019a). Respondent 9 noted that before one is able to establish such a fund, one needs to look at the feasibility thereof, in terms of the sources of revenue, and whether the municipality will be able to motivate downstream water users and beneficiaries to contribute financial resources to the fund.

Part of that feasibility process will be to investigate whether some of the catchments are worthy of further work. Once that feasibility study is done, one can see if it is viable, and whether the model in its current form will work for Durban. There is a whole process linked to that. Also, some of Durban's catchments, in terms of water security, are quite transformed, so you need to investigate whether investing in the ecological infrastructure will generate the yields, from a water security perspective. (Respondent 9)

According to the TNC, the rationale behind a water fund is the utilisation of nature-based solutions, or catchment conservation, to improve a city's water security.

A 'water fund is a mechanism for downstream users to directly or indirectly compensate upstream users for activities that deliver water benefits to the payer. Public and private water users, including businesses and local governments, invest collectively in conservation of catchments from which they source their water. Many cities are coming to see these funds to minimize treatment costs and reduce the chance of water shortages in the future' (TNC 2019b).

This quote indicates that the municipality conceptualises water security in a specific way, in that it relates only to water quality and quantity. What is also noticeable is that a water fund is a positivist and instrumental way of improving water security.

A water fund, therefore, is a financial mechanism that is used to address the water quantity and quality problems faced by local governments and the private sector. In this context, water users can invest in five strategies to reduce pollution and improve water security, namely, forest protection, reforestation, agricultural best management practices, riparian restoration and forest fuel reduction (TNC 2019b).

As has already been mentioned (p. 9), the feasibility study that Respondent 9 refers to is in the form of a cost-benefit analysis, which utilises empirical data and which is, by default, positivist planning (Meissner 2015a) or development that is framed by positivism. We can draw a similar conclusion from the River Horse project that Respondent 5 spoke about. The empirical data, therefore, will not only relate to the feasibility of funding, but also in terms of the potential impact it could have on the rivers that Durban depends on for its water resources.

When it comes to initiatives, such as the establishment of a water fund, Respondent 11's opinion about eThekwini's enterprises is as follows:

> eThekwini won the Greening award again this year, which is an indication that impressive work has been done. While there are pockets of excellence, much more needs to be done, on a significant scale, to contain global warming and to reduce greenhouse gases, transport costs, water and energy consumption and waste generation. "Sexy Projects"... get votes and 'appear' to address service delivery, while, in reality, they tend to deal with the symptoms and with unsustainable reactive measures. What is the cost-benefit of floating wetlands for removing nutrients and converting polluting chemicals, compared to that of properly managed exotic aquatic-weed control?

Here again, there is talk of cost-benefit thinking and the reduction of measurable elements relating to the environment, such as greenhouse gases and transportation costs, as well as a reduction in energy consumption. Respondent 11 quoted Duncan Hay at length, with regard to the water quality problems encountered in the uMngeni River basin. In 2017, Hay said that:

> What is abundantly clear is that our river basin [the uMngeni] is in a very unhealthy and steadily-deteriorating ecological state - it is being subjected to slow, but intense, violence... In order to reverse this state, an immediate, integrated, multi-agency response on a large scale is required. For this, technical and financial resources are required. These resources are available either in-house, in the various management agencies, or as grants and loans. A lack of resources is not the issue. We need inspired political and technical leadership in order to mobilise the state, civil society and the business sector to solve the problems. We are all in this together.

This quote also contains normative elements, namely, political and technical leadership and collaboration, with the response to this activity being framed in positivist language. Hay recommends that collaboration should occur in a structured and in triangular fashion involving three actors or groupings, namely, the state, civil society and business (Meissner and Jacobs 2016).

On a completely different matter, when I asked him about the social and financial costs and benefits of green infrastructures, Respondent 12 indicated that:

> With regard to financial aspects, I think we have a lot of financial support for operational matters like research, assessments and staffing. The problems come in when we deal with the capital aspects. I have been saying it for four years. All the international support we are getting has to do with operational support for research and assessments. What we need more funds for is largely for the implementation of capex.

Respondent 12's concern was more for the maintenance and upgrading of capital expenditure, especially the physical infrastructure, such as water reticulation systems and SuDS. His answer suggests that eThekwini relies heavily on research and assessments to inform its policies and plans, and that it spends a lot of financial resources on such endeavours and too little on capital expenditure.

Furthermore, he believes that the huge migration to the city results in burgeoning urbanisation. This migration is not only from other parts of South Africa, but also from the rest of Africa, which overburdens Durban's infrastructure. Hence, there is a need to shift the expenditure from research and assessments to capital expenditure projects.

Respondent 16 spoke at length about various technological options, including nature-based solutions, for dealing with the solid waste from wastewater treatment plants. He gave a specific example when he said that:

> There is a decentralised waste water system in Newlands, Kwamashu, which serves a block of houses. The water goes through a wetland and this wetland supplies water to a market garden where they grow crops. The solid matter can be turned into pellets and there is a lot of technology available to refine the solid waste material, like sludge, and turn it into useable stuff. Then there is the black soldier fly project. We use the larvae that are grown on the sludge...to extract oil from them. Therefore, if you deal with the sludge, at the same time you get products from the larvae. The next step would be to do it on a large scale. This means that you have all these baskets of technologies that can deal with this resource, namely, the solid waste..., which contains incredibly valuable phosphates. I think it is important that you have an ecological infrastructure, a built infrastructure and a circular economy that meshes them all together. The black soldier fly larvae would also be part of the ecological infrastructure, as they are part of a nature-based process.

With positivism in mind, the respondents framed their recommended action through the lens of 'Who governs and who benefits?' (Hobson and Seabrooke 2007, 2009). This analytical organising question assumes that there is a hierarchical order (Table 2.1). This grading consists of the state actors, who govern at the top, and the non-state entities or individuals who benefit (Hobson and Seabrooke 2007) at the bottom. We see this manifesting in the responses of Respondents 5, 7, 9, 11, 12 and 16. Respondent 5 said that the eThekwini Municipality should do more to educate people and communities about alien vegetation clearing and rainwater harvesting, as this would benefit the people and communities. Respondent 7 recommended more studies into the benefits of green and ecological infrastructures for communities and the wider public. Respondent 9 also mentioned the feasibility of the water fund, while Respondent 11 noted that the municipality should govern more for the benefit of everybody in the uMngeni River basin, and Respondent 12 argued that the municipality should invest more in 'capex projects' that will, by implication, benefit Durban's residents. We also see this in Respondent 16's lengthy quote about a decentralised wastewater treatment plant and the use of the black soldier fly for treating solid waste from the plant. This enterprise will result in the Newlands community benefitting from the municipal initiative.

In all the recommended actions above, the respondents argued that the eThekwini Municipality, as a structure of rule, as well as part of a capitalist economy (Hobson and Seabrooke 2007, 2009) is the entity that is responsible for improving the water and human security of both the municipality and its inhabitants. For instance, Respondent 9 specified how the establishment of a water fund could solve the municipality's water security issues, while Respondent 12 spoke of the 'incredibly valuable phosphates' being a source of income.

The prime empirical focus in all the respondents' recommended actions is the supply of order and welfare maximisation (Hobson and Seabrooke 2007, 2009) by the municipality and, to a certain extent, the TNC. In this regard, the locus of agency is top-down (Hobson and Seabrooke 2007, 2009), which means that the municipality and the TNC are, or should be, the change agents that address the various problems (i.e. the lack of education, water insecurity and pollution in the uMngeni) experienced by the stakeholders (i.e. the people, communities, the private sector) within the municipality's boundaries and beyond (i.e. the uMngeni River basin).

Moreover, a prevalent structuralist ontology (Hobson and Seabrooke 2007, 2009) is present in the respondents' recommended actions. This relates to the municipality as a structure of rule and the notion that, as a state entity, it is more powerful for constituting change than individuals are. People view the municipality as an entity with vast financial and knowledge resources, and as being the sole enabler and enforcer of bylaws and other legislative aspects, like SuDS. From the discussion on the potential formation of a water fund, we see that Respondent 9 also places a lot of emphasis on a feasibility study to ascertain the viability of the fund. This shows that the ethics around the water fund and its feasibility study is of a positivist nature. Ethics is defined here as the relationship between the subject (altered river catchments) and the researcher (municipal officials and/or UKZN researchers), together with the influence of research (feasibility studies) on the population (eThekwini's residents and private sector entities) (Lincoln et al. 2011, 2018).

Furthermore, the 'hegemony' of research in the answers given by Respondents 9 and 11 is a positivistic sign. Lincoln et al. (2011, 2018) describe research hegemony as the influence of researchers and their research on others. For Respondent 9, the purpose of the assessments (feasibility studies) is clear, namely, to determine the financial costs and benefits of the water fund. This also goes for Respondent 6's explanation of the planned study to investigate the benefits of the River Horse project. For Respondent 11, Duncan Hay is a voice of authority and reason on matters pertaining to the poor water quality of the uMngeni River.

The recommended actions are, moreover, in line with the positivist view, namely, that objectivity should guide the municipality's actions to bring about change in Durban's water security aspirations. In other words, the dictum of 'science speaking truth to power' (Eriksson 2014) (see pp. 33–34) is of relevance. According to Eriksson (2014: 96) '[t]his notion of "speaking truth to power" expresses a strong belief in the enlightenment function of scholarship [or research] and implies a traditional fact/value distinction'. Put differently, the research should be 'speaking truth' to the municipality in the formulation of policies around the water fund, the River Horse project, community participation and the deplorable state of the uMngeni River.

This is also applicable to Respondent 7's recommended action, when he spoke about the role of the epistemic community in investigating and communicating the benefits to a wider audience. He is asking researchers to conduct more studies on the social benefits of green infrastructure initiatives, thereby invoking the role of the epistemic community to find the 'truth' about this and to communicate it to the municipality. Such research would give policy-makers an explanation of how people perceive green infrastructure projects. The public could then, through research that is expressed as 'truth', convey their wishes for projects that benefit eThekwini's residents directly. We see that his recommended action is to gain an increased knowledge of the benefits of green infrastructure initiatives by interpreting how individuals that could benefit would perceive and interact (Lincoln et al. 2018) with municipal officials, via the epistemic community.

3.4.2   Theories

The dominant theories expressed in the interviews are presented in Fig. 3.16. Two perspectives, namely, the hydrosocial contract theory and neo-liberalism, had exceptionally high scores, with 316 and 289, respectively. Five theories followed, namely, social constructivism, risk society, interactive governance, modernity and the social learning and policy paradigms, with scores of 148, 147, 141, 140 and 128, respectively. A theory that stands out among the lower scored perspectives is the Regulatory International Political Economy (RIPE), with a score of 98. I will lump together interactive governance, the RIPE, as well as the social learning and policy paradigms, under neo-liberalism. Since these theories fall under neo-liberalism, I will relabel them as liberal intuitionalism (more on this later). In the following section, I will discuss the manifestation of these theories in the interviews and interpret the significance of their presence.

Why did the hydrosocial contract theory receive such a high score? Is it because my study focuses on green and ecological infrastructures in promoting water security, or is it because water plays such a fundamental role in eThekwini's green and ecological infrastructure policy arena? I would say that it is a bit of both.

When I designed the study, my purpose was to investigate how green infrastructures contribute to Durban's water security. The questions that I put to the respondents always revolved around the notion of green infrastructures and their link to water security. This implies that the questions that had water security as their central focus (p. 84), guided the respondents to think about the concept and 'forced' them to talk about water security aspects within eThekwini's green infrastructure policy landscape.

Even though this was the case during the first two interviews, the respondents mentioned the concept of an ecological infrastructure, which I had initially overlooked in the original design. During and after these face-to-face encounters, I started asking the follow-up question concerning the difference between green and ecological infrastructures (p. 84). The value of an open-ended analysis shows itself here, in that the researcher is not guiding the research along a certain path. The respondents also played this role by way of their ontological awareness. Nevertheless, by asking this follow-up question, I realised that although the main aim of a green infrastructure is not to promote water security, but biodiversity protection and climate

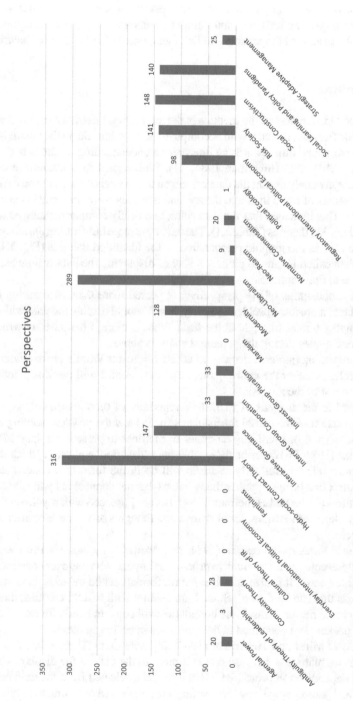

Fig. 3.16 Perspectives present in the analysed interviews (derived from Meissner 2017)

change adaptation, an ecological infrastructure places the concept of 'water security' at centre stage. The uMngeni and Palmiet Rivers are, in this sense, important ecological infrastructure components that link them, ontologically, to water security.

3.4.2.1 Matrices

To discuss the 'appearance' of the theories in the interviews, I needed a practical and communicable framework that would accomplish the '…scientific, in the positivist's sense of the word, and aim towards an interpretive understanding, in the best sense of that term' (Miles and Huberman 1984: 21). With regard to communication, I realised during my analysis that the data contained in the interviews were comprised of the assumptions of more than one theory, and that they were, in some instances, simultaneous. This indicated that the data cannot be neatly compartmentalised into each of the identified theories (Table 2.1). The data were more fluid during the analysis than was the case in the graphic presentation of the identified theories (Fig. 3.16). This dynamism called for me to pioneer a way of displaying the data from which I would be able to draw conclusions.

Since the propositions of the perspectives appeared more than once during the interviews, I had a practical problem, namely, how I should display the data without repeating lengthy quotes throughout the text? What is more, I had already written down elaborate quotes during the paradigm analysis above.

To avoid repeating quotes, I decided to utilise Microsoft Word's cross-reference function to refer to either the entire quote, or parts thereof. I will use this practical method in the rest of the analysis.

With regard to the analysis of PULSE[3]'s repertoire of theories (p. 61), as well as the interpretation of the causal mechanisms (p. 58) and the problem-solving and critical theories (p. 63), I will rely on tables or matrices to present the data. Miles and Huberman (1984: 26) describe these tables as a 'descriptive matrix', which they found 'uncommonly fruitful'. My matrices will be in the form of a conceptually clustered-matrix that brings together the variables that are connected by the theoretical assumptions (Miles and Huberman 1984). Before I proceed with my theoretical conceptually clustered matrices, I will share a few thoughts on the development and use of matrices.

According to Miles and Huberman (1984: 26), 'Matrix formulation is, in our experience, a simple, enjoyable and creative process, as anyone who has ever constructed a dummy table knows. It is also decisive: matrix formats set boundaries on the type of conclusions that can be drawn'. Since I am dealing with several theories, theory classes and causal mechanisms, these boundaries will come in handy for organising the densely packed data contained in the numerous interview quotes.

Furthermore, Miles and Huberman (1984: 26) note that: 'There is a catch, of course [in using matrices], because one is limited to the data in the display. Thus, a great deal depends on the core [data] that have been selected from the field notes [interviews and photographs], and how far they are aggregated or abstracted'. It is for this reason that I will rely on Microsoft Word's cross-reference function to confirm the presence of theoretical assumptions, theory classes and causal mechanisms in the interviews. Before I do so, I would like to mention a methodological challenge.

3.4.2.2 Methodological Conundrum

When identifying the perspectives contained in the respondents' answers, I was confronted by a peculiar methodological challenge. It is unlikely that the respondents have heard of theories like 'agential power' or 'social constructivism'; however, the elements of these perspectives that are represented in Fig. 3.16 above are known to me, and I identified them when analysing the respondent's answers by using PULSE[3]. Since I can identify the theoretical elements, and the respondents are unaware of them, there is no exact match between them and the respondents' answers as they appear in scholarly texts. How then can one relate the respondents' answers to a scholarly text? Eisner (2017: 4) gave some ideas on this when he asserted that it is quite important to emphasise 'voice' in qualitative research, not for the mere sake thereof, but because it 'serve[s] epistemological interests'. He went on to explain that:

> What we look for, as well as what we see and say, is influenced by the tools we know how to use and believe to be appropriate. The language of propositions, that language fundamental to the empirical sciences…cannot take the impress of the life of feeling. For feeling to be conveyed, the "language" of the arts must be used, because it is through the form a symbol displays that feelings is given virtual life. The point, therefore, of exploiting language fully is to do justice to what has been seen; it is to help readers come to know… Arnheim believes that most knowledge is visual in nature and that propositions and visual art are two ways of representing what has been conceptualized. Goodman argues that there are as many worlds as there are ways of describing them, and that the worlds we know, are the worlds we make.

I would therefore like to argue that the theoretical elements that were identified during the analysis of the respondents' answers, serve a particular epistemological interest: they contain the thinking process and the relationship between what the respondents know and what they see (Lincoln et al. 2018). For me, the theoretical elements that I have identified over the many years of reading theory are the most appropriate tools, and I believe that they are useful for analysing the respondents' answers. The scholarly texts are a language of propositions that are fundamental to the empirical and social sciences. Theoretical reflection is therefore relevant, in its broader social purpose, (Selby 2018) for explaining and understanding the world around us. Furthermore, I believe that theories are like paintings, in that they give us a glimpse of the world that is captured within a bounded frame. The utterances of the respondents during the interviews are also theoretical frames, since they describe their world when they work with green or ecological infrastructures.

There is, therefore, a common denominator at play here, namely, the theoretical frames that are contained in the analysed answers and those represented by PULSE[3]. Both the PULSE[3] theories and the respondents' answers, which conform closely to the theories, are conceptualisations of the many worlds that we know, whether it is my PULSE[3] analytical framework world or the world of the respondents' work relating to green or ecological infrastructures.

In the section that follows, I will present the 'voice' of the respondents and how they conform to, or are like, the PULSE[3] theories. By expressing their voices, the nuances of their world of working in the green or ecological infrastructure domains come to the fore. The section will unfold as follows: I will first give a brief description of the theory, as described in scholarly texts, and then I will capture the respondents' answers that contain elements of the theory, in order to explain how the answers conform to the theoretical elements.

3.4.2.3   Hydrosocial Contract

The hydrosocial contract theory, which was first mooted by Prof Jeroen Warner from the Wageningen University, emphasises the natural resource and its structural and normative constituting role. The hydrosocial contract theory gives an insight into how water resource management practices originated (Meissner and Turton 2003) and how they evolve in society by means of two transitions. It is an unwritten contract between the public and the government (Turton and Ohlsson 1999), and the transitions permeate into a Hobbesian contract, which characterises the first transition, and the Lockean and Rousseauian forms, which typify the second transition (Warner 2000a, b, 2004; Turton and Meissner 2002). In the following pages, I will explain the features of the theory, before identifying its elements in the interviews.

The first transition

The first transition takes place when a social entity encounters water scarcity that is caused by a major event like a drought. Other societal, economic, environmental, political and institutional changes can also act as independent variables of this transition, a typical example of which is the industrial and mining pollution of existing water resources. These events culminate into a Hobbesian form of the hydrosocial contract. During this time, a bipolar configuration appears between the government and the water-consuming public, which closely resembles the writings of Thomas Hobbes (1651), who argued that the state takes the final responsibility for securing water provision in all aspects of society, either through dams, pipelines and hydropower installations, or the hydraulic mission (Turton and Meissner 2002; Warner 2004).

The desire to control nature as a norm constitutes the philosophical foundation of the hydraulic mission. This norm is evident in the writings of Francis Bacon (1561–1626) and René Descartes (1596–1650) (Turton and Meissner 2002). The positivist paradigm, and particularly Newtonian physics, with has the longing to control nature, underlines this norm (Turton and Meissner 2002; Meissner 2017). In this regard, Bacon (1620) argued that humankind can use 'the noble discoveries' of science to 'renew and enlarge the power of the human race itself over the Universe' (Kitchen 1855: 129, cited in Turton and Meissner 2002). The subsequent work of Descartes (1637) supported Bacon's thesis when he argued that:

[I] saw that one may reach the conclusions of great usefulness in life, an[d] discover a practical philosophy [the natural sciences]...which would show us the energy and action of fire, air, and stars, the heavens, and all other bodies in our environment and [we] could apply them...and thus make ourselves masters an[d] owners of nature. (Anscombe and Geach 1954: 46, cited in Turton and Meissner 2002)

The norm of controlling nature is still relevant in the natural sciences today and, in the case of the hydraulic mission, it is manifested in hydraulic engineering (Turton and Meissner 2002). As already mentioned in Chap. 2, this philosophical foundation influences the way in which scientists construct knowledge and, in turn, it has a bearing on the way in which we interpret information (Turton and Meissner 2002; Meissner 2017). Robinson (1934: 236) sums up the special place of engineering and the engineer in the following way:

In our control of physical nature we are served by a special class of men called engineers, who are rigorously trained, not only in the practical tricks that can be used in the harnessing of physical forces, but also in the mathematical and experimental sciences necessary for straight thinking about the physical world. Within their own sphere of activity the engineers consti-tute, in the finest sense, an élite. It is generally accepted that the outstanding peculiarities of present-day western civilization are principally the contributions of the engineer.

As a response to dynamics, such as drought and pollution, which result in a scarcity of water, water conservation, through the construction of reservoirs and irrigation schemes, becomes an act that takes water from a water body and supplies it to areas in need (Turton and Meissner 2002). This first transition, therefore, constitutes adaptive behaviour through engineering solutions, or a coping strategy to control the natural environment (Turton and Ohlsson 1999).

As mentioned above, the first transition characterises a Hobbesian-like contract, which is manifested in a bipolar arrangement between government and society (Warner 2000a, b). Power relations creep into water management. The state and engi-neers take the responsibility of supplying water through engineering projects, like the Inanda Dam (Fig. 3.13), which grants the engineer a privileged position (Robinson 1934). The contract gives the government a mandate to assume and execute its responsibility, and acts as the foundation for the development of institutional arrange-ments, such as government departments and catchment management agencies. The contract also indicates to the public what fair and legitimate practices are like, for example, sustainable development. Politicians and engineers, therefore, dominate the first transition, with the government being the custodian of water resources and engineers determining the sanctioned discourse. Social instability could result if the state cannot deliver on its promise; the state can either ignore society's plea for water provisioning, due to capacity constraints or corruption, or it may supply water to a select few, as was the case during Apartheid (Turton and Ohlsson 1999).

Second transition

The second transition occurs when water deficits become visible, in the face of engineering solutions (Turton and Meissner 2002). The result of this transition is a Lockean type of hydrosocial contract, which is characterised by a triangular config-uration between the government, the public and interest groups, or other elements in

Fig. 3.17   Businesses close to the River Horse wetland

civil society. John Locke emphasised the triangular role of civil society in politics, as opposed to Hobbes's bipolar configuration (Warner 2000a, b; Turton and Meissner 2002).

During the second transition, two elements are vital. The first is that the implementation of engineering solutions is costly, in that government and local authorities find it increasingly difficult to finance water infrastructure programmes when no other source of water exists.

Secondly, as social conscience surfaces in the form of environmentalism and sustainable development, civil society becomes prominent. This psyche argues that supply sided solutions are environmentally damaging. Within an urban context, this awareness transcends from one where large cities with complex patterns of interaction move from an agenda of demand-supply to 'multi-criterion models' that take into consideration factors like the impacts of climate change, differing settlement patterns, human vulnerability, as well as resource optimization, in an effort to balance the society-ecology relationship (Ghosh and Kansal 2019: 59).

The hydrosocial contract theory postulates that if society does not handle this transition in a sound manner, it can cause social instability, like political unrest and varying levels of civil disobedience (Turton and Ohlsson 1999). By lobbying their activities, interest groups become the major actors in this transition (Meissner and Turton 2003; Meissner 2004, 2005, 2016; Warner 2012). Warner (2012) proclaimed that, 'Lockean liberalism stresses freedom, diversity, tolerance and interdependence... For a Lockean, the prime mover for releasing and harnessing ingenuity is not the state, but civil society: the key social relation is horizontal, not vertical'.

Through the normative elements mentioned by Warner (2012), these actors mobilise support for water infrastructure projects, such as large dams (Meissner 2005) (Fig. 3.23), and they demand that water resource development becomes more

Fig. 3.18 Municipal workers cleaning municipal solid waste in the wake of the devastating April 2019 storm. Photo courtesy of Douw Steyn, Plastics SA

Fig. 3.19 View across the uMngeni Vlei—conceptualised as an ecological infrastructure. Photo courtesy of Duncan Hay

Fig. 3.20   uMngeni Vlei Note the snow-capped hills and Drakensberg Mountains in the background. Photo courtesy of Duncan Hay

Fig. 3.21   Midmar Dam spill way.   Photo courtesy of Duncan Hay

Fig. 3.22 oHlanga Estuary, at Umhlanga Rocks, north of Durban

sustainable. They also voice their concern against large-scale water pollution. Then they become new members of the discursive élite, with the sanctioned discourse embracing the principles of sustainable water utilisation (Turton and Ohlsson 1999; Meissner 2005). Therefore, interest groups often enter into an agreement with government entities (Meissner and Turton 2003), in order to gain a place in the negotiation of water management principles.

Warner (2004) suggested a third type of hydrosocial contract after he observed conflicts over the privatisation of water and infrastructural conflicts, which highlighted a crisis for the Lockean contract. Water privatisation is an abomination to the Rousseauian social contract (Rousseau 1762; Warner 2004), since Jean Jacques Rousseau, the original social contract theorist, argued that a government should look after the rights of people and social equality. He dismissed the notion of private property and saw government and morality as having a strong bond with each other. If a state wants to lose its legitimacy, it only has to become immoral (Warner 2004). An argument that is rooted in Rousseauian philosophical foundations would argue that local communities and citizens are the best protectors of water and, therefore, must be respected as equal partners with the government in regulating and protecting water resources (Warner 2004).

Green and Ecological Infrastructures through the Hydrosocial contract theory lens

In the next section, I will highlight the transitions that I identified during the interviews in Table 3.3.

In Table 3.3, we see that the first transition is characterised by several features or themes. eThekwini is situated in a disadvantageous geographical position in the uMngeni Catchment; it is the last urban centre to receive water by means of the natural hydrology of the uMngeni River. In order to supply the municipality and its residents with water, national government and the municipality have implemented a number of coping strategies over the years, in the form of large dams (Figs. 3.13 and 3.23), aqueducts (p. 112) and water purification plants. This was, and still is, out of a necessity to supply water to a growing urban centre around the port of Durban, with an accompanying growing population and industrialised centre.

Because of the eThekwini Municipality's downstream position in the uMngeni Catchment, with its poor water quality properties (p. 95), it has started negotiating with other upstream municipalities to address pollution problems (see Figs. 1.2, 1.3, 1.5, 1.6 and 3.14) as a water security facilitation mechanism (Meissner et al. 2018a). According to Respondent 4, the municipality is doing this in an impressive manner and has also championed the UEIP to facilitate the promotion of an ecological infrastructure and to assist in securing better quality water for its residents and for the industrial complex. It is therefore not impossible to conclude that eThekwini's downstream location has forced it to change its discourse, as indicated by Respondent 15 (p. 112).

However, what is absent from this discourse is the role and function that estuaries might play in eThekwini's hydraulic mission, as expressed earlier by Respondents 1 and 2 (p. 95). This indicates that the municipality's downstream position is forcing it to look upstream, while it ignores the role of estuaries in the interaction between freshwater resources and the marine environment (Meissner et al. 2018a, b). eThekwini, as a popular tourist destination, relies on the water quality of the Indian Ocean, directly offshore from its beaches, as a source of tourist revenue. The ocean and eThekwini's beaches are, after all, part of its conceptualised ecological infrastructure landscape.

Furthermore, the ecological infrastructures, together with the political structures (i.e. UEIP) constitute an extended coping mechanism to further its water security aspirations and goals. This is the case, because SANBI, with its normative definition and instrumental policies, played an influential role in the establishment of the UEIP (Respondent 4). SANBI, therefore, forms part of the élite that informs a combination of grey with ecological infrastructures. We should also note that eThekwini employed engineers who contracted with SANBI in a bipolar configuration and, in a way, co-opted a concept that is closely aligned with the engineering profession (p. 5).

A casual observation would place this under the second transition. However, I would argue that SANBI's role and involvement constitutes a Hobbesian-type of contract, since the organisation and eThekwini formed a bipolar, instead of a triangular Lockean-type configuration. That being said, the structure constitutes a specific hydrosocial contract type. As Respondent 4 mentioned earlier, the purpose of the concept 'ecological infrastructure' is to accommodate engineers and economists (p. 90), so that they can play a meaningful role in promoting not only ecological

Table 3.3 The hydrosocial contract theory

Theory	Assumptions	Interview data exemplifying assumptions	Themes or a case of
Hydrosocial contract	First transition	'[W]e do create artificial wetlands and detention ponds and retention ponds.... So, what we did in the past was to use sports fields and parks to mitigate flooding downstream. Then, you divert the excess storm water into them and they fill up. You store that water and then you release it slowly. Instead, the engineers at the moment like to get rid of everything quickly, which is sometimes not the right thing... For me, that would be an ecological infrastructure.... It is like a series of ponds. We need those flood plains, right down to the estuaries...and we need to protect them and get them to work properly. Estuaries are heavily impacted—we've got sand mining in our rivers, which impacts the estuaries. We've got pollution, which is a big issue. We've got quite a lot of alien invasive species, which are in the estuaries because of nutrient levels, and the nutrient levels are high because of sewerage pollution in the main' (Respondent 1)

'Estuaries are quite a good example....they are huge natural assets [Fig. 3.30] and they offer some of that protection. But their functioning is dependent on the amount of water coming in, or on sediment movement, whatever the natural dynamics are. But these have already been upset in some way...due to water abstraction. I think it just becomes quite complicated when it gets down to the coastal [green infrastructures] along the coast, especially when you look at the city [of Durban]' (Respondent 2)

'Then we get to the Umhlangane [River project]...it is Geoff Tooley's project. In their minds, it is much cheaper and more effective to use river and stream systems to manage water than it is to use storm-water systems. Building storm-water drains to regulate water flow i.e. for flood mitigation, is much cheaper for rehabilitating a river or stream system than to go and build a bunch of storm water systems that gets knocked out with every flood. The rehabilitation of environmental systems is linked to job creation. I personally think it is a wonderful project, so long it is institutionalised—through policy and investment by the municipality and not so much with external funding. It needs to become a normal practise and not a temporary project' (Respondent 4)

'Apart from negotiating with Umgeni Water and [Department of Water and Sanitation] (DWS), which Durban is doing very powerfully, because they receive about 76% of the water that Umgeni Water produces. It is by far the largest consumer of Umgeni Water's water, which puts them into a powerful position, and it uses this to leverage consistency of supply, as well as other things. As it relates to nature [ecological infrastrucres], they recognise that their water comes from upstream and they are at the end of the pipe and, together with SANBI, they were instrumental in setting up the UEIP.... So, this might not be a formal policy, but there is a strategy to engage upstream. Where it is problematic in terms of them engaging upstream, it is difficult, but not impossible, for a municipality to invest outside its geographical area. It is not impossible, it can happen... So, it can make a direct investment by championing the UEIP and by deploying staff to be part of it' (Respondent 4)

'When eThekwini initiated D'MOSS [Durban Metropolitan Open Space System] 25 years ago, 'the idea was to have natural areas that could support water supply and clean water (Respondent 8)'. With respect to the [Working for Ecosystems] WfE, Respondent 8 furthermore noted that: 'They clear the land and they look after it for the ecosystem's benefit and specifically for water or fire. Biodiversity is the main emphasis. It is the clearing of alien invasives... The idea is to work for ecosystems by mobilising people and their understanding and then care about the environment and protect water' (Respondent 8) | Water management practices through engineering

Ecological infrastructure as part of the hydraulic mission

Political structures informed by Hobbesian thinking to involve communities and citizen science (WRC and UKZN)

Pollution and environmental changes due to human influences |

(continued)

Table 3.3 (continued)

Theory	Assumptions	Interview data exemplifying assumptions	Themes or a case of
		'Durban has realised that the uMngeni River is a bit like a sewer, taking all the challenges of [KwaZulu-Natal] KZN down to Durban. There is the huge pressure of water security, and I would say that all the initiatives are integrally concerned with water security, both sorting out leaks and wastage and so to trying to protect the catchment supplying the water...' During the flood of 2017, 'It was interesting because the N2 [highway] flooded and lots of people lost their lives and hundreds of cars got washed away. One of the main reasons that it was so bad was that the culverts and drains were full of litter. Therefore, human behaviour made the flood so much worse. Just south of Durban there is a citizen science project at Amanzimtoti, and they've been doing community action for five or six years, which has resulted in no flooding on the N2. So, within a 10-kilometre radius, you have extreme flooding, where there was no community action, and less flooding, where there was community action...' (Respondent 8)	
		'These are natural systems that almost provide a natural engineering function, whether it is flood attenuation, purification, and so on. This can mean anything, from our natural ecosystems through to things like our green roofs, but then moving more towards certain technologies and the design of infrastructure, like sustainable urban drainage systems' (Respondent 9)	
		'Within that setting there is obviously the securing of our riparian zones and the fact that there should be enough plants and matter that will clean the water as it goes by. We are always thinking about how to harvest rainwater for use in the ablutions or minimise the use of municipal water at this site [Buffelsdraai Landfill]' (Respondent 5)	
		'SuDS is looking at the actual engineered infrastructure to manage storm-water flow, whereas water-sensitive urban design looks at the whole water cycle and it is more inclusive of the water cycle' (Respondent 6)	
		'A green infrastructure, in my understanding, does not incorporate natural systems, but it refers largely to constructed natural systems, like artificial wetlands' (Respondent 10)	
		The Botanical Garden's 'high water table' could be a source of water for the gardens '…we have water as a resource, and we try to look at ways of trying to utilise that water. It is also situated on the slope of the Berea, so we have gravity feeding the system. We pick up storm water from the storm-water system to use for irrigation. There is a lot of potential we can still explore. The work has started, and it is being managed by Geoff Tooley's department. In the first phase, they installed the take-off pipe from the storm-water system and the next part involves the design of the constructed wetland, which still needs to happen. Once those tests have been completed, it will determine the design or the extent of the design of the constructed wetland. Before that, they need to test what comes out of that system. We would like to build a wetland in the garden that is connected to the storm-water system. The lake in the garden is a source of water, since it discharges through the nursery. It is a source of irrigation. It gets fed by the ground water, which means that the aquifer is quite shallow. The water bubbles up through the ground. Geoff was saying that there might be an amount of storm water that percolates through the system into the ground and then feeds the lake. There would then be some continual flow. We want to collect the storm water system for use in the Botanical Gardens because we are using a lot of potable water in the garden for irrigation purposes. Litter in the storm-water drains is a major issue because it clogs the storm-water system during big rain events and then you get some erosion from the overflow of water…' (Respondent 13)	

(continued)

Table 3.3 (continued)

Theory	Assumptions	Interview data exemplifying assumptions	Themes or a case of
		'I don't think that they have policies directed towards [green infrastructure and water security], but they do have game strategies related to it. They do have a strategy that relates directly to the ecological infrastructure, but it does not link the ecological infrastructure directly to water security. However, there is an indication of management showing the role of ecological infrastructure in the management of floods…. They have projects like the Sihlangimvelo Project that looks at rivers and rehabilitating them [to] improve water security… I don't think that these are…directly linked to their policies in terms of water. I think it is slowly getting to that point because that project now has the Water and Sanitation Department of eThekwini and the climate change project' (Respondent 14)	
		'We are always thinking about larger elements because they are easier to quantify with regard to providing water services. The volume of water that they can give or the quality of water that they can create is more about the dimensions of the infrastructure. [A]lien vegetation, in terms of ecosystem goods and services in the context of South Africa, is something very negative, in terms of water usage—water hyacinth has infested areas and you cannot get access to reservoirs (Respondent 15)	
		Linked to eThekwini's geographical position, its negotiations with Umgeni Water and its UEIP involvement, Respondent 15 also said that: 'I think their activities and their discourse are shifting because they realise that they need the [uMngeni] catchment for their water needs. It is not only about telling Umgeni Water that they are the biggest user and biggest payer and that they must be serviced first. There is more of an acknowledgement and preparedness to engage with the catchment and the elements throughout the catchment. Come to think about it, my conversation about water and the catchment have come mainly from the thinking of the 'climate change people' in eThekwini. The debate is around water resources within the uMngeni Catchment. Even so, I've never heard them discuss estuaries, which is obviously a big element. The estuaries are a big problem. The UEIP is not about the estuary or the marine environment, at least not while I have been involved in the last two years. The Palmiet is linked to that WRC uMngeni project around [ecological infrastructures], which is quite high on the agenda'	
		'The reason why [the Palmiet River rehabilitation] project happened was because they had three proof of concept projects for the UEIP… The idea was that if you build the reconstructed wetland downstream of the informal settlement then all the water sort of dis-surface. What is happening at the informal settlement gets cleaned by the wetlands by the receiving waters of the uMngeni. So, the uMngeni is the focus and not the communities, to ensure that the uMngeni is not filthy… As the focus changed to the community's relationship with the river, it became a core part of that and you would sit by the river and see it take our refuse away and think that it provides us with a service… Through the UEIP and the Western Aqueduct system we managed to reach up into the full catchment approach' (Respondent 16)	
		'For instance, the water that we re-treat is of a much better quality that the water we receive in the uMngeni catchment flowing into the Inanda Dam because of the malfunctioning wastewater treatment works its water is much worse than our treated water. What people don't understand is that the water they drink from the tap has been treated and the river water is much worse. Then there is the cost of the treatment. The solution 'from toilet to tap' is very important from a water security point of view, but it is also about how you manage your system and the catchments and the ecological infrastructures therein. It is also about your built infrastructures and how you operate them because some of the systems are so old. Do you now go and rebuild your sewer systems and centralised water treatment works, or do you now go and decentralise your water treatment works?' (Respondent 16)	

(continued)

Table 3.3 (continued)

Theory	Assumptions	Interview data exemplifying assumptions	Themes or a case of
		'From our perspective in the climate change strategy, we focus on water security, namely, the provision of water as well as the reduction of flooding, two very important aspects of ecological infrastructure. There is a very clear reference to green infrastructures and the need to conserve ecological infrastructures and water quantity and quality provision. The Palmiet Catchment initiative is part of the UEIP that has fallen on us. Originally it was driven by water and sanitation and to create wetlands below the Quarry Road informal settlement. One of the problems was that the municipality has a very difficult relationship with the Quarry Road informal settlement. When Neill retired and we took over the mantle of that project, we did a rethink straight away and we started working in the informal settlement with Cathy [Sutherland] and her research team. We started a stakeholder process and it became a much more stakeholder-driven process and a catchment-based project. We developed an action plan for the catchment rehabilitation.... So, there was a complete change, but it is still around ecological infrastructure' (Respondent 16) 'These programs are context-specific and they have differences in where they are implemented. However, there are [several] broad principles that they work on. First, they are based on the premise that you are looking after your natural assets, be it in a catchment, or part of the catchment, because those assets will then look after you, if they are well-managed. If they are not well-managed and they are full of alien invasives, or if they are full of solid waste and they block up your infrastructure, you have rat infestations and black mambas coming into feed on the rats' (Respondent 16)	

(continued)

Table 3.3 (continued)

Theory	Assumptions	Interview data exemplifying assumptions	Themes or a case of
	Second transition	'[Green infrastructure] is an important question or "buzz-word"…' (p. 92)'. [A] parallel intervention to a "green Infrastructure" is required to reverse ongoing environmental degradation, that of "retro-engineering" storm-water and waste-water systems, so that they function effectively and not destructively, and so that they contribute positively, by providing the goods and services that nature previously provided, particularly along the water paths. 'Besides encouraging the right vegetation in appropriate places on a large scale, it is vital that the root causes of environmental degradation are acknowledged and that the policy is revised, with revised maintenance regimes; and the retro engineering where past planning, design, construction and maintenance practices are the major causes of environmental degradation' (Respondent 11). 'Why on your and my watch do we have the following: industrial pollution? fresh-water-pipe bursts? sewage pollution? poor storm-water management (erosion, riverbed scouring, banks collapsing)? solid (plastic) waste disposal? invasive alien species—feral cats, birds, fish, invertebrates and plants? land fragmentation and misuse? and the poor application of laws?' (Respondent 11) 'Beyond the master drainage plans, there is a need for effective city-wide storm-water management projects to be implemented that address the root causes of poor water quality and poor river health (River Health Assessment will begin in September 2018), namely: a. The reduction of stream and river "high-flow" (storm-water flow-quantity and energy management to acceptable non-destructive levels of flow); or b. The increase of stream and river "low-flow" (detention, attenuation, wetland and water harvesting); or c. Improved and acceptable river-health management; or d. Acceptable stream and river water-quality management' (Respondent 11) 'The graphs and images of the data collected by the Palmiet River Watch show that, like many other rivers in eThekwini, the Palmiet River is highly polluted and that the environmental health of the valley is in dire straits (Pollution Events, *E. coli* Counts and River Health Assessments)' (Respondent 11) 'Regarding the Palmiet, littering and flooding are big issues. Those are two top themes. There is also a huge water quality issue. People who do their washing in the Palmiet River develop rashes when coming into contact with the water. The conservancy higher up is of relatively good quality, but then it degrades quite drastically in the last couple of kilometres' (Respondent 11)	Social conscious from an interest group and what is needed to tackle the Palmiet River's pollution problems

Fig. 3.23 Umkomaas Estuary, south of Durban

Fig. 3.24 An egret hunting fishes in the oHlanga Estuary

infrastructure practices, but also the concept of an ecologhical infrastructure. We see this in the press release when eThekwini launched the UEIP (p. 5).

This means that eThekwini is not only utilising grey infrastructure (p. 8) for water security purposes (e.g. water quality and quantity provisioning, flood attenuation and storm-water drainage), but it is also tapping into green and ecological infrastructures as part of its hydraulic mission. Moreover, storm-water utilisation for the Botanical Gardens (Respondents 6 and 13) is a typical example of a grey hydraulic mission that serves a green infrastructure. In this respect, we can identify a grey hydraulic mission working in tandem with a green and ecological infrastructure hydraulic mission.

When considering the specific use of green and ecological infrastructures, Poff et al. (2016) noted that, in the face of climate and hydrological uncertainty, humanity is required to manage its freshwater resources in a novel way. Although water practitioners view the rehabilitation of an ageing infrastructure and new water supply projects as a solution for reducing future climate risks, attaining the comprehensive goals of freshwater sustainability requires an expansion of the current water management paradigm. This new worldview should incorporate socially valued ecosystem functions and services and move away from the prevalent narrow economic criteria of traditional water management practices (Poff et al. 2016).

From the discussion on liberal institutionalism below, we will see that, in the face of climate uncertainty, eThekwini has embraced sustainability as a normative direction-finding beacon for informing its policies and practices, with respect to green and ecological infrastructures, that are linked with water resources management. In this regard, eThekwini is practicing a form of participatory co-engineering that is intentional and purposeful (Daniell et al. 2010).

In their research, Daniell et al. (2010) moved away from the mechanical notion of engineering and stress that it is a creative process that synthesises and implements knowledge and experience to enhance the welfare, health and safety of individuals and communities, while taking the environment and sustainability into consideration. Even though eThekwini is practicing co-engineering, it is only doing so to a certain extent. eThekwini places a lot of emphasis on knowledge and experience, both within and outside of the municipality, but what is absent from it's particular style of co-engineering is the analysis of social interactions, conflicts and negotiations (Von Korff et al. 2012) among the project teams, as well as the particular green and ecological infrastructure processes that are linked to stakeholder participation.

In Table 3.3, we see how the Hobbesian and Lockean forms of the hydrosocial contract exemplify a limited notion of co-engineering. Respondent 16 noted this when he said that, through the Palmiet River project, which is part of the UEIP, there is a noticeable ecological infrastructure engineering component involving the constructed wetlands downstream of the Quarry Road informal settlement. The focus is on the uMngeni River and not the community. Later, that focus changed and the relationship of the community with the river became the core part of the project (p. 112). It is Hobbesian, since the river and the pollution problems are the central concern, and it involves a Lockean-type contract, since the municipality is conscious of, and informed by, its sustainability. The community was conscious that it was living on the banks of a polluted and flood-prone river, but it did not play an influential bottom-up role. Stated differently, the municipality governed, while the community benefited (Hobson and Seabrooke 2007, 2009).

According to Respondent 16, the municipality has had a difficult relationship with the Quarry Road informal settlement residents, and after the previous Head of Water and Sanitation Department retired, the Durban Municipality involved Dr. Sutherland from UKZN to rethink the project. In other words, a member of the epistemic community suggested that the municipality should change its vision for the project and involve the informal settlement community. This means that the municipality did not contract with the community at first, but that the epistemic community informed it of how its practices around the uMngeni River affected the Quarry Road settlement. Therefore, the first transition was not only characterised by a hydrosocial contract between the local government and the engineer, but other members of the epistemic community became involved in the changes within an engineer-influenced project.

That said, the Hobbesian form of the hydrosocial contract informed the top-down political structure, which only later involved the community. We also see this influence in other ecological infrastructure projects, such as the River Horse project, where the municipality plans a corporate social responsibility (CSR) initiative with the business associations (p. 130) (Fig. 3.18), and where the Small, Medium and Micro Enterprises (SMMEs) contracted to clear alien invasives from rivers and streams (p. 128), the water fund (p. 133) and the Amanzimtoti Citizen Science project (p. 111). The municipality bases these projects on the premise that if you look after the environment, the environment will look after you (Respondent 16). To a certain extent, this entails the personification of nature, which translates into the moral responsibility that people should have towards the environment (Koch 2020).

The PRW's criticism of the concept 'green infrastructure' and the municipality's role in the continued degradation of the river (Respondent 11) are typical characteristics of the second transition. In this regard, the PWR has a clear consciousness about the environmental degradation and protection issues (Schnaiberg 1994). The interests of the different actors, for example, those who seek to utilise the environment for economic purposes and those having an interest in preserving it, have been central in the fostering of environmental consciousness (Rannikko 1996). Environmental consciousness always has an intellectual and agential, or affective, element. This means that a knowledge of the environmental problems, awakens a knowledge of the environment (Rannikko 1996).

Environmental consciousness also has an action component, which determines how individuals operate and function in 'real' situations. That said, environmental consciousness has a role to play in the establishment of environmental movements and interest groups and in the orientation of their activities. Consciousness always relates to a subject that is either human or inanimate (Rannikko 1996), and in this case, it relates to the green infrastructure policies and the poor state of the Palmiet River. The concrete situation surrounding the Palmiet River could have fostered a spatial consciousness (Brand 1999) in the PWR, as an interest group.

From the interview transcript of Respondent 11, we not only see a spatial consciousness in action, but also an epistemological awareness. This type of consciousness relates to the use of the 'green infrastructure' concept. An episte-mological awareness usually comes through reflection when we communicate with others on a matter (Huber 2010). Respondent 11 has interacted with Dr. Sutherland from UKZN, as well as with municipal officials working on green and ecological infrastructures, and it is within this context that Respondent 11 questions the legit-imacy of using the concept 'green infrastructure' to gain access to funding, instead of solving the myriad of problems that the environment is facing.

Respondent 12 made a similar argument with regard to the capex investments of the municipality (p. 100). For Respondent 12, therefore, the concept does not live up to its instrumental purpose of mitigating environmental degradation, but it has a more 'sinister usage', namely, to access financial resources. Because of this, he proposes that a grey epistemology and ontology be incorporated, in order to realise the aspirations of a green infrastructure and its positive effect on an ecological infras-tructure. The consciousness of Respondent 12 plays a systemic constructivist role (Finnemore and Sikkink 2001) in eThekwini's green and ecological infrastructure policy landscape.

Whereas the hydrosocial contract theory focuses its attention on the structural and normative elements of water, liberal institutionalism emphasises the actors and their organisational and non-structured relationships. It is to this theory that I will now turn.

3.4.2.4   Liberal Institutionalism

Although I refer to neo-liberalism in Fig. 3.16, in the following text I will use the concept of 'liberal institutionalism'. Neo-liberalism has many labels, since it is a wide-ranging theoretical perspective of politics and international relations that highlights states and organs of state, such as local and national governments, as well as non-state entities and the structural relations between, and within, these actors. Observers of the IR theory also label neo-liberalism as 'liberal pluralism'. The theory embraces several perspectives, such as complex interdependence, regime theory, liberal internationalism, idealism, liberal institutionalism, neo-liberal insti-tutionalism, neo-idealism, functionalism, neo-functionalism, interest group liber-alism, transnationalism, interactive governance, integration theory and sociological liberalism. On these varying perspectives, Jessop (2002) noted that:

> Liberalism rarely, if ever, exists in pure form; it typically coexists with elements from other discourses, strategies and organizational patterns. Thus, it is better seen as one set of elements in the repertoire of Western economic, political and ideological discourse than as a singular, univocal and internally coherent discourse… Likewise, it is better seen as a…significant principle of economic, political and social organization, in a broader institutional config-uration, than as a self-consistent, self-sufficient and eternally reproducible organizational principle. Thus, the meaning and import of liberalism can vary considerably.

Because of varying and complex forms of liberalism, it is not a straightforward exercise to define this perspective (McCarthy and Prudham 2004). I will use the concept 'liberal institutionalism' because my research investigates individuals who work in an institutionalist setting, namely, the eThekwini Metropolitan Municipality. Within this organisational structure, strategies and discourses play a prominent role. In fact, the information that I uncovered was replete with these reproducible elements. The 'label' liberal institutionalism speaks to the 'organizational patterns' that Jessop (2002) refers to. What is more, liberal institutionalism does not exist in a pure form; we can see this in eThekwini, where collaboration and policy critique happen simultaneously, as mentioned by Respondent 11 above.

Liberal institutionalism rests on several assumptions. For example, it assumes that states consist of citizens, interest groups, local authorities and government departments that continuously compete with one another (Viotti and Kauppi 1999) to receive the scarce resources emanating from society and the economy. However, the state and its governing apparatus, is not the only significant player in (world) politics (Viotti and Kauppi 1999), because non-state entities also have autonomous preferences (Stone 1994), or an inclination to act independently, or free from external control or influence.

With the plurality of actors in mind, ideologically, liberalism claims that economic, political and social relations should be organised through the free choices of unrestricted and rational actors. These role players do so to advance their own material or ideational interests within an autonomous institutional arrangement (Jessop 2002). To recognise a range of actors, including non-state actors, liberal institutionalists substitute sovereignty, which is associated with the state system, with autonomy, as a settled norm (Heywood 1997). The theory accepts that the relationships between plentiful actors resembles a cobweb of relations, with the numerous performers on the political stage being linked to each other in diverse ways, based on their highly complex and multiple interdependent relationships (Heywood 1999).

Economically, the self-regulating market is one of the central features of the theory. The market is self-regulating because it has a wide geographical scope and a comprehensive governing mechanism for allocating all goods and services. This market is also important for organising and evaluating institutional performance; in other words, everything needs to be commodified, even the natural resources, in order for the market to do all these things (McCarthy and Prudham 2004).

Politically, collective decision-making involves both a constitutional state (Jessop 2002; Ikenberry 2009) and its governing apparatus. These have limited 'substantive powers of economic and social intervention' while, at the same time, they are committed to increasing the formal freedom of actors and stakeholders in the economy, as well as the liberty of actors in the public sphere (Jessop 2002).

With respect to these highly complex relationships, there is no clear distinction between domestic affairs and international politics. The two realms of political action are interdependent, with the one influencing the other (Moravcsik 1997; Ikenberry 2009), and the actors have simultaneous roles to play in both domains. What is more, cooperation between the actors in both systems is quite normal, since international relations experts perceive the systems as being free (Stone 1994; Ikenberry 2009).

Green and Ecological Infrastructures through liberal institutionalism

In this section, I will discuss eThekwini's green and ecological infrastructure initiatives through the lens of liberal institutionalism. Table 3.4 is an elaboration of the theory's assumptions and how they are manifested in the respondent's answers. From Table 3.4, we can see that several themes or topics stand out, namely, collaboration, networking and interdependent domestic and international relations, the commodification of nature, competition, domestic politics and policies, functionalism, interest group pluralism and organs of state.

Collaboration, networking, domestic and international relations

Advisory body

In Table 3.4, Respondent 2 indicated that the existence of an advisory body institutionalises the collaboration in which she and Respondent 1 participate to advise the municipality on coastal matters. Here we get the first hint that the municipality invites experts to advise it on matters pertaining to the local government's knowledge capability.

This form of collaboration, together with the cooperative relationships that eThekwini has formed with other local and international research bodies over the years, relates to the theory of interactive governance. On interactive governance, Kooiman et al. (2008: 1) noted that:

> Governance in its broad sense suggests that not only the state but also market and civil society have prominent roles in the governing of modern societies, from local to international levels. Interactive governance highlights the interactions between entities belonging to these social parties.

In an earlier publication Kooiman et al. (2005, cited in Mahon 2008) defined interactive governance as '...the whole of public, as well as private, interactions that are initiated to solve societal problems and create societal opportunities'. A related concept to interactive governance is governability, or the 'governance status of a societal sector or system such as a...coastal region, as a whole' (Kooiman et al. 2008: 1). We recognise three features of interactive governance and governability throughout eThekwini's green and ecological infrastructure policy landscape, namely, the system that is to be governed, its governing system and the governance interactions (Kooiman et al. 2008; Mahon 2008). The typical features of a governance system include diversity, complexity and dynamics (Mahon 2008). The overall system to be governed are eThekwini's green and ecological infrastructures, with, for instance, the advisory body governing matters pertaining to the coast, as well as the governance system, which manifests itself in the form of municipal officials who collaborate with researchers from other entities. It is within this arrangement that we see a diverse and complex set of local and international actors, and a dynamism that spans not only the actors' relations but also the natural processes and international boundaries (see below).

Contracted research and DRAP

The eThekwini municipality has also established research relationships by contracting scholars and consultants to investigate specific issues and to develop

Table 3.4 Liberal Institutionalism and eThekwini's green and ecological infrastructure policy landscape

Liberal institutionalism	Collaboration, networking, functionalism and interdependent domestic politics and international relations	Respondent 2 indicated that she works closely with Respondent 1 on coastal engineering matters. Respondent 2 also 'sits on the coastal working group of the municipality…that forms their advisory body for a lot of things that come up, in terms of coastal policy'	Inter-institutional collaboration between the municipality and a research institute (ORI) on coastal matters
		'eThekwini will sub-contract specialists to do different pieces of work for them. The CSIR have previously been involved in SEA [a Strategic Environmental Assessment], which we are doing. They work quite a lot with Cathy Sutherland [and] Margaret McKenzie (Urban Earth). Anton Cartwright (a Cape Town-based resource economist) has done some work for them. Anton did the first piece of work with Christine Colvin on water funds here in the uMngeni River basin… eThekwini has a pool of specialists on whom they draw for the research side of things' (Respondent 10)	Contractual collaboration between the municipality and various research institutions and independent researchers on specific research endeavours
		'Another partnership that we are engaging in is the Durban Research Action Partnership (DRAP) which has to do with global environmental change to provide evidence-based decision-making. It has a…strong partnership with the UKZN. UKZN's full flagship programme is the African City of the Future, which is looking at what Durban wants to be in the future' (Respondent 9) DRAP '…has a good partnership between the university [UKZN] and the municipality; it generates a lot of research and insight into protecting important spaces of the ecological infrastructure in the city' (Respondent 3) 'DRAP is quite big—it is not only this site… I coordinate whatever the researchers do on this site…. The communities are heavily involved in reforestation projects' (Respondent 5)	An institutionalised partnership with UKZN to investigate a specific global governance issue to inform policy
		'DRAP is funding the research of the students. The last contract was with UKZN and a lot of them were from the Environmental faculty of the university in Pietermaritzburg. The Community Reforestation Programme (CRP) is funded by the DBSA [Development Bank of Southern Africa] and involves the African Conservation Trust and the Buffelsdraai community. WESSA did a lot of environmental education and is aligned to the work that we do here' (Respondent 5)	DRAP as an institutionalised partnership involving funders and other epistemic stakeholders

(continued)

Table 3.4 (continued)

'We also work with the IPCC and we are the lead authors of specific reports' (Respondent 9). In the Paris Agreement in 2015, there was an organisation that was established, called C40—it is a group of 90 cities that come together in a forum, which is funded by Marko Bloomberg, to discuss climate change and to exchange and share information and knowledge. They have a few programmes. Our Mayor, Ms Gumede , is the Vice President of C40, so Durban plays a prominent role in it. One of the C40 projects was to disaggregate what the Paris Agreement meant for cities. C40 then commissioned the research called Deadline 2020, from which there were a number of offshoots, one of them being that certain cities were chosen as pilot cities in which to develop climate implementation action plans. These action plans cover thematic areas of mitigation and adaptation and I am certain that water security features highly in those plans, at least from an adaptation point of view. Over and above that, we are busy with a C40 programme right now which involves new buildings in the city. Initially, the core element of this programme was the carbon element' (Respondent 12)	Participation and knowledge production on a specific global governance issue translating into specific local government objectives and programmes (interactive governance at international level)
The municipality will sub-contract smaller SMMEs to clear alien invasive plants and promote the spread of indigenous vegetation (Respondent 5). eThekwini '…work[s] a lot with private companies, and the SMMEs that have been established through the [WfE] are now being hired by eThekwini to implement their projects' (Respondent 8)	Contractual collaboration between the municipality and Small, Medium and Micro-enterprises (SMMEs) and communities for specific ecological infrastructure services
'A lot of the work will be community-based and require public participation. Very seldom will you find the municipality going out and doing things unilaterally. I don't think it is part of our ethos or culture, as the eThekwini Municipality, to do things unilaterally. So, stakeholder involvement and consultation is critical. Even the Durban Climate Change Strategy [DCCS] was not a desktop strategy with somebody sitting behind a screen and writing the strategy. We have been through processes that were very iterative, and where we went through the stages of consultation with NGOs, other government departments, private citizens and academia. So, I would say that stakeholders are consulted. Now the next part is to get the neighbouring municipalities on board' (Respondent 12)	
'We are also working on a coastal project with Cathy [from UKZN] and we have put in an application to WIOMSA [the Western Indian Ocean Marine Science Association] to partner with Mombasa regarding the Western Indian Ocean, in the form of a learning exchange partnership' (Respondent 9)	Epistemic local and international collaboration within an institutionalised ocean governance structure

(continued)

Table 3.4 (continued)

Collaboration between eThekwini and DUCT [Duzi Umngeni Conservation Trust] exists concerning the River Horse Project. The purpose of this is to establish a Corporate Social Investment (CSI) initiative to implement the recommendations of the consultants on the maintenance of the wetland (Respondent 6). Respondent 6 said that he wants to get the CSI off the ground, together with the River Horse Valley Business Management Association. '[T]hey, (the business association) have done the investigations for us and the feedback that we've had is that Unilever is one of them, and there are a number of companies that said they are excited about contributing to this, especially because the wetland is literally on their doorstep. The companies said they usually fund things that are kilometres away, and now we have the opportunity to fund something that is literally on our doorstep'	specific interest groups (DUCT and the business association) with a specific interest group (business association)	
The '…former head of eThekwini Water and Sanitation' established and championed the UEIP. The aim is '…to coordinate action around improving the state of [ecological infrastructures] in the uMngeni Catchment, in terms of water security. There is…a whole governance component that coordinates action across research institutions and government entities and NGOs [non-governmental organisations], and it is quite complex. There is a strong research component to that' (Respondent 9). eThekwini, therefore, is a key partner in the UEIP, which has several associates, for example, '…the Msunduzi Municipality, Umgeni Water,… the KZN Provincial Department of Environment and Tourism, the Department of Water and Sanitation, the WRC, KZN Wildlife (Ezemvelo), UKZN (Sabine and Cathy, and the other person is Graham Jewitt)—so they are very well-represented on the UEIP. CSIR is also part of the partnership. [T]here are a few [NGOs], for example, WESSA, Wildlands, which is a KZN-based NGO, Birdlife South Africa and SASA [South African Sugar Association], as well as quite a few municipalities, for example, the uMgungundlovu District Municipality (Respondent 7). According to Respondent 16, the UEIP consists of 23 stakeholders. There is inter-municipal collaboration with upstream neighbouring municipalities, as mentioned by Respondent 7, climate change compact partnerships and communities, particularly in the Palmiet Catchment, as well as associations representing the sugar industry and forestry The Palmiet Catchment initiative is part of the UEIP, and it has fallen on us. Originally, it was driven by Water and Sanitation to create wetlands below the Quarry Road informal settlement' (Respondent 16) eThekwini collaborates with various stakeholders, such as Umgeni Water, WWF [World Wide Fund for Nature] and other interest groups (PWR) and the epistemic community (DUT and UKZN) when implementing green and ecological infrastructure initiatives (Respondents 7 and 9) Other initiatives by the Municipality are the Sihlangzimvelo Project, on the Umhlangane River, and the Green Corridor initiative. The latter is under the auspices of eThekwini's Economic Development Department, which has an agreement with the DUCT. The initiative involves the clearing of alien invasive species and alien management in the uMngeni River (Respondent 6). [The Green Corridor] was a subsidiary of DUCT before they decided to make it independent. Now it is an independent entity that is mainly funded by eThekwini. It is like a parastatal NGO, because it is very dependent on eThekwini for funding. And then there is [WfE], which is becoming very famous, and which is a Durban project…' (Respondent 8)	Establishment of a collaborative structure to promote the municipality's water security, considering its downstream status A cobweb of relations involving various actor types, although a local government-centred partnership constituting specific green and ecological infrastructure initiatives	

(continued)

Table 3.4 (continued)

	'There is also a joint venture project with the Japanese, DUT and Durban' (Respondent 7)	Transnational cooperation involving the municipality, a local epistemic entity (DUT) and an international donor government (Japan)
	Respondent 9 was involved in signing a memorandum of understanding with The Nature Conservancy concerning the water fund. '[T]he water fund, by its very nature, will involve other stakeholders and will never be run by the Municipality. It is a multi-stakeholder entity' (Respondent 9)	Collaboration between the municipality and a transnational interest group and funder to establish a regulatory economic structure
	'To go back to that Compact partnership: the African model these days is that you have these compacts around Africa, but not on the scale of a single city; rather, you have a driving city in the middle and you have these compact partners because they are using each other. We have now one that started in Mozambique, and Lusaka is busy establishing one as well. They were established after having done learning exchanges with them and having looked at the community-based ecosystem approaches, water management approaches and managing landfills. So, we have this network of African cities that are developing around these good principles of adaptation. This would be another form of partnership on a much bigger scale. It is called the Durban Adaptation Charter Hub and Compact approach' (Respondent 9) What we have is the Durban Adaptation Charter [DAC], which is one of the outcomes of COP 17. This Charter rests on the compact model. This means that here in KZN we have the central KZN climate change compact, which is a forum of municipalities that get together to exchange information, knowledge and best practices. This is one step in the direction of dealing with other municipalities. So, you use eThekwini, perhaps, as the anchor, and if you go up the east coast of Africa, there are two other compacts. One is in Mozambique, while the other is in Dar es Salaam (Tanzania)' (Respondent 12)	Institutionalised twinning between eThekwini and other cities in Africa on climate change adaptation

(continued)

Table 3.4 (continued)

	'We've got a partnership with our sister city, Bremen [Germany]; they sent out a team to see the issues that our city is facing, and we…went to Bremen…which is an old city; they've got problems and have refurbished portions of the city…. So, they were looking to us, as an expanding new city, and testing out new stuff to see what works, and we are learning from them in terms of the challenges that development has caused in an old city. When they came to SA [South Africa], we asked, what do we show them, and where do we start? So, we focused on the Umhlangane River… We have all the land uses, we have industrial, commercial, high-, medium- and low-density residential uses, we have large spaces of open land, and we've got wetlands… And then we asked, "What are the challenges this city faces in the catchments?" So, we started what we call the Lighthouse Project on the Umhlangane… The Bremen partnership is a political partnership because it was approved by council. It started with a lot of Durban NGOs partnering with Bremen and it has now become a full partnership between the two cities. An agreement has been signed between the two cities… GIZ [*Gesellschaft für Internationale Zusammenarbeit*] is the implementer of the Bremen partnership and it is also linked to the C40 initiative' (Respondent 6)	City twinning across international borders that are developed from a non-state actor initiative
	'I would say that Durban is one of the leading cities in the world for these matters. It is a C40 city [see page 120]. So, sustainability and sustainable development goals are on their front burner. That is why so many cities in the world are twinning with Durban. Bremen has just joined them. I think there are 34 cities around the world that have twinned with Durban in the last eight years. The purpose is to sustain sustainability practices and link them to the C40 city movement. C40 is all about the most sustainable cities on the earth, now there are about 93 cities around the world and the objective is to sustain sustainability practices, like those we've just described' (Respondent 8)	
	Debra Roberts conducted research on the D'MOSS in the late 1980s and early 1990s, and her research culminated in the adoption of D'MOSS as a spatially-explicit policy, which looks at metropolitan open spaces that generate a whole lot of benefits to us humans; for example, flood retention and flood mitigation. So, a lot of those open spaces would have been along rivers and streams. Either implicitly or explicitly, there has been a policy that has addressed these issues… Therefore, there have been many years to refine, establish, and get buy-in' (Respondent 4). It has been developed through a systematic conservation planning tool, so it is quite a scientific way of helping to identify key biodiversity or ecological assets and then working to protect them' (Respondent 9)	Policy continuity constituting an authoritative local government organ that utilises science to develop a policy
	'[W]hat I understand is that the Environmental Planning and Climate Protection Department is a very strong department in the Municipality. It has done a lot of work on ecosystem services and community-based ecosystem adaptation, using ecosystem services to adapt to climate change to support development in the city, and now, more recently, to address water security. D'MOSS is the first layering of this [policy], which is why the idea of linking biodiversity to ecological infrastructure to ecosystem services is important. So, D'MOSS is the…important first layer that defines it in the city and, of course, it is in the spatial development framework. They've been clever to combine existing legislation to support [D'MOSS]' (Respondent 3)	

(continued)

Table 3.4 (continued)

On policies relating to estuaries, Respondent 4 said that: '…in the abstract [Respondent 1] spoke about the estuarine component…they have policies in place that relate to estuaries, because they have water treatment works along their estuaries and they have policies that regulate the quality of water going into the estuaries. They are also directed by a whole bunch of national policies, as they relate to estuary management. Remember, estuary management is a national competency that has not been correctly delegated to local governments. It has been delegated, …but in law it is a problem. Durban is, therefore, directed to manage their estuaries under national law directives'		Interdependence between national and local government policies
On the knowledge produced through DRAP, Respondent 5 noted that: 'The research will be used by the municipality. It was in the early 2000s that the first initiative around this research was introduced. One of the first things they focused on was the KZN grasslands and they had a whole lot of the students doing research on that. The second was the reforestation initiative. This was the community reforestation initiative at this site [Buffelsdraai], the Paris Valley Nature Reserve, and the Inanda Mountain Project, which is being kicked off now'. Concerning the sewer bylaws, Respondent 6 said that: "There are also sewer bylaws and they talk a lot about managing your sewers and sewer discharge, so you don't create pollution. There is coverage from that point of view as well. The sewer bylaws don't incorporate green or ecological infrastructures directly, but there has been work by the city and the university to investigate alternatives for sewerage management that would, in certain cases, include engineered wetlands'		The institutionalisation of the science-policy interface
'We have the storm-water management policies (storm-water bylaws) that are in the process of going through a public participation process. They will then go through for approval and then Gazeting. The storm-water policies essentially give details about the type of materials and pipes. They will also deal with new developments in that they will provide the storm-water lens for such developments. We don't specify how, but we specify that we need to maintain the previous development rate flows. Our bylaws don't presently cover quality issues. Cape Town has incorporated quality issues in their bylaws, and we are hoping to learn from them. One of the challenges they have are the standards…' [W]e are not prescriptive on how they implement storm-water systems. We run several courses and the WRC is running a number of courses around the country looking at the best practice regarding water-sensitive design. Architects and engineers are now becoming more sensitised about water-sensitive design in the use of such practices. There is now a commission looking at developing guidelines for the SuDS applications'. (Respondent 6)		Partially prescriptive storm-water policies and bylaws

(continued)

Table 3.4 (continued)

	'[T]he Durban Climate Change Strategy (DCCS)…is not an adaptation of a climate change mitigation strategy, but more a climate change response strategy. It includes adaptation, mitigation and some over-arching thematic areas. Water security is very prominent there' (Respondent 12)	A strategy aimed at a global governance issues (climate change)
	'From our perspective the climate change strategy focuses on water security, so the provision of water and the reduction of flooding are two very important aspects of [ecological infratsructures]. There is a very clear reference to [green infratsructures], the need to conserve [ecological infratsrutucres], as well as water quantity and quality provision. The biodiversity theme is also a very good example of the policy there. The Climate Change Strategy outlines the policies regarding [green infratsructures]' (Respondent 16)	A structure that deals with water security aspirations is defined as water provisioning and flood reduction
	'I don't think that they have policies that are directly related to [a green infrastructure] and water security. They do have a strategy that relates directly to the ecological infrastructure, but it does not link the ecological infrastructure directly to water security. However, there is an indication that management is showing the role of [ecological infrastructure] in the management of floods. The Sihlangzimvelo Project is the only project I know of that links green infrastructure to water security. The others deal with biodiversity conservation, but there is no direct link with water security' (Respondent 14)	Policies that indirectly deals with the problem of water insecurity
Commodification of nature	'From a coastal productivity point of view, the estuaries are unbelievably productive [see Figs. 3.22, 3.23 and 3.24]. These are the sort of things that I would pick out as ecological assets or green assets' (Respondent 1) Estuaries are quite a good example because they are huge natural assets and they offer some of that protection' (Respondent 2)	Natural elements defined within the discourse of economic productivity and as an asset
	'[P]eople can relate to the need for the conservation of wetlands…or the value that they add. Whereas the coast is still considered to be an economic resource…in terms of the city, it is the biggest tourist attraction. It is managed for it's socio-economic and financial benefits. However, maybe the understanding as to the natural value of ecological infrastructures is not necessarily there. The other part is that the coast is extremely developed. I think there is an understanding of the value' [see Fig. 3.25] (Respondent 2)	An ecological infrastructure is an economic commodity that produces revenue for the city

(continued)

Table 3.4 (continued)

Competition	'However, in the traditional authority areas, it is a whole new big challenge. There is a very interesting situation in eThekwini, where you've got the geo-governance areas; it has all changed. Durban is a very good case study for planning these different systems and integrating them into the environment. In hybrid planning, the environment comes in strongly. In traditional authority areas, like Mnini, coastal buffers have been put in place. So, in policies and bylaws, you've got these coastal protection areas, but the traditional authority are allocating land in these buffers—new houses have been put into these zones. The application of bylaws, policies and legislation is therefore difficult in those situations. It changes, because you are not controlling the allocation of land… On the city scale, they are doing these bigger pieces of management, but it is hard in the context of dual governance [traditional leadership authorities versus the municipality]. The [KZN] Provincial Government is very important in terms of the traditional authorities because the Province has the mandate and influence over it. They have different guidelines that they've put out for the *Amakhosi* to use, to try and protect water buffers and wetlands. There are all these guidelines, but the application is quite difficult. The *Amakhosi* see these guidelines as imposing on their traditional authority and their right to make decisions over land' (Respondent 3)	Discordant governance over land allocation of free and rational actors (*Amakhosi*)
Interest group pluralism	When asked why there is 'the poor application of laws', Respondent 11 was critical of the municipality and how it formulates and implements its policies. He also noted that 'the D'MOSS does not protect eThekwini from the insidious erosion of the goods and services caused by "development"'. To solve some of the problems that Respondent 11 identified, with respect to the city's storm-water management (see his responses under the hydrosocial contract theory section above, for an elaboration), he noted that: 'Achieving these requires an integrated approach. While funding and resources tend to be spent on sector- or 'silo'-specific hard infrastructure, and, at best, localised unsustainable reactive interventions and activities. This is largely due to the silo structures, poor communication and a limiting relationship and connection between the authorities and the landowners'	An interest group critical of a fundamental local government policy aimed at protecting biodiversity and other environmental issues, such as climate change adaptation and mitigation

evidence-based policy instruments, such as SEAs, and resource economics investigations (p. 122). Therefore, it has a pool of experts on whom it can rely to do such contractual work. This group of experts also extends to the universities, such as the Durban University of Technology (DUT) and the Durban- and Pietermaritzburg-based UKZN.

eThekwini has institutionalised such a contractual partnership through the DRAP, which focuses on global environmental change issues, including biodiversity loss and global climate change, to provide evidence-based research for policy development and implementation (p. 122). Through UKZN's flagship programme, the African City of the Future, Durban has contracted researchers to investigate future scenarios related to global change matters. DRAP fosters the mutual benefits for the municipality and university, while the municipality gains valuable insights into global change matters, and the university generates post-graduate studies, which are valuable, from a government subsidy perspective. The research produced by the students investigates specific matters pertaining to eThekwini's open spaces as green and ecological infrastructures. What is noteworthy in Table 3.4, from a natural and social science philosophy perspective, is that most of the students hail from the university's Environmental faculty.

DRAP is not only a partnership between the municipality and UKZN, but also with the DBSA (p. 122), of which the municipality is a client (DBSA 2020). This bank has funded the community reforestation programme, which also involves the African Conservation Trust (ACT) and WESSA.

ACT started its work as a volunteer organisation in 2000, with the vision and aim of assisting environmental, conservation and heritage efforts in the Southern African Development Community (SADC). Since 2003, it has focused on South Africa as its area of operation and has implemented various initiatives, ranging from food security to erosion control (ACT 2020).

The aim of WESSA is to create '…and support high-impact environmental and conservation projects to promote participation in caring for the earth' (WESSA 2020a). The organisation focuses much attention on environmental education and works with schools and teachers, throughout the country, to promote several local and international programmes that have to do with learning about the environment and educating learners on how to take environmental action in their own areas. Sustainable development is the theoretical foundation that underpins WESSA's work (WESSA 2020a).

eThekwini, therefore, links itself not only to universities, but also to organisations that could play a pivotal role in realising its DRAP-inspired initiatives. The identity of the various organisations involved in the partnership is defined by their role within the enterprise, with ACT being identified with conservation, food security and heritage, and WESSA playing the role of an environmental educator.

IPCC and C40

eThekwini's Environmental Planning and Climate Protection Department (EPCPD) is active in the IPCC (p. 122), with some of its officials having co-authored sections of the IPCC reports (Revi et al. 2014a, b).

Debra Roberts, founder and former manager of the EPCPD, and currently the Head of eThekwini's Sustainable and Resilient City Initiatives Unit, is also co-Chair of the IPCC's Working Group Two. She shares the chairpersonship with Hans-Otto Pörtner, who has affiliations with the Alfred Wegener Institute in Bremerhaven and the University of Bremen in Germany (IPCC 2019). The IPCC, which is a body of the United Nations (UN), assesses science that is related to climate change, or more specifically, that 'determines the state of knowledge on climate change' (IPCC 2020). According to the IPCC (2020), it was '…created to provide policymakers with regular scientific assessments of climate change, its implications and potential risks, and to also put forward adaptation and mitigation options. It identifies where there is agreement in the scientific community on topics related to climate change, and where further research is needed'.

C40 (p. 122), on the other hand, is a global network of the world's mega-cities. It is closely linked to matters pertaining to climate change, and '…supports cities to collaborate effectively, to share knowledge and to drive meaningful, measurable and sustainable action on climate change' (C40 2020). The fact that the former Mayor of eThekwini, Councillor Zandile Gumede, was the Vice President of C40 indicates the role that the municipality plays in the organisation and on the issue of climate change adaptation and mitigation in cities. By being the co-Chairperson of the IPCC, it is possible that Debra Roberts played a pivotal role in elevating Ms Gumede to this position.

At the time that I conducted the interviews, Gumede was the Mayor of eThekwini. However, a year later she resigned from her mayoral duties, but later revoked the notice. Finally, the ANC announced Gumede's departure from the municipality's Executive Committee. Nevertheless, she has remained on as an ordinary councillor. According to the ANC, Gumede's ousting was more about the poor administration of the local government than it was about the criminal charges brought against her for alleged fraud and corruption (Singh 2019). Gumede faces charges of attempting to influence the award of a tender of about R208 million for a Durban Solid Waste contract in 2016 (Naidoo 2019). It is not impossible to assume that the circumstances surrounding Gumede's ousting as mayor has tarnished the reputation of Durban as a high-profile C40 city, especially since it involved an environmental issue, such as solid waste removal.

Small, medium and micro enterprise contracting

Speaking of contracts and tenders, another way in which eThekwini 'collaborates' with business is by awarding contracts to SMMEs for specific ecological infrastructure service delivery, particularly the clearing of alien vegetation along rivers, streams and around wetlands (p. 123). According to eThekwini (2020), 'Ecosystem restoration, sustainable development and the provision of green jobs are key factors in achieving community ecosystem-based adaptation to climate change'. In 2006, the EPCPD initiated an ecosystem programme, which was based on the philosophy 'to build a holistic and positive interaction between local communities and the environment' (eThekwini Metropolitan Municipality 2020a, b). The municipality uses the programme to create training and employment opportunities that help in the management of environmentally important areas. The aim is to achieve the self-sufficiency of communities (i.e. poverty alleviation) (WESSA 2020b) and to allow communities to be more adaptable to natural disasters that may occur in the future, due to climate change. WESSA is the implementing agent of the initiative (eThekwini Metropolitan Municipality 2020a, b), since the programme encompasses a prominent environmental training component. Since WESSA plays this role, the EPCPD provides guidance with respect to the management and selection of the sites (WESSA 2020b).

Marine science and governance

The collaborative network between the municipality and the local and southern African epistemic community also includes marine science and governance.

Regarding local epistemic relationships, Durban's geographical location along the Indian Ocean coastline informs this epistemic association with Dr. Cathy Sutherland from UKZN (p. 122, 123). Sutherland is a Senior Lecturer at the university's School of Built Environment and Development Studies. She is a trained geographer whose research interest lies '...at the interface between social and environmental systems, with a focus on sustainable development' (UKZN 2020). Furthermore, her research interests are concerned with the relationship between society, space and the environment and how these three aspects shape environmental politics and policy-making (UKZN 2020b). Since she is a geographer working on environmental issues and their influence on politics and policy-making, the EPCPD sees her as a relevant epistemic partner in its green and ecological infrastructure programmes and policies. Sustainable development, which is the theory that underpins her knowledge, further strengthens this association. Therefore, the EPCPD and Sutherland are not strange bed-fellows, so to speak, when it comes to complementing each other's work.

In the southern African context, Durban is actively involved in the Western Indian Ocean Marine Science Association (WIOMSA). This organisation was established as a regional, non-profit membership entity in 1993. It dedicates itself to promoting the educational, scientific and technological development of all research of marine science throughout the Western Indian Ocean. WIOMSA consists of 10 countries, namely, Somalia, Kenya, Tanzania, Mozambique, South Africa, Comoros, Madagascar, Seychelles, Mauritius and Réunion, and it focuses on the sustainable development, use and conservation of the marine resources in the area. An interest of WIOMSA is to link the knowledge emerging from research with governance issues that influence marine and coastal systems (WIOMSA 2020). Since South Africa is a WIOMSA member, and considering that Durban is located along the Indian Ocean coastline, it is logical that eThekwini would be involved in WIOMSA activities (p. 123). In addition, WIOMSA's philosophy of sustainable development links it with, and possibly informs, Durban's vision of sustainable marine development matters.

An inextricable link exists between freshwater security and the marine environment. Since Durban is located along the Indian Ocean, storm-water systems carry solid waste and sewage pollution into the sea during storm events. The waste not only has a negative influence on the coastal and marine environment, but it also has an undesirable influence on human health, especially for those using Durban's beaches for recreational purposes. The main concern here is the high level of pathogens (Meissner et al. 2018b), such as the *Escherichia coli* (*E. coli*) bacteria after stroms. These bacteria normally reside in the intestines of healthy humans and animals and are harmless, or they could cause relatively brief diarrhoea; however, certain strains, like *E. coli* O157:H7, can be responsible for severe abdominal cramps, bloody diarrhoea and vomiting. Contaminated water and food could expose humans to the bacteria. After rainfall events in the city, human and animal faeces may pollute the surface water (Mayo Clinic 2020) and may end up in the sea along Durban's beaches, where *E. coli.* could infect recreational bathers and surfers (Meissner et al. 2018b). This could have a potentially damaging influence on Durban's local tourism industry.

Corporate social responsibility

Corporate social responsibility (CSR) (p. 123) is the collaboration capability between corporations and the community in which businesses are located. This type of cooperation aims at the mutual benefit of the partners: communities and companies (Boehm 2002). Certain arguments exist that CSR is embedded in liberal institutionalism, while others state that this type of cooperation is incompatible with liberal institutionalism. Kinderman (2012) argues that CSR and related initiatives are complementary to liberal institutionalism and can act as a substitute for institutionalised social unity. Furthermore, Kinderman (2012: 30) goes on to say that:

> As a quid pro quo for a lighter regulatory touch, CSR provides compensation for some of the social dislocations that result from neo-liberalism. Moreover, it appeals to the businesspersons' moral sensibilities, thereby helping to legitimate their conduct among themselves and vis-a-vis society in a way that purely instrumental rationality cannot.

Likewise, Kinderman (2012) notes that CSR is associated with neo-liberalism in that it reinforces neo-liberal institutions, which provides fertile ground for a rise in CSR. Richter (2010: 625) also explains that the 'conventional reasoning on corporate social responsibility…is based on the assumption of a liberal market economy in the context of the nation state'. The theory assumes that the responsibility of businesses is directed towards a national society in the liberalist reasoning of their corporate roles (Richter 2010).

In this context, society contracts, or charters, 'give life' to corporations through the state's governing apparatus, and then they regulate their behaviour through local, provincial and national government structural devices, such as the planned River Horse corporate social initiative, which is mentioned by Respondent 6 (p. 123). Society, on the other hand, depends on business for employment and income, while, at the same time, it supports corporations by buying their goods and services (Fitch 1976). During this cyclical relationship, and particularly when businesses manufacture goods and services, problems could arise, such as pollution and other ills that affect the natural environment negatively. CSR is an attempt by businesses to 'solve' problems (Fitch 1976), by focusing on the ethical aspects of business to achieve a balance between profitability and social responsibility (Balza and Radojicic 2004).

Although the preceding brief discussion of the philosophy and substance of CSR focuses its attention on the link between corporations and the different governing apparatuses of the state, such as local governments, it is not uncommon to find non-state entities, like DUCT, promoting CSR (p. 123) (Balza and Radojicic 2004; Spodarczyk and Szelągowska-Rudzka 2015). After all, the various activities of NGOs not only promote a respect for human rights, but also for the care of the environment within the business world (Balza and Radojicic 2004).

Linked to the ethical credence of CSR, part of DUCT's vision states that communities within the uMsunduzi and uMngeni River systems will 'show respect for the rivers and will take ownership and responsibility for the condition of the rivers, seeking to preserve their natural function and beauty' (DUCT 2020). Part of the preservation function is the conservation of biological diversity. For the NGO, government, business, the scientific community and civil society must cooperate by providing resources for the effective management of river systems (DUCT 2020).

uMngeni Ecological Infrastructure Partnership

In February 2013, several organisations (see Respondent 7 in Table 3.4 for a list of the entities) signed a Memorandum of Understanding (MOU) to formally establish the UEIP (p. 5). At the time, these organisations were '...committed to finding ways of integrating ecological infrastructure solutions to support built infrastructure investments in addressing the challenges of water security in the uMngeni Catchment' (SANBI 2020) (see Figs. 3.19 and 3.20). Currently, the UEIP consists of over 30 organisations that hail from local, provincial and national governments, business and academic institutions, as well as civil society. As a co-founder of the UEIP, SANBI has been central in coordinating the partnership since it was established (SANBI 2020). SANBI's notion of an ecological infrastructure is the key for exploring the role of this type of infrastructure and for improving the water security of the uMngeni Catchment. For instance, Respondent 3 indicated that:

When the UEIP was established in [2013], Debra Roberts and I wrote an opinion piece on the formation of the UEIP and took up the concept of an ecological infrastructure. So, it has been framed like that for a long time because of SANBI's influence...

This quote is interesting, in the sense that the establishment of a collaborative partnership between a few independent actors prompted the two researchers to define an ecological infrastructure in relation to an institution's epistemology and ontology (see the SANBI (2019) definition on p. 5). The quote by Respondent 3 does not only allude to a highly complex and interdependent relationship between several actors within the UEIP structure, but also to an epistemic relationship between the collaborative partners and the epistemic function of the concept 'ecological infrastructure'. In this sense, the concept acts as a direction-finding beacon for the UEIP partnership to anchor its work in an empirically formulated model. Furthermore, this shows that the actors working on green or ecological infrastructures have acted in a 'rational' manner to adopt a realistically fashioned concept to formulate the configuration of the UEIP. It is from these concepts that eThekwini has developed 'rational' policies to systematically order its ecological infrastructure policy landscape.

When considering water security as a concept, SANBI (2020) defines it within the ambit of population growth and economic development in the catchment, which has resulted in the uMngeni River system being unable to meet the current water demands in the area.

The UEIP has several objectives that are aligned to ecological infrastructure, governance, institutional capacity, policy, knowledge generation, collaboration and co-learning. Firstly, it sets out to establish strategic investments in the ecological infrastructure that will contribute to water security enhancement in the catchment. Secondly, it would like to improve governance in order to slow down the degradation rate of the ecological infrastructure. Thirdly, the partnership wants to strengthen the institutional capacity to rehabilitate, maintain and protect the ecological infrastructure. A fourth objective is to foster an enabling policy environment for investment in the said rehabilitation and management activities and to improve the knowledge base that informs the policy and practice on ecological infrastructures. Lastly, the collaboration between the various partners will focus on 'effective' collaboration, coordination and co-learning, which enables the partnership to consolidate, grow and demonstrate its value (SANBI 2020).

Since its inception seven years ago, the UEIP has initiated three demonstration projects, namely, the Baynesspruit and Palmiet River Rehabilitation initiatives and the Save Midmar Project. Three local governments manage the respective projects, namely, the Msunduzi Local Municipality, the uMgungundlovu District Municipality and the eThekwini Metropolitan Municipality (SANBI 2020).

Msunduzi is a UEIP partner in the Baynesspruit Rehabilitation project, and its efforts in the Baynesspruit, a tributary of the Msunduzi River, focus on water quality deterioration due to pollution from 'illegal industrial waste discharge, solid waste and contamination from dilapidated storm-water and sanitation infrastructure. Activities to address these problems include alien vegetation control, riverbank rehabilitation to reduce soil erosion, citizen science and school projects to clean-up the river' (SANBI 2020). The overall aim of the project is to restore the existing ecological infrastructure areas (SANBI 2020).

The uMgungundlovu District Municipality partners with the KwaZulu-Natal Department of Economic Development, Tourism and Environmental Affairs on the Save Midmar Project. They coordinate their activities to restore the ecological infrastructure in important areas around the Midmar Dam (Figs. 3.11 and 3.21), which is a major reservoir that stores water for use by Pietermaritzburg and Durban and the surrounding areas. The dam also hosts the popular annual Midmar Mile sporting event. Sewerage, industrial pollution and upstream agricultural activities have been responsible for the deteriorating water quality. Save Midmar focuses on wastewater infrastructure construction and repair, wetland rehabilitation, as well as community education and outreach (SANBI 2020) involving WESSA's education and learning efforts (p. 129) (Taylor 2020).Researchers from UKZN assist eThekwini in the Palmiet River project, which aims to address the river's deteriorating water quality (p. 123). Furthermore, the initiative focuses on improving governance by fostering 'good relationships' with catchment stakeholders, in order to understand the socio-economic and political dynamics, especially in informal settlements. Pollution, which is caused by solid waste in informal settlements, sewer leaks and 'illegal industrial effluent discharge', is particularly challenging in this regard (SANBI 2020).

We can conclude from this that the relationship between eThekwini and the epistemic community shows an over-reliance on the internal (eThekwini) and external (SANBI and UKZN) epistemic communities to infuse eThekwini's green and ecological infrastructure as water security-related policies informed with scientific evidence. This over-reliance points to a structural-functional reasoning.

Functionalism is an explanatory problem-solving device (Callicot 1998). We see it operating in the framing of problems, such as adaptation and biodiversity, through green and ecological infrastructure initiatives and, more particularly, ecological infrastructures in the UEIP. Wellstead et al. (2013) argue that this 'structural-functional' reasoning in eThekwini, '…presents a false sense of what is possible, and proper, in terms of policy processes and outputs'. The logic presents management and policy innovations as 'quasi-automatic responses' to climate change- and water security-related challenges and do not sufficiently acknowledge the political factors that impact the policy-making and policy results (Wellstead et al. 2013). A fallacy of structural-functionalism is that its 'input-output' methodologies disregard the propensity of environmental governance systems to support the status quo in management options and policy outcomes (Wellstead et al. 2013). This is evident in the three projects, where instrumentalism, in the form of the rehabilitation of natural systems, is the order of the day. However, there is no mention of the governance structures that are responsible for the deplorable state of the rivers and the dam, in the first place. Wellstead et al.'s (2013) study found that positivism dominates within a structural-functionalist discourse. In other words, a positivist-orientated liberal intuitionalism infuses the logic behind the UEIP's ecological infrastructure initiatives by ignoring the policy processes in their entirety and the possible influence of traditional leaders (*Amakhosi*) (p. 127) who want to maintain the current status quo.

Green Corridors

Green Corridors (p. 124) is an organisation that aims to see communities thriving and in balance with their surrounding habitats. The organisation describes itself as a 'social-purpose and impact-focused' entity. The organisation was established in the uMngeni River Valley in 2009 and it later fostered a partnership with eThekwini. Green Corridor establishes hubs across rivers, the ocean and the ecosystems within Durban. These centres act as incubators for social and environmental initiatives. The projects emanating from Green Corridors highlight youth and sport development, open space management and restoration, as well as eco-tourism and learning. Furthermore, these projects focus on local sustainability, balanced habitats and community resilience. The philosophy that underpins Green Corridors rests on eco-system thinking and the community's interdependency with nature (Green Corridors 2020), which relates to WESSA's mission and objectives.

Water funds

The TNC is an American organisation that is funded by philanthropy and that has a global reach that establishes water funds throughout the world. At the time of the interview in 2018, the organisation was trying to set up a water fund in Cape Town. It was also discussing the establishment of such a fund (Respondent 9) with the eThekwini Municipality (p. 124). As mentioned earlier, eThekwini plans to establish the water fund with TNC, which will allow the Municipality to 'protect important water source areas outside of its municipal boundary' (p. 98).

According to Respondent 9, 'One of the Nature Conservancy's big agendas is water security and investing in [ecological infrastructures] for water security. In the work we are doing with the Nature Conservancy, we have engaged and partnered with eThekwini Water and Sanitation, because it is also involved in the UEIP. There are also other departments working in that space'.

Water funds are governance and financing mechanisms that have the purpose of linking downstream users with upstream landowners, in order to generate sustainable streams of finance through that relationship. If it is industrial or agricultural entities that benefit from a sustainable and secure water supply downstream, they may choose to contribute financial resources to the fund. The mechanisms linked to this philosophy enable decisions to be made on where to invest that money in the ecological infrastructure and how to maximise the benefits for water security. Water funds have also been established elsewhere in Africa, like Kenya (Respondent 9).

The planned water fund in eThekwini speaks to the self-regulating market feature of liberal institutionalism and indicates an intrinsic and inextricable link between liberal institutionalism, environmental change and environmental politics. These linkages show that liberal institutionalism is not only responsible for the environmental consequences we are witnessing, but that they can also change the social relations with the natural environment (McCarthy and Prudham 2004); in this case, it would be for the ecological infrastructure to improve water security. At the time of the interview, eThekwini was planning to establish the water fund with TNC and to link it to water security, which shows that the local government pays serious political and ideological attention on the social regulation (McCarthy and Prudham 2004) of ecological infrastructures that are linked to water security. These concerns are manifested through the advancement of the self-regulating market that is constituted by the fund.

The water fund is also a feature of the RIPE a '...standard regulatory approach—found especially within liberal institutionalism, neorealism and, more recently, systematic constructivism...' (Hobson and Seabrooke 2007: 5). The organising question that is fundamental to RIPE is 'who governs, and how is...order regulated?' (Hobson and Seabrooke 2007: 5) (see Table 2.1). The actors that are central to a RIPE epistemology and ontology are its structures, institutions or élites (Hobson and Seabrooke 2007), such as transnational corporations (Meissner and Ramasar 2015) and, in this case, TNC and eThekwini. The theory places little importance on actor identity and has a tendency to view '...actors' preferences as being aligned with their material self-interest' (Hobson and Seabrooke 2007: 13). Systematic constructivism, which discusses how actors and norm entrepreneurs diffuse their identities, view RIPE and its material self-interest manifestation as an internalised obligation, rather than a source of everyday actors (Seabrooke and Sending 2006; Hobson and Seabrooke 2007), such as the *Amakhosi* and Durban's wider public. This obligation is capital-centric and views the capital or financial resources as the 'agent' of change (Herod 2007). Therefore, the water fund could act as a catalyst for eThekwini to realise its water security aspirations through ecological infrastructure projects.

The hub and compact approach

eThekwini signed the DAC in 2011, and as the Charter Secretariat, it drives the implementation of the DAC, which has resulted in the development of the hub and compact approach (Respondent 9) (p. 124) for the implementation of the charter. This approach is a network of those networks that are involved in collaborative climate change adaptation action. The strategy started with the foundation of a partnership between Durban and Fort Lauderdale in the United States of America (USA). The two cities began with a series of city-to-city exchanges that enhanced peer-to-peer learning outcomes within the areas that had a mutual adaptation interest (DAC Secretariat 2020).

As mentioned by Respondent 12 in Table 3.4 (p. 124), the DAC is an outflow of the United Nations Framework Convention on Climate Change (UNFCC's) Conference of the Parties (COP 17), which was held in Durban in November 2011. The Charter is linked directly to the IPCC's call '…for more extensive and rapid adaptation than is currently occurring to reduce vulnerability to climate change' (DAC 2011).

The Charter contains several clauses that are directly related to climate change, as well as adaptation and mitigation measures. The first of these articles commits the signatories to '…climate change adaptation as a key consideration in all local government development strategies and spatial development frameworks' (DAC 2011: 2). The link between D'MOSS (p. 125) and climate adaptation is obvious in this clause. Secondly, the Charter urges members to 'understand climate risks through conducting impact and vulnerability assessments' and, thirdly, to 'Prepare and implement integrated, inclusive and long-term local adaptation strategies designed to reduce vulnerability' (DAC 2011: 2). Furthermore, the signatories need to link the adaptation and mitigation strategies and to promote the use of adaptation, which recognises the needs of vulnerable communities and ensures sustainable local development. Clause 6 states that they should, 'prioritise the role of functioning ecosystems as core municipal green infrastructure' (DAC 2011: 2). In Clause 9, the Charter promotes multi-level and integrated governance and advocates for sub-national and national government partnerships on climate action, while Clause 10 envisages the promotion of partnerships at all levels, as well as city-to-city cooperation and a knowledge exchange (DAC 2011).

In this respect, the Charter promotes institutionalised twinning between eThekwini and other cities in Europe, North America and southern Africa. It is also linked structurally to global climate change governance, at a local government level, through biodiversity protection, and more specifically, through green infrastructure initiatives. The Charter, therefore, is the structural expression of an empirical reality that climate change exists and that cities must improve the situation through organisational cooperation and by using international and operational innovations at a local level.

Twinning

City twinning is a form of international friendship that has been institutionalised by agreements between municipalities. Cities are actors in the international relations system who play an important part in defining foreign policies (Vion 2007). By twinning, cities can cultivate an identity that transcends state borders and, in this way, they can enter the arena of world politics through cooperative endeavours (Joenniemi and Jańczak 2017).

Increasing numbers of cities are pursuing policies that combine the local and the international through twinning. Cities have freed themselves, by establishing various ways of twinning (Joenniemi and Jańczak 2017) with other like-minded city governance structures. City twinning is a European phenomenon that has spread to all continents (Joenniemi and Jańczak 2017), with agreements being signed between cities from one continent and those located on another. The twinning between Durban and Bremen, as well as Durban and Fort Lauderdale, are typical examples of how the partners have facilitated twinning through the River Horse project and the Climate Adaptation Charter, respectively (p. 125). Twinning does not only occur through charters or agreements, but also through the initiation of specific projects. This shows that it is a global practice for eThekwini (Joenniemi and Jańczak 2017), particularly with cities across Europe, North America and southern Africa (p. 122 and 124).

As already mentioned, city twinning, as an institutionalised practice, transcends the local or domestic political environment, as well as the sphere of world affairs, with cities increasingly operating as links between local politics and international relations (Joenniemi 2017; Joenniemi and Jańczak 2017). Cities entering into twinning agreements, accompanied by structural and operational arrangements, are not as restricted by the policies of the states (Joenniemi and Jańczak 2017) that must keep diplomatic protocol and sovereignty in mind, when conducting international relations. Because of this emancipatory function of twinning, cities can surpass a variety of normative (autonomy) and geographical boundaries. It increases the regulative competency of local governments in ways of their own choosing. This makes a difference to, and has an impact on, national sovereign boundaries (Joenniemi 2017; Joenniemi and Jańczak 2017: 427) and national interests.

What is more, institutionalised and operational activities are fountainheads of a variety of political, social and cultural innovations, which are less restricted by the numerous accessories of modern politics, such as sovereignty, nationhood and national cultures (Joenniemi and Jańczak 2017). It is in this normative notion that we see autonomy substituting sovereignty as the settled norm (Heywood 1997) (p. 118). City twinning adds to a transworld notion and, as an ontological concept, it contributes to something that is akin to a civilisation of international relations (Joenniemi and Jańczak 2017). All in all, the practice plays both an affective and instrumental role (Joenniemi 2017); it is affective, in that it brings about another view of international relations, while it also plays the role of an instrument, to foster closer relationships between cities on functional matters, such as climate change adaptation.

The commodification of nature

Green and ecological infrastructures are natural assets that can be commodified in terms of performing a certain function or service, such as flood attenuation and water purification at an estuarine level (p. 89 and 127), in order to benefit human well being. These functions have either a direct or imbedded economic value; they are direct, when the economic value of the infrastructure can be measured, and they are imbedded, when these resources decrease the risks and create economic opportunities. Therefore, the productive function of the infrastructure relates to the economic realm of sustainability, since these purposes have a certain market value; for instance, green and ecological infrastructures provide agricultural products, such as food, animal feed, fibre, biofuel and medicinal resources. Sustainability occupies the intersection of the environmental, economic and social pillars (Lovell and Taylor 2013), and in this intersecting realm, green and ecological infrastructures provide opportunities for delivering environmental, social and economic benefits (Mell 2008; Wright 2011). It is within the ambit of sustainability thinking that we find a green and ecological infrastructure becoming an economic value proposition to society (Roe and Mell 2013). Benedict and McMahon (2002) asserted that a green infrastructure is the ecological framework that a society needs for environmental, social and economic sustainability. Dunn (2010) makes a similar argument when he said that, 'Green infrastructure is an economically and environmentally viable approach for water management and natural resource protection in urban areas'. The ontological foundation of these arguments is grounded in the negative impact of urbanisation on the socio-economic and environmental conditions that affect not only the health of city dwellers (Dunn 2010), but also their economic well-being, which manifests itself in poverty. From an economic perspective, green and ecological infrastructures have an economic function, as they provide an attractive and economically useful environment that entices business, tourism, an improved quality of life and increased property prices (Roe and Mell 2013). It is within the ambit of sustainability thinking that Respondents 1 and 2 conceptualised their definition and purpose of green and ecological infrastructures (p. 89 and 127).

With regard to the 'need for' and 'value of' wetlands, Respondent 2 refers not so much to their economic need and value, but to the need to protect them against risks, since wetlands purify water through natural, instead of technological, processes. Furthermore, ontologically, wetlands provide direct and indirect benefits to people, in the form food and water. Often, people take these benefits for granted and promote the technologies that successfully provide water, sanitation (Horwitz and Finlayson 2011) and food through a grey epistemology (p. 3). However, when compared with the coast, wetlands are perceived as being less of an economic resource than the beaches, which attract tourists (Fig. 3.18). The development of the coast also relates to technological advances through the construction of so-called hard infrastructure, such as buildings, roads, walkways and storm-water systems, (Fig. 3.25) and it is within this 'hard infrastructure' ontology that people place estuaries and their role in purifying water, before it enters the sea.

Not only are green and ecological infrastructures performing certain market functions that are linked to sustainable development, but governments and business

Fig. 3.25 Luxury apartments on the beach front at Umhlanga Rocks

can also invest huge sums of money in these infrastructure types. Such investments finance green infrastructures, with the argument that they are necessary to sustain economic growth and improve living standards (Chen and Warren 2011). It is here where green and ecological infrastructures can attain the target of economic prosperity (Roe and Mell 2013).

Not only do green and ecological infrastructures perform economic functions, but the normative arguments linked to this ontology state that governments and business must invest their financial resources in these infrastructure types, in order to realise their social, environmental and economic functions. We see this happening with eThekwini's proposed water fund and the CSR initiative of the River Horse project. Within the view of liberal institutionalism, municipal officials hold an epistemological and ontological view on green and ecological infrastructures, which commodifies nature. Epistemologically, these infrastructure types perform certain

Fig. 3.26 Small fish in the oHlanga Estuary north of Durban These fish are Glassies (*Ambassis natalensis*), which is one of few fish species that completes its life-cycle in South African estuaries, whereas most are species that spawn at sea, recruit into the estuaries as young juveniles and use the systems as nurseries (Weerts 2019)

economic functions and have an economic value. Ontologically, eThekwini views the liberal institutionalist types of intervention, such as the water fund and CSR, as mechanisms that give expression to sustainability thinking, and the economy plays a subsidiary role (Figs. 3.22, 3.23 and 3.24).

Competition

From what Respondent 3 said about traditional authority (p. 127), we can see that local governments are not just simple or complicated structures of rule but, in certain instances, they are normatively quite complex when it comes to the overlap between different governance systems.

The socio-economic and political form of governance over the protected coastal areas and, by implication, green or ecological infrastructures, is often not harmonious (Rhodes 1996; Meissner 2015b; Meissner et al. 2017a, b). According to Respondent 3, this friction occurs when the municipality is not controlling the allocation of land everywhere in the 'buffer' zones. In the dual governance areas, the *Amakhosi*, under the KwaZulu-Natal Provincial Government's authority, have the right to distribute land. Therefore, the complexity does not relate only to the (empirical) number of

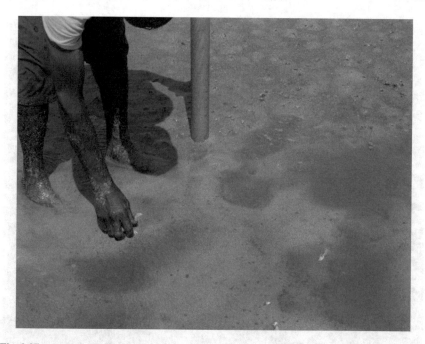

Fig. 3.27 An angler collecting sand prawn *Callichirus kraussi* for bait from the oHlanga Estuary at low tide (Weerts 2019)

actors interacting and competing with one another, but that the (normative) forms of authority are also different.

In this sense, Alcock and Hornby (2004: 28) noted that the '*Amakhosi* are not elected leaders and cannot be held accountable through a vote. However, they are subject to powerful mechanisms of accountability, recourse and communication within the tribe', such as the *ibandla*, or any gathering of older men for the purpose of discussing social issues. The *ibandla*, therefore, is influential at a communal governance level (Alcock and Hornby 2004). According to Respondent 3, the Provincial Government, which is another structure of rule, together with the *ibandla*, also has authority over the *Amakhosi*. According to Respondent 3, we see that municipal bylaws, policies and legislation do not have a significant role to play in protecting the land in buffer zones, particularly in dual-governance areas. In this sense, the policy environment is characterised by diffusion, which indicates that the constitutional state and its governance apparatus (Jessop 2002) have limited substantive economic and social powers. The *Amakhosi* seem to act more freely in the public sphere, while influencing an economic resource (i.e. land).

This can be explained by the fact that there has been an increase in traditional rule and customary practices since the introduction of liberal democracy in 1994. Liberal democracy, as an ideology, is not the only contextual factor at play, as there are numerous institutions that have an indigenous base and that are founded on customary practices, with the *Amakhosi* being only one example (Beall et al. 2004).

Fig. 3.28   An angler with collected sand prawn *Callichirus kraussi* from the oHlanga Estuary (Weerts 2019)

In the case of land allocation, we see discord between municipal structures of rule and the traditional customary practices pertaining to land apportionment. This indicates that the municipality is not the only important actor and that significant autonomous preferences (Viotti and Kauppi 1999; Stone 1994) emanate from the *Amakhosi's* traditional structures of rule with regard to the social construction of the importance of green or ecological infrastructures.

Because of the perceived discordant relationship, I argue that green or ecological infrastructures are a Western concept of value that is at odds with the traditional customary practices of land allocation. We see this in Respondent 3's response to the situation in the Quarry Road informal settlement, where their research is about 'making a difference in people's lives'. This speaks to the impact that research is supposed to have, or the perceived value of knowledge, which links back to the liberal institutionalist assumption of the 'economic' value of things, which includes an ecological infrastructure and knowledge. In this respect, Respondent 3 argues that the implementation of an ecological infrastructure that is underpinned by sound research could make a difference in the lives of the informal residents. This stands in contrast to the situation that has been created by the apparent disharmony of land allocation by the *Amakhosi*. Furthermore, Respondent 3 argues that the implementation of the provincial guidelines is difficult because of the *Amakhosi's* attitude towards them; they view them as a threat to their authority. That being said, it appears as if

Respondent 3 perceives the *Amakhosi* as being a hindrance to the implementation of green or ecological infrastructures, despite the fact that research suggests that implementing such infrastructures can improve people's lives. In other words, research produces knowledge about ecological infrastructures, and it is constructive in that it brings about positive change, whereas discord produces the exact opposite.

Moreover, we see several organs of state operating within the green and ecological infrastructures issue area. The one that stands central is, of course, the eThekwini Metropolitan Municipality. However, an organ of state, which is not easily categorised as such, includes the *Amakhosi*. Nevertheless, the Traditional Leadership and Governance Framework Act (No. 41 of 2003) (RSA 2003) stipulates the role and functions of traditional leadership and individual leaders, and therefore the role of the *Amakhosi*, in the political processes. The Act defines a traditional leader as '...any person who, in terms of customary law of the traditional community concerned, holds a traditional leadership position, and is recognised in terms of this Act' (RSA 2003: 8). The Act also defines a 'senior traditional leader' as '...a traditional leader of a specific traditional community who exercises authority over a number of headmen and headwomen in accordance with customary law, or within whose area of jurisdiction a number of headmen or headwomen exercise authority' (RSA 2003: 8).

If we follow Easton's definition of politics (1965 and 1985, cited in Meissner and Ramasar 2015) as the authoritative allocation of resources in society, we see that the *Amakhosi* have such an authoritative role, especially with regard to the allotment and distribution of land in buffer zones. However, as mentioned above, because the participation of the *Amakhosi* is linked to provincial government structures, we recognise that the relationship between the organs of state, particularly between the local government and traditional leaders, is not always harmonious. This is an intrinsic characteristic of the practice of governance in society, which is not always, and under all circumstances, in agreement. This trait not only reveals the quality of the governance practices concerning an ecological infrastructure, but it also shows that governance hierarchies have no clear boundaries and that one government actor is not always at the top (Knill and Lehmkuhl 2002; Rosenau 2006, 2008; Meissner and Ramasar 2015).

What is more, collaboration and contestation entail the involvement of dissimilar actors, resulting in the multi-varied nature of governance systems (Funke and Meissner 2011) and their enmeshed hierarchical nature (Meissner and Ramasar 2015). The fact that various governance authorities are involved in the implementation of green and ecological infrastructures shows that the issue is not always a clear-cut local government function, which is something that decision-makers should take into consideration in their planning. A further feature that characterises the contested nature of ecological infrastructures is the notion of autonomous preferences that give rise to the *Amakhosi's* choices in allocating land in the buffer zones.

Where liberal institutionalism focuses on collaborative, or non-harmonious, structures and actors (e.g. in allocating land), social constructivism highlights the normative aspects that are operational in the structures constructed by various role players. I will turn to this social theory in the next few pages (Fig. 3.25).

3.4.2.5 Social Constructivism

Social constructivism has a different view of knowledge and power than what liberal institutionalism and the hydrosocial contract theory teach us. Guzzini (2000: 147) argues that: '…constructivism is epistemologically about the social construction of knowledge and ontologically about the construction of social reality'. Social constructivism highlights interpretivism in social science and puts it centre stage in the analysis of social reality. This philosophy refers to the levels of observation and action, as well as their relationship to each other. At the level of observation, we find epistemological constructivism, which was developed out of social constructivism's critique of empiricism and positivism. Then there is the dogma of methodological intersubjectivity, or sociological constructivism (Guzzini 2000), and hence the term 'social constructivism'. This methodology sprouts from the critique of the rational choice approach. Lastly, there is also the concept and analysis of power in constructivism that '…functions as the reflexive link between observation and action' (Guzzini 2000: 156). A central theme of power is the 'art of the possible', which is challenged by social constructivism, and, by doing this, the theory presents a wider definition of politics, since meta-theories matter, both empirically and politically (Guzzini 2000). Since PULSE[3] identifies and gives meaning to social structures through a meta-theoretical analysis, social constructivism is apt to produce a deeper comprehension of the wider political nuances that are discernible in eThekwini's green and ecological infrastructure policy landscape.

Social constructivism attempts to explain and make sense of social relations by describing the construction of the socio-political world through human practice (Du Plessis 2000), which is constituted by ideational elements. Guzzini (2000: 149) '…understands *constructivism in terms of both a social construction of meaning (including knowledge), and of the construction of social reality*' (emphasis in the original). As such, it is '…an approach to social analysis that deals with the role of human consciousness in social life' (Finnemore and Sikkink 2001: 391). Social constructivism argues that ideational factors primarily shape human interactions. This means that it is not only the material factors, or visible actors, that play a role in relations. Furthermore, the approach notes that the most significant ideational influences are widely shared intersubjective beliefs, which we cannot, through other perspectives like the hydrosocial contract theory and liberal institutionalism, reduce to the individual level (Finnemore and Sikkink 2001). In this light, social constructivism asserts that positivist theories have been exceedingly materialist and agent-centric (Hobson 2000), with social constructivism taking the middle ground (Adler 1997; Guzzini 2000), so to speak, between the rationalism and postmodernist theories (Risse and Wiener 1999). Hence, the focus on the ideational elements constitutes social relations. Moreover, the shared beliefs create the actors' purposive interests (Finnemore and Sikkink 2001), which are translated into action.

To reiterate, from a political perspective, social constructivism focuses its attention on the core ideational elements of intersubjective beliefs (ideas, concepts, assumptions, rules and procedures) that people share (Ruggie 1998; Risse and Wiener 1999; Jackson and Sørensen 2003). Norms (p. 31) construct the identities of the actors, states and their governing apparatuses, and they define their interests. Norms have an important function in that interests change, and as norms reconstruct identities, it leads to changes in policy. In short, norms channel actors along certain socially dictated conduits of appropriate behaviour, and they are therefore autonomous. On the contrary, positivism sees norms as the intervening variables that are situated between the basic causal variable (power actors) and the political consequences (Smith 1997; Price 1998; Du Plessis 2000; Hobson 2000).

Social constructivism, like liberal institutionalism, treats the domestic and international spheres as two facets of a single social and political order (Reus-Smit 2001). The theory is concerned with the dynamics of political change (Reus-Smit 2001). For instance, interest groups exist as a political engagement community in politics. They have a meaningful impact through networks that teach governments what is appropriate to pursue in politics (Price 1998; Meissner 2005).

Norms create agents out of individuals by giving them an opportunity to act upon the world. To be sure, agents use all the means at their disposal to achieve their goals, including norms. On this, Onuf (1998: 4) commented by saying that: 'These means include material features of the world. Because the world is a social place...rules make the world's material features into *resources* available for agents' use'. According to Wendt (1995: 73), '...social structures include material resources, like gold and tanks... [C]onstructivists argue that material resources only acquire meaning for human action through the structure of shared knowledge in which they are embedded... Material capabilities, as such, explain nothing; their effects presuppose structures of shared knowledge, which vary and which are not reducible to capabilities'. All in all, the medium of norms and practices develops the actors' relations and understanding. If norms were absent, power and action would be meaningless. Actors specify their actions through norms that will lead other actors to realise their identity and respond to it appropriately (Hopf 2000).

Green and ecological Infrastructures through the social constructivist lens

Respondent 4's 'sex doll' analogy (p. 90) is an epistemological construction of these infrastructure types. He is critical of the construction of these terms in that natural scientists invented the concepts to change power relationships. Green and ecological infrastructures provide economists and engineers with a stake in this constructed knowledge system. Without economists and engineers, the implementation of green and ecological infrastructure initiatives would be perceivably impossible, since local governments require financial resources, as well as an engineered solution, to carry out such projects. This means that green and ecological infrastructures have become widely shared ideational elements so that we can involve those disciplines that will enable the seemingly successful implementation of green and ecological infrastructure projects.

For Respondent 4, the notion of nature, and nature-based solutions, is more encompassing because both scholars and the public share a common understanding of what they mean. It is not impossible that South African natural scientists observed European and American (p. 143) green and ecological infrastructure projects and concluded that we need to follow the same rational choice route. It is scientifically acceptable to follow the construction and implementation of the infrastructure types of these developed countries. Green and ecological infrastructures have 'resurrected' nature (Guzzini 2000: 157) to show that, through empiricism and rational choice, we will be able to make it work, and this legitimises the green and ecological infrastructure discourse and projects in the eyes of the public.

Therefore, green and ecological infrastructures consist of shared epistemological beliefs about what they both mean and, in so doing, they are purposeful concepts for the decision-making actors. It is no coincidence that most of the people that were interviewed shared SANBI's definition of what an ecological infrastructure is, and that some, like Respondent 3 (p. 90), have moved away from green infrastructures, in preference for ecological infrastructures. Both concepts give meaning to the epistemic community, both inside and outside of the municipality, by relating a knowledge of green and ecological infrastructures as an adaptation initiative to climate change and as an empirical reality. The treatment of this specific ontology provides a reason for their actions (James 2007), namely, to use green and ecological infrastructures as adaptation tools. In other words, the social construction of the concepts, when linked with purposeful actors and actions, constitutes the various practices that are manifested in competent performances (Adler and Pouliot 2011), which can be measured empirically. The conceptualisation of the infrastructure types has the capacity to inform the ideas and identities of social relations, networks and institutions (Carpenter 2003; Grant 2018) that are found in eThekwini's policy landscape.

From Table 3.5, we can see that there is an agential constructivism (Grant 2018) at play in eThekwini's green and ecological infrastructure policy landscape. Grant (2018: 256) described this type of constructivism as the '...rationalist variant of constructivism that is particularly apposite in providing the conceptual framing for understanding not only how norms influence state (and non-state) behaviour but also how these actors influence the dynamics of...norms'. In other words, norms influence non-state and state actors, and *vice versa* (Grant 2018).

From the definitions of green and ecological infrastructures (pp. 3–4), we can see that the concepts are widely used in other parts of the world. It is, therefore, safe to say that the concepts, as well as the discourse around them, constitute a transnational norm, although we can observe differences in their conceptualisation. As with the standards of appropriate behaviour, green and ecological infrastructure projects are practices that outline how local governments must adapt to climate change and bring about positive problem-solving techniques for urbanisation and the separation of people from nature in urban centres.

Returning to the concepts of green and ecological infrastructures as the shared beliefs of purposeful actors, the agential constructivism that emanates from the conceptualisation and use of the two infrastructure types, shows that rational actors, as the developers and implementors of such projects, shape the political, institutional and social environments in which they operate (Grant 2018). That said, the structures within which they operate, such as the EPCPD and the UEIP, are not less important

Table 3.5 Social constructivism and eThekwini's green and ecological infrastructure policy landscape

Theory	Assumptions	Interview data exemplifying assumptions	Themes or a case of
Social constructivism	Widely shared ideational elements	The 'sex doll' analogy of a green or ecological infrastructure (Respondent 4) (p. 90). 'That is [my] opinion of an ecological infrastructure. The terms that I like are nature and nature-based solutions, which are what the Europeans and Americans like. Therefore, it is nature-based solutions for water security. Nature-based solutions, conserving your wetlands, building artificial wetlands. So, that is how I engage outputs in our discussion' (Respondent 4)	The concept 'nature' as a widely shared term to critically engage with the notion of an 'ecological infrastructure'
	Shared beliefs constitute purposeful actors	'People use these terms interchangeably. Sometimes, 'green infrastructure' is used…more in the policy field or the applied field' (Respondent 3) (p. 90). 'In our work and engagement here in eThekwini, we predominantly use the term 'ecological infrastructure', (Respondent 3) (p. 90). 'Green infrastructure is to me the more popularised policy term that gets used' (Respondent 3) (p. 90). 'I am not sure what SANBI's intention was in terms of ecological infrastructure; in other words, whether they had…considered green infrastructure as a component' (Respondent 10) (p. 97)	Green infrastructure as a purposeful policy concept. SANBI's ecological infrastructure definition constitutes a shared belief. An ecological infrastructure institutes the purposefulness of the epistemic community and epistemic municipal officials (e.g. Debra Roberts). Green infrastructure as a discourse, with political implications

(continued)

Table 3.5 (continued)

Theory	Assumptions	Interview data exemplifying assumptions	Themes or a case of
	Norms creating interests and identities	'Because, I think it takes people quite a bit of time, and that is where I think eThekwini is very good in understanding the difference between biodiversity, ecosystem services and ecological infrastructure. To me, the closer relationship is between an ecological infrastructure and ecosystem services, because it goes into that ecological infrastructure framing. So, the infrastructure provides you with the services. Biodiversity is part of that, but it is also different to it. So, you can give up on some biodiversity, in order to get ecosystem services in the ecological infrastructure' (Respondent 3) 'So, terminology is important, and I would say that there is a difference in terms of terminology. Green infrastructure for experts and professionals means one thing, and for the layman it means another thing. What I perceive as a green infrastructure, at least in eThekwini, I think that when people think and talk about a green infrastructure, they are perhaps talking about pavements made from recycled tyres, energy-efficient buildings or those buildings that generate their own electricity. An ecological infrastructure is obviously the natural environment, as opposed to the built environment. That is how I perceive the differentiation' (Respondent 12)	eThekwini as an organisation that understands how biodiversity, ecosystem services and ecological infrastructure relate to each other interdependently Green infrastructure is used by actors with different interests and identities (municipal officials and the public)

(continued)

Table 3.5 (continued)

Theory	Assumptions	Interview data exemplifying assumptions	Themes or a case of
	Global change elements	'Come to think of it, my conversations about water and the catchment come mainly from the thinking of the climate change people in eThekwini. The debate is around water resources within the uMngeni Catchment. Even so, I've never heard them discuss estuaries, which is obviously a big element. The estuaries are a big problem. The UEIP is not about the estuary or the marine environment, at least, not since I have been involved in the last two years.... You know the idea about constructing all those parks in the city is connected to well-being and the quality of the urban environment. As far as I know, they do link with the green infrastructure' (Respondent 15)	The EPCPD influencing the view of a member of the epistemic community in her framing of ecological infrastructure Climate change played a role in the establishment of UEIPs
	Rules constituting resources	'I think there will be differences [between the definitions of green and ecological infrastructure]. So, an ecological infrastructure is, by implication, more about the natural systems that have ecological processes running through them' (Respondent 16) (p. 87) 'These programs are context-specific and they have differences in where they are implemented. However, there are [several] broad principles that they work on. First, they are based on the premise that you are looking after your natural assets, be it in a catchment or part of the catchment, because those assets will then look after you, if they are well-managed' (Respondent 16)	Epistemological rules defining what constitutes a green and ecological infrastructure Manage your resources well, and they will look after you

(continued)

Table 3.5 (continued)

Theory	Assumptions	Interview data exemplifying assumptions	Themes or a case of
	Agents and structures	'It depends on how you interpret a green infrastructure, so I will come at it from an ecological infrastructure perspective because that's the field I am more familiar with and I sit within a biodiversity department' (Respondent 9) (p. 87) 'Coming from an environmental planning department, the biodiversity that is closest to my heart is a green infrastructure, for its intrinsic value. However, it is also about the services it provides to humans. So, when we communicate about it is as a green infrastructure, it is all about your natural ecosystems, but also created ecosystems by humans, so that is immediately different to the biodiversity side. All the natural infrastructures, e.g. the ecological process or physical processes like filtering water, that is green infrastructure and the services that they provide to humans and how we communicate, quantify and use services and the need for the management of green infrastructure to advance job creation' (Respondent 16)	Normative concepts defining the structures in which agents operate

than the actors, because they '...and environments both constitute and provide the contours for the interactions of [the] transnational [and local] actors' (Grant 2018).

From what I have said thus far, we should also consider the 'technological dimension' of green and ecological infrastructures'. From a consciousness perspective, which exemplifies the relationship between the ideational and material, Doherty et al. (2006) asked whether technology determines human practice (technical determinism), or whether people construct technology through human agency (social constructivism)? This question relates to the topic in hand, since green and ecological infrastructures share, to a certain extent, the same space as technologies in open spaces and the uMngeni River. From the perception of social constructivism, our understanding of technology is social, by its very nature, in that people devise and use technology, instead of technology reflecting its capabilities (Doherty et al. 2006) in the absence of human activity. The foundation for this argument is that the application of identical technologies, in a similar context, can have a dissimilar influence on organisations (Doherty et al. 2006) and society at large.

Within the realm of technology and human agency, situated cognition or 'that knowledge and the conditions of its use, [that] are inextricably linked' (Tam 2000: 55) play an important constitutive role. Knowledge is 'situated', which means that it is 'in part a product of the activity, context, and culture in which it is developed and used' (Brown et al. 1989: 32). What social constructivism emphasises is the role that 'others' play in our knowledge construction efforts. Learning is, therefore, a social process. 'Explicit here is the belief that individuals bring implicit theories and perspectives derived from the cultural milieu..., and that inter-psychological aspects of knowledge creation themselves assist in the formulation of this very cultural context' (Adams 2006: 249). This learning operates within eThekwini's green and infrastructure policy landscape when municipal officials interact with the epistemic community and NGOs, such as SANBI, DUCT and WESSA, to create an epistemological and ontological consensus, or meaning, of the infrastructure type.

Part of the knowledge that is generated in the formulation of the cultural context is the notion of risk and a risk society. In the next section, I will now turn to this theory to explain the role of risk in generating knowledge about eThekwini's green and ecological infrastructure policy landscape.

3.4.2.6 Risk Society

The concept of a risk society began with the premise that pre-industrial societies were at the behest of natural hazards, such as famines, floods and earthquakes. With the advent of an industrial society, the power of natural disasters was diminished by institutional interventions (Mythen and Walklate 2006), which is similar to what the hydrosocial contract theory has to say about the hydraulic mission. Industrial modernity has been so successful that it has developed into a new, or reflexive, modernity of a risk society (Tuathail 1999). The new modernity and its associated technologies have the characteristics of a risk society (Beck 1992, 2009; Ritzer 2000), namely, a society that is troubled by comprehensive and globalised techno-scientific risks (Tuathail 1999) or 'dangers produced by civilization, which cannot be socially delimited in either space or time' (Beck 1996: 1). Risks are, to a large extent, the side-effects of modernisation (Beck 1992; Bulkeley 2001). Industrialism and the incessant accumulation of wealth produced by industrial processes are the main sources of risks in modern society (Mythen and Walklate 2006; Meissner 2004).

Consequently, the new modernity era is not less risk prone than the pre-industrial phase was. Through the process of capitalist modernisation, society has created a series of volatile manufactured risks, or "side-effects", of techno-economic development, which points to a movement away from a 'regimented industrial society to an unplanned and chaotic risk society' (Mythen and Walklate 2006: 383), which is often seemingly out of control.

This perspective conceptualises a risk as having a large negative impact on various environmental and social systems, and that it has a small probability of occurring (Björkman 1987). In this instance, Blowers (1997) identifies several nature-based risks on a grand or global scale, such as biodiversity loss, climate change, desertification and ozone depletion. Temporal and spatial determinants are unable to delimit these risks, but they have greater temporal and spatial mobility (Mythen and Walklate 2006) across the large land surfaces of the entire world. With respect to these 'grand risks', we are living in an industrial and risk society. In other words, we are in a transitional period between these two civilizations, with the society in which people live having elements from both worlds (Beck and Ritter 1992; Ritzer 2000). Not only that, risks pose a 'greater potential for harm in the modern age with risks not only spanning the globe, but also producing "irremediable effects"'. Modern technologies that multiply the risks, instead of remediating them, create a 'new Pandora's box' (Mythen and Walklate 2006: 384). There is another worrying dimension of a risk society that is not located in an industrial society. Mythen and Walklate (2006: 384) summarise this element by saying that:

> ...[T]he destructive energy of manufactured risks shatters existing principles of insurance. In industrial society, hazards are predicted and managed through insurance policies, welfare practices and legal constraints. By contrast, in the risk society, the ferocious force of "worst imaginable accidents"...disempowers existing methods of institutional regulation. These radical transformations in the nature of risk engender fundamental transformations in the nature of politics and patterns of social distribution. In industrial society, the general public pressed political parties to ensure adequate distribution of "goods", such as income, health and housing. Conversely, in the post-needs risk society, individuals become preoccupied by protection against social "bads", such as pollution, crime and terrorism. Since nobody craves ownership of bads, the logic of the risk society is no longer based on possession but avoidance....

Modernity is the reason why individuals and citizens are operating more independently, as society becomes more classless. Risks can either be reduced, redirected or, as stated above, avoided. Safety is of the essence and, in this sense, citizens look towards the defensive goal of an institution or structure that spares them from danger (Cohen 1997; Ritzer 2000). The dream of a risk society is that everyone should be secure from 'poisoning', and that the slogan 'I am afraid' should encapsulate the driving force in the risk society (Beck 1992).

The creation and accumulation of wealth is producing risks in modern society (Ritzer 2000; Beck 2009), which implies that there are those producing risk, those who are affected by it (Beck 2015) and those who enjoy the benefits of it, but who do not have to bear the costs (Beck 2009). This creation of wealth fosters a culture of increased economic affluence, the disregard of natural resources, which are seen as economic assets, and it has negative consequences for the environment (Selin 1987).

Speaking of wealth and its distribution, a connection exists between risk and class, with risks being associated with class patterns, although not in the way that we think. Wealth is the norm at the top of the class structure, while risks are the standard at the bottom. With poverty come risks. However, the wealthy (in terms of their income, power or education) have the means at their disposal to purchase safety and freedom from risk (Beck and Ritter 1992; Ritzer 2000; Beck 2015) in the form of insurance and assurances, with money being a cushion against risk.

This implies that two types of risk exist, namely, individual and collective. Individuals face risks when they are exposed to the probability of 'getting hurt' or dying. I would go so far as to argue that humans do not feel individual risks on such a grand scale as the risks that Blowers (1997) speaks of, but that they are more localised in their nature and extent, since such risks have an individual nature, as opposed to one that is collective. Collective risks touch more than one individual and do not spare large communities, states and even humanity (Selin 1987). At the time of writing, the global coronavirus disease of 2019 (COVID-19) pandemic is a good example of a collective risk.

As mentioned earlier, the source of risk alleviation is contained in modernity itself. The risks produced by modernisation also have the capacity to provide the reflexivity that is needed to think about modernity and its created risks. People are the main reflexive agents, especially those affected by risks. They observe their environment and gather information on the risks and their effects. Through observation, citizens become experts on risks. In this sense they critically look at the pace of modernity and the danger that it holds for society (Ritzer 2000). Individuals can also solve environmental problems and address the collective risks produced by modern society through functional and rational discourse (Eder 2000).

In certain instances, the advancement of knowledge is taking place outside government structures, not by standing in opposition to it, but by ignoring it (Beck and Ritter 1992). This has led to an erosion of the state and its governing apparatus, as well as its main power base, which is called the 'unbinding of politics' (Beck 1995; Bulkeley 2001), with discourse playing a central role in this unravelling process. This means that politics is no longer only the government's responsibility; rather, individuals, industry, interest groups, the wider epistemic community, NGOs, rural communities and other private sector institutions also have a responsible role in politics (Meissner 2004). This does not mean that the state and its governing apparatus have lost all political power (Beck 1995; Bulkeley 2001), but it is dispersed by a range of individuals and collective actors across numerous social structures.

Green and ecological infrastructures through the risk society lens

Table 3.6 elaborates on the risks that were identified during the interviews. The first

Table 3.6 Risks identified in the interviews

Risk	Expressed during the interview	Individual risk	Collective risk	Characteristic of a risk Society
Flooding and damage to property	'[O]ne of the big issues for us is that guys just want to build on the flood plains. We try to keep them out because they are part of the green infrastructure. We need those flood plains, right down to the estuaries' (Respondent 1)	X	X	**Initiated by:** Wealth generation and accumulation, as well as urban migration, resulting in people constructing more formal and informal dwellings **Affecting:** Property owners and those living in informal settlements near rivers and streams **Temporality and spatiality:** During summer rainfall events, flooding occurs with damage and harm to properties close to river systems and those who have constructed their homes on or near flood plains
Negative impacts on estuaries	'Estuaries are heavily impacted' (Respondent 1) (p. 110)		X	**Initiated by:** Sand mining for building material, pollution from wastewater treatment plants and other sources, and alien invasives, such the common water hyacinth (*Eichhornia crassipes*) and water lettuce (*Pistia stratiotes L.*) **Affecting:** Aquatic fauna and flora (Fig. 3.26), humans using estuaries for various activities, like collecting bait for angling (Figs. 3.27 and 3.28) and bathing **Temporality and spatiality:** Both the temporal and spatial dimensions of risks to estuaries are constants, since sand mining and alien invasives are present within the spatial dimensions of estuaries (river systems). Pollution has an irregular character since pollution from wastewater treatment plants and other sources, such as sewage spills and industrial pollution, are unpredictable as no one can determine when they will take place
No 'protection' from natural hazards	Natural habitats or natural resources provide 'ecosystems that function for the purposes of protection. Estuaries are quite a good example' of this (Respondent 2) (p. 110)		X	**Initiated by:** Not enough water in river systems to provide estuaries with water so that they can, in turn, function naturally and optimally **Affecting:** The natural environment occurring in estuaries. Should an estuary not function properly, it will not be able to offer some protection to the natural environment and human settlements and the infrastructure **Temporality and spatiality:** Water abstraction is a constant temporal and spatial feature in the uMngeni River, especially upstream from eThekwini. The same can be said for sediment movement, where sand mining is taking place in the uMngeni and its tributaries

(continued)

Table 3.6 (continued)

Risk	Expressed during the interview	Individual risk	Collective risk	Characteristic of a risk Society
Human encroachment onto coastal protection areas	'Traditional authorities are allocating land in these coastal protection areas' (Respondent 3) 'You can't develop within a certain distance from the sea and to ensure that infrastructure does not get damaged by tidal and storm events. You might not have to worry about tsunamis, but you'll have to consider tropical cyclones, which all fall within your disaster management context' (Respondent 4)	X	X	**Initiated by:** Decisions by traditional leaders at odds with the prevailing Provincial governance system on land allocation rights. In other instances, where traditional leaders are not involved, wealth generation and accumulation, as well as urban migration, constitute such risks **Affecting:** Coastal land resources, like dunes and estuaries, as well as the fauna and flora occurring in such bio-physical systems **Temporality and spatiality:** Temporally, the allocation of land in buffer zones by the *Amakhosi* is happening constantly, since the governance system cannot halt the allocation of land by the traditional authorities. Spatially it is only happening along the coastal areas like Mnini. For instance, the KwaZulu-Natal Province has influence over that and it is the Provincial Government's mandate. The Provincial Government has also put out 'different guidelines for the Amakhosi to use, to try and protect water buffers and wetlands' (Respondent 3). Spatially, one should not build too close to the sea, because one could run the risk of property and infrastructure damage from ocean storm surges
Flooding	'The D'MOSS looked at the benefits of metropolitan open spaces to humans, like flood retention and mitigation' (Respondent 4) 'It was interesting because the N2 [highway] flooded, lots of people lost their lives and hundreds of cars got washed away. One of the main reasons that it was so bad, was that the culverts and drains weres full of litter, so human behaviour made the flood so much worse. Just south of Durban there is a Citizen Science project at Amanzimtoti, that has been doing community action for five or six years, and there was no flooding on the N2. So, within a 10 kilometre radius, where there was no community action, you got extreme flooding, and less flooding where there was community action. That flood cost the insurance companies R600 million and the community projects at Amanzimtoti cost R500 000 per year. So, by investing half a million a year will mobilise people to care about water, litter and solid waste, and then you won't have the big floods that cost you R600 million per year to fix. That is the real cost. If you don't have a good system to sort out solid waste, you are going to pay a lot of insurance money. It is going to be very expensive for society. It is not paying a few people to clear away stuff. It is the risk of flooding and the loss of life that goes with it'. (Respondent 8)	X	X	**Initiated by:** Durban's high rainfall and people building near rivers, streams and floodplains. The lack of, or poor, maintenance of storm-water drainage systems on the N2 and surrounding areas. Litter that is washed into storm-water drainage systems clogs the infrastructure **Affecting:** People, property and built infrastructure **Temporality and spatiality:** Temporally during summer rainfall events and localised to streams, rivers and floodplains. Spatially, the N2 is a road, so the flooding occurred in a specific location

(continued)

Table 3.6 (continued)

Risk	Expressed during the interview	Individual risk	Collective risk	Characteristic of a risk Society
Inadequate water supplies	'You need to split off water security, in terms of flood mitigation, with water security, in terms of supply'. Durban's policy on flood mitigation is 'one thing' 'because they get lots of rain. Their policy on water supply and ecological infrastructure is something else, because Durban gets its water from elsewhere' (Respondent 4)	X	X	**Initiated by:** eThekwini's downstream status in the uMngeni River system, from which it gets most of its water supply **Affecting:** The entire population of eThekwini **Temporality and spatiality:** Flooding might occur during the summer rainfall season. However, the potable water supply is a constant issue because it is sourced from upstream. In addition, eThekwini's water security, in terms of quantity and quality, is spatially dependent on what happens (geographically) upstream in the uMngeni River system
The dumping of household refuse into rivers and alien vegetation along river banks, where criminals hide	'One of the big issues is the dumping of garbage, combined with alien invasives; when there is a flood, all that junk ends up in the harbour and they have to stop the tugs from operating because the garbage and trees gets caught in their cooling intakes. Every time there is an event, it costs them R2-R3 million to clean up the system. If you clear those areas that choke the place, there are other benefits too—criminals hide in those areas. In such choked areas give criminals free reign because they can hide easily' (Respondent 4)		X	**Initiated by:** The behaviour of people, namely, by littering, as well as poor household refuse removal services in certain areas of the municipality and the proliferation of introduced alien vegetation along the banks of rivers and streams **Affecting:** The Durban Harbour, as well as the logistics sector that connects to Durban from beyond the South African borders (foreign freight shipping companies) and those logistics operators that transport goods from the harbour to various parts in South Africa (Fig. 1.5 and Fig. 1.6) Criminals targeting eThekwini residents **Temporality and spatiality:** The dumping of household refuse is a constant issue because people do it every day. Alien invasives grow over a longer period, sometimes years, and 'push out' the indigenous forests incrementally. The 'junk' that ends up in the harbour happens during heavy rainfall events. Spatially, these risks are localised to river systems and the Durban Harbour Criminality is difficult to predict, in terms of time and space—one does not know when criminals will operate and from which specific area
Sewage spills are difficult to remediate	'What we have discovered up here in [Pietermaritzburg] is that the sewer lines that run along streams are choked with alien vegetation, and when there is a spill, you can't get to the blockage because of the vegetation. So, it is critically important to keep the servitudes clear of junk, so that you can monitor and manage your hard infrastructure. So, there are a whole bunch of benefits that are not just about flood mitigation and getting rid of the storm water and things like that'. (Respondent 4)		X	**Initiated by:** Blocked sewer drains, as well as the inability of municipal workers to reach those drains to unblock them, due to the choking vegetation **Affecting:** The natural environment including aquatic fauna and flora and potential animal and human health **Temporality and spatiality:** Alien vegetation grows slowly and pushes out the indigenous forests incrementally. Sewage spills are difficult to predict—one does not know when and where sewage spills will take place The municipality should conduct routine management of 'hard infrastructure' and along the entire sewer line to prevent or minimise sewage spills from happening

(continued)

Table 3.6 (continued)

Risk	Expressed during the interview	Individual risk	Collective risk	Characteristic of a risk Society
Erosion of riverbanks	'The municipality is getting rid of alien invasives and it is sometimes physically restoring the topography so that when the thing floods, you don't get huge erosion' (Respondent 4)		X	**Initiated by:** Alien invasives 'kill' the indigenous undergrowth along riverbanks. When workers clear large alien invasive trees and shrubs, it exposes the barren riverbank **Affecting:** The flora of rivers and streams **Temporality and spatiality:** Alien invasive clearing is a constant process across large tracts of land. 'The Palmiet River is an example of this' (Respondent 4)
Global climate change, global warming and biodiversity loss	'The big focus for us was to restore the former sugarcane areas as a carbon offset for the 2010 Soccer World Cup. Now we are moving towards restoration ecology' (Respondent 5) 'While there are pockets of excellence, much more needs to be done on a significant scale to contain global warming, to reduce greenhouse gases, transport costs, water and energy consumption and waste generation' (Respondent 11) 'I think they have some stuff that protects natural resources, but their intention is not necessarily from a water security perspective. So, I think the intention is around biodiversity conservation and things like that, in terms of their mandate under Section 24 of the Constitution, where people have the right to a clean environment' (Respondent 10) 'The Climate Change Strategy outlines the policies regarding green infrastructure' (Respondent 15)		X	**Initiated by:** The release of carbon dioxide and other greenhouse gases, like methane, into the atmosphere through various carbon-based fuel combustion activities, such as wood burning and fuel consumption in internal combustion engines. The processes responsible for these releases include, but are not limited to the industrialisation of agriculture, economic growth and increased energy consumption (Bulkeley 2001) **Affecting:** eThekwini's entire human and animal population, as well as the natural fauna and flora **Temporality and spatiality:** Anthropogenic global climate change is a long-term issue that is affecting the entire planet, while in the case of eThekwini's 'restoration ecology', its temporal scale is much shorter and more localised to specific areas within it's municipal boundary (e.g. Inanda) (Respondent 5)
Sewage pollution	'There are also sewer bylaws and they talk a lot about managing your sewers and managing your sewer discharge, so you don't create pollution' (Respondent 6)		X	**Initiated by:** Improper sewerage discharge by not following governance mechanisms (e.g. bylaws) and no, or improper, maintenance of sewer infrastructure **Affecting:** Aquatic ecosystems and, potentially, human and animal health **Temporality and spatiality:** Periodic sewage spills are a constant threat to rivers and streams and, by default, human and animal health. The spatial coverage of sewer spills, in terms of bylaws, covers the entire municipality
Development influencing water quality in rivers and streams	'Everything you do in the catchment has an impact on water quality, and the water quality for us is an indication of what is happening in the catchment. Whether there is silt, or a little bit of pollution or a lot of pollution, everything ultimately links back to the management of the catchment' (Respondent 6)		X	**Initiated by:** Development initiatives and objectives of various magnitudes within a river's catchment **Affecting:** Water quality and humans and the bio-physical environment that is dependent on the water **Temporality and spatiality:** Temporal development is a constant, with localised (spatial) impacts where it is taking place

(continued)

Table 3.6 (continued)

Risk	Expressed during the interview	Individual risk	Collective risk	Characteristic of a risk Society
Fire hazards	'They clear the land and they look after it for the ecosystem's benefit and specifically within the water or fire programme. Biodiversity is the main emphasis. It is the clearing of alien invasives (Respondent 8)		X	**Initiated by**: Alien invasives and human behaviour, as well as natural processes that start fires. **Affecting**: Human settlements and fauna and flora. **Temporality and spatiality**: Temporally, clearing alien vegetation is a constant process to minimise the fire hazard in specific locations where alien vegetation is prolific
Failure of modern communication technology	'We have been using the Citizen Science and the GRDK system as an app to monitor where things are going wrong. And when the big flood happened on the N2 and those cars got washed away, if you used your Garmin or Google device, it would have steered you onto the N2 because it would have said there is not traffic on the N2 at all. So, that was the place to go if you were trying to get home through that flood. Ironically, all the instruments on that road were flooded—all the phones on that road stopped working' (Respondent 8)	X	X	**Initiated by**: Various techno-scientific mishaps, like disruptions in electricity supplies. **Affecting**: Human society in various forms and with various consequences. **Temporality and spatiality**: Temporally, modern communication equipment operates under normal condition, such as when there is a constant electric power supply and no other disruptions to the infrastructure system. Spatially, modern communication technology failures could take place at an individual, household, communal, area, municipal or national level
Environmental damage	'[B]y reversing the accumulated environmental damage caused by past planning, design and maintenance regimes that have impaired, if not obliterated, the natural environment, so that it no longer provides goods and services' (Respondent 11) 'Vast natural forests (not plantations), large bodies of water, the earth's atmosphere and its crust all provide vital goods and services, which mankind has seriously impaired in the name of economic and social development: with little, if any, regard for the environment' (Respondent 11)		X	**Initiated by**: Various human activities. **Affecting**: Humans and the bio-physical environment. **Temporality and spatiality**: Temporally, environmental damage caused by past planning, design and maintenance regimes can be either short or long term, or they can be either localised, within a specific geographical area, or extensive
Water security	'While there are pockets of excellence, and great bulk water supply and water disposal systems, neither a "green infrastructure", or "ecological infrastructure" has been committed to, or implemented, on a scale that is required to make any significant contribution to the improvement of water security in a sustainable manner across the eThekwini Municipal Area' (Respondent 11)	X	X	**Initiated by**: Various human activities that influence water resources, especially their quality and quantity, as well as human interaction with the natural environment, to conduct water-related activities (Meissner et al. 2018a, b, 2019). **Affecting**: Humans and the natural environment. **Temporality and spatiality**: Temporally, water insecurity can be a short-, long-term or a permanent condition on a spatial scale ranging from the individual to the country levels

(continued)

Table 3.6 (continued)

Risk	Expressed during the interview	Individual risk	Collective risk	Characteristic of a risk Society
Loss of goods and services	The D'MOSS does not protect eThekwini from the insidious erosion of the goods and services caused by "development"' (Respondent 11)		X	**Initiated by:** Human activities that have an adverse influence on the natural environment when humans over-exploit ecosystem goods and services alike **Affecting:** Both humans and the bio-physical environment alike **Temporality and spatiality:** Temporally loss of goods and services can either be immediate or incremental, for instance, when alien invasives take over an indigenous forest. Spatially, this can be localised or widespread
Poor water quality and poor river health	eThekwini is implementing ineffective city-wide storm-water management projects (Respondent 11) 'People do their washing in the Palmiet River and develop rashes when coming into contact with the water. The conservancy higher up is of a relatively good quality, but then it degrades quite drastically in the last couple of kilometres' (Respondent 15) 'The idea was that if you build the reconstructed wetland downstream of the [Quarry Road] informal settlement then all the water sort of dis-surface and flows into the receiving waters of the uMngeni. So, the uMngeni is the focus and not the communities, to ensure that the uMngeni is not filthy' (Respondent 15) '[P]eople throwing nappies down the sewer lines. In the Aloe River, they have the nappy project which collects nappies by providing skips for them. They collect tens-of-thousands of nappies' (Respondent 15) 'Our mean interest in the UEIP is the fact that we are downstream from these municipalities' dysfunctional wastewater treatment works. For instance, the water that we re-treat is of a much better quality that the water we receive into the Inanda Dam in the Umgeni catchment because of the malfunctioning wastewater treatment works' (Respondent 15)		X	**Initiated by:** Littering (people disposing of soiled nappies in rivers and streams), sewer pollution from blocked sewer lines and dysfunctional waste water treatment works, as well as the littering of general household waste in rivers and streams (Fig. 3.14) **Affecting:** Humans, with adverse impacts on human health (skin rashes), as well as eThekwini, since it is located downstream from the pollution sources **Temporality and spatiality:** Temporally, these have a more long-term impact because the project has a lead time of months or years. Spatially, should the municipality realise one system across Durban, it could have ramifications across the entire city
Poor service delivery	'"Sexy Projects": The current projects gets votes and "appear" to address service delivery, while they tend to deal with the symptoms by using unsustainable reactive measures' (Respondent 11)	X	X	**Initiated by:** By a range of local governance and political aspects like mismanagement, corruption, skills shortages, the shortage of financial resources and other capacity constraints **Affecting:** The poorer communities of eThekwini **Temporality and spatiality:** The temporal and spatial dimension of risks emanating from 'sexy projects' depend on their temporal length (days, weeks, months or years) and the geographical extent

(continued)

Table 3.6 (continued)

Risk	Expressed during the interview	Individual risk	Collective risk	Characteristic of a risk Society
Pollution	'What is the cost benefit of floating-wetlands to remove nutrients and convert polluting chemicals, compared to properly managed exotic aquatic-weed control?' (Respondent 11)		X	**Initiated by:** Pollutants released into water systems that have an impact on human activities **Affecting:** The natural environment, particularly aquatic ecosystems, as well as human and animal health **Temporality and spatiality:** The temporal and spatial dimension of risk reduction from floating wetlands depends on their temporal length (days, weeks, months or years) and the geographical extent of its implementation
Urban migration and water (in)security	'My concern, however, is that as South Africa is moving to the metros, and perhaps in a complex adaptive system, as you solve one problem, you create another. So, my one concern is that if we ensure that our cities are water secure, or our country is water secure and the rest of Africa is experiencing drought, what is going to happen? Where are people going to migrate to? So, us being water secure and our infrastructure being secure can have these negative repercussions. Sometimes you need a team to sit and think carefully about these issues. So, if our systems are successful, you'll find that people will come to Durban or Cape Town because we have mitigated some of these negative impacts'. (Respondent 12)	X	X	**Initiated by:** Human migration patterns from rural areas in South Africa and other southern African countries, due to people searching for a better livelihood **Affecting:** People and the natural environment **Temporality and spatiality:** Temporally, migration is a constant process, while spatially, it takes place from one place to another, where the migrant believes there are better living conditions and infrastructure
Alien infestation	'During the conversation, someone mentioned that alien vegetation, in terms of ecosystem goods and services and in the context of South Africa, is something very negative in terms of water usage—water hyacinth infests areas and you cannot get access to the reservoirs' (Respondent 15)		X	**Initiated by:** High nutrient levels in water bodies and the proliferation of water hyacinth **Affecting:** Aquatic fauna and flora **Temporality and spatiality:** Temporally water hyacinth occurs in water bodies throughout the year, due to Durban's sub-tropical climate. Spatially, water hyacinth is a prolific water weed that occurs in almost every water body in and around eThekwini
Rodents and snakes	'If they are not well-managed and they are full of alien invasives, or if they are full of solid waste and they block up your infrastructure, you have rat infestations and black mambas coming into feed off the rats. Some people get bitten by the mambas and we've lost two people in [the] Quarry Road [informal settlement]. So, we've started a program of snake monitors in the settlements. We have about ten inhabitants in settlements that act as snake monitors and tell people about snakes. If there is a snake around, they will call the local snake handler to come and remove it' (Respondent 15)	X		**Initiated by:** Urbanisation and the proliferation of informal settlements where no, or limited, refuse removal services occur **Affecting:** Humans and other animals, like dogs, living in informal settlements where rodents like rats (*Rattus*) (Fig. 3.31) attract black mambas (*Dendroaspis polylepis*) (Fig. 3.32). According to the African Snakebite Institute, the black mamba is the largest venomous snake in Africa, with the African Snakebite Institute (2019) classifying it as 'very dangerous' **Temporality and spatiality:** These are constant threats, especially where refuse removal is inadequate. Spatially, rats and black mambas occur naturally in and around Durban, due to its sub-tropical climate (Evans 2019)

Fig. 3.29 Alien vegetation and erosion near Durban

column states the specific risk, while the second column contains the direct quote that expresses the risk. I then classify the risk as either individual or collective, before stipulating its characteristics. These features show the origins of the risk, who is affected by it, as well as its temporal and spatial scale. It is important to classify the risks in this way, since it interprets the functionality and rationality of the discourse contained in the quote. From the table, we see that most of the risks that I have identified conform to the collective type. The reason for this relates to the system that the municipal is using to implement green and ecological infrastructures. The natural environment is all-encompassing in that we live in the environment and we utilise the resources emanating from it. This implies that the risks emanating from the environment and human-constructed structures, such as the economy, are of a systemic nature and that they have the propensity to affect more than one individual at a time. Since Durban is, in certain parts, densely populated, the environmental risks are widespread, both spatially and temporally (Figs. 3.29, 3.30, 3.31 and 3.32).

3.4.3 Causal Mechanisms

In Table 3.6, we see that there are several causes of risks. In this section, I will discuss the causal mechanisms that I identified in each interview, in order to further highlight

Fig. 3.30 Anglers at the Umkomaas Estuary

the epistemological and ontological embodiment of causation in the interviews. These mechanisms are not only related to risks, but also to the overall green and ecological infrastructure policy landscape, or the context (p. 55).

Figure 3.33 indicates the total number of causal mechanisms. When considering what Gerring (2010) noted about them (p. 60), I would say that I identified micro-level (microfoundational) explanations for a causal phenomenon, as well as easy-to-observe causal factors, in the form of keywords, that describe it. This will become apparent in the following explanation. The mechanisms that I identified the most were material causal mechanisms (582), followed by structural mechanisms (336), agential mechanism (332), and lastly, ideational mechanisms (270). This distribution of causal mechanisms across the interviews is not surprising, since the study is about green and ecological infrastructures, or material causal mechanisms. The questions that I asked the respondents, therefore, highlighted such material causal mechanisms because the questionnaire focused on green and ecological infrastructures that the respondents viewed as natural resources, in terms of their materiality. Likewise, two of my research questions dealt directly with policies, programs, plans and strategies, which are all structural causal mechanisms. What is interesting to note about agential mechanisms is their close count with structural mechanisms, which implies that agents and structures have a close affinity to each other. In all the interviews, the respondents spoke about the activities and actions of multiple actors who were involved in green and ecological infrastructure initiatives, such as

Fig. 3.31   A rat (*Rattus*) along the banks of the uMngeni River near the Quarry Road informal settlement

Fig. 3.32   A black mamba (*Dendroaspis polylepis*). Photo courtesy of Johan Marais, African Snakebite Institute

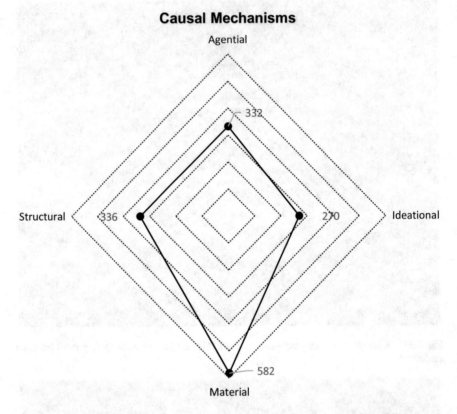

Fig. 3.33 The sum of causal mechanisms in the analysed interviews (after Sil and Katzenstein 2010; Meissner 2017)

'alien invasive clearing' (Respondent 5), pollution control (Respondents 1, 6 and 11) and twinning (Respondent 8). Lastly, the ideational causal mechanisms scored the least, for example, the concepts 'green and ecological infrastructure' (Respondents 1–16), 'mimicked' (Respondent 15), 'sex doll' (Respondent 4), and 'sexy projects' (Respondent 11).

In the following pages, I will highlight each respondent's causal mechanisms in a separate graph. In previous research conducted on water security in the Sekhukhune District and eThekwini Metropolitan Municipalities, we also developed a capability matrix, which indicated short- and long-term and positive and negative causal mechanisms (Meissner et al. 2018a, 2019). This capability matrix will be presented in the Appendix, at the end of the book.

In Respondent 1's causal mechanism graph (Fig. 3.34), he emphasised material causal mechanisms more than any others. These causal mechanisms are linked to the context of his career development within the municipality. His vocation, as a civil engineer, brings him into contact with various material resources, such as rivers,

Causal Mechanisms

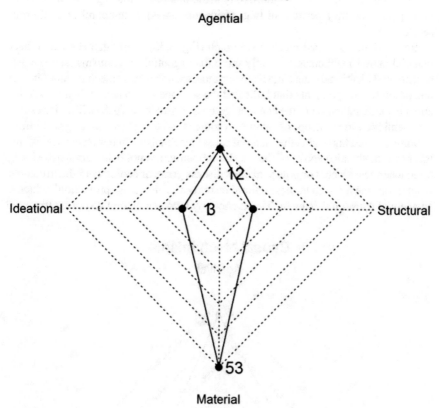

Fig. 3.34 Respondent 1's identified causal mechanisms (after Sil and Katzenstein 2010; Meissner 2017)

storm-water systems, estuaries, percolation areas, wetlands and reed beds (p. 89). These are natural systems, as well as various observed 'engineered' material causal mechanisms, such as artificial wetlands, detention and retention ponds, in the form of sports fields and public parks. The municipality used these recreational areas in the past to 'mitigate flooding downstream' (Respondent 1). Respondent 1 also alluded to the current thinking within the Engineering department, which constitutes an ideational causal mechanism, where 'engineers would like to get rid of everything quickly' (Respondent 1). This would therefore constitute a cognitive ideational mechanism that shapes the preferences of engineers, whereby they frame problems in a specific way. For Respondent 1, the context is the green and ecological infrastructures that are of cardinal importance in attenuating floods, with an eye on the prevention of flooding. Within this specific flood prevention context, Respondent 1 emphasised an agential causal mechanism when he said that 'We try and keep them out [of the flood plains], which are part of the green infrastructure' (Respondent 1). Here, Respondent

1 talked about people who would like to settle and build housing structures on the flood plains, be they permanent (e.g. residential areas) or temporal (e.g. informal settlements).

Respondent 2's causal mechanism graph (Fig. 3.35) shows that she emphasised material causal mechanisms slightly more than agential mechanisms, and she also accentuated ideational causal mechanisms more than structural mechanisms. She was one of the few respondents that indicated the salience of 'context' when I asked her about her understanding of the concept 'green infrastructure' (p. 89). She also explicitly mentioned the concept of 'natural resources' when talking about green infrastructures, which highlights the material causal mechanisms in her responses. Within this context, she also spoke of the coastal infrastructure not being emphasised very often when researchers or policy-makers discuss green or ecological infrastructures. In addition, she referred to these natural resources as affording 'protection', which is an agential causal mechanism. This dichotomy between natural resources (material

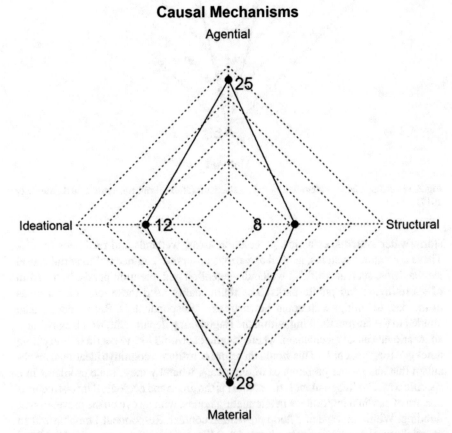

Fig. 3.35   Respondent 2's identified causal mechanisms (after Sil and Katzenstein 2010; Meissner 2017)

mechanisms) and protection (agential mechanisms) was mentioned frequently during the interview. Respondent 2 spoke of several ideational causal mechanisms when she told me about like Transnet's sustainable initiatives after the catastrophic flooding of October 2017. In this regard, she talked of 'sustainable initiatives' and the establishment of a 'green approach…, [which gives] an understanding that it affords better protection than engineering' (Respondent 2). Her low score for structural causal mechanisms reflects her limited knowledge of eThekwini's policies, plans, bylaws or standards for implementing green infrastructures.

The causal mechanism graph of Respondent 3 (Fig. 3.36) shows that she highlighted ideational causal mechanisms more frequently than the other mechanisms. Her score for these was 36, with agential mechanisms coming in second at 30, and the material and structural mechanisms scoring 25 and 24, respectively. The reason why ideational mechanisms scored so high could be because she spoke at length about the interpretation of the green and ecological infrastructure concepts (p. 89). For

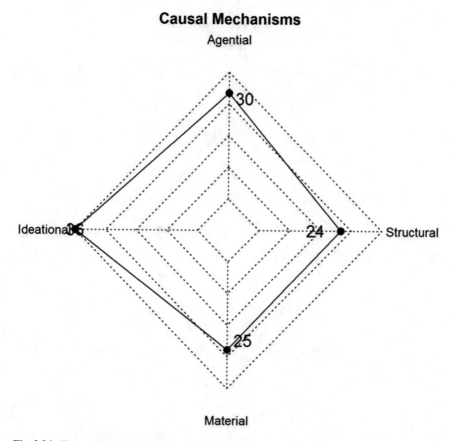

Fig. 3.36 Respondent 3's identified causal mechanisms (after Sil and Katzenstein 2010; Meissner 2017)

instance, she said that she has 'framed' the concept 'as an ecological infrastructure' and that a green infrastructure is 'in my mind...when city planners are thinking of putting in green open spaces to try and provide a whole lot of ecosystem services'. In this regard, Respondent 3 constructed the concept of green and ecological infrastructures within an academic or scientific context. She noted that: 'I've learned about it [an ecological infrastructure] from a science point of view' (Respondent 3), which is in line with the SANBI definition and the writings of Graham Jewitt. She also spoke about 'band-wagoning' when people use the term 'green infrastructure' in the '...battle for the environment...' (Respondent 3). Some of the structural mechanisms she mentioned are Durban's policies and bylaws that protect coastal buffers, and she also mentioned the 'dual governance context' between traditional authorities and the municipality, which constitutes a structural causal mechanism (Respondent 3).

Respondent 4, just like Respondents 1 and 2, highlighted material causal mechanisms the most, with a score of 51 (Fig. 3.37). Agential and ideational mechanisms

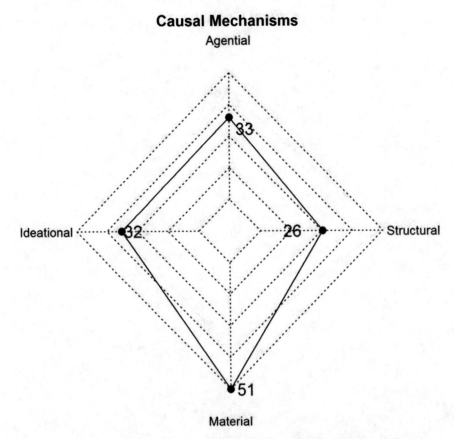

Fig. 3.37 Respondent 4's identified causal mechanisms (after Sil and Katzenstein 2010; Meissner 2017)

came in second, with scores of 33 and 32, respectively, while structural mechanisms scored the least, with 26. The explanation for the high number of material causal mechanisms relates to Respondent 4's position in the INR, which conducts research on environmental matters. He indicated that he works in a water context and that a green infrastructure is designed to improve the quality and quantity of water. What is more, when I asked Respondent 4 about his understanding of an ecological infrastructure, he indicated that it means 'nature'. He also spoke about manufactured wetlands and floating islands that are used to purify water. Ideational causal mechanisms transpired during the conversation when he started using analogies to describe ecological infrastructures (p. 90). He also argued that, 'when we talk about an ecological infrastructure, we become quite mechanistic about the way we think about nature' (Respondent 4). He emphasised agential mechanisms when he said that nature has all '…the passion, all the mystery, all the nuances, everything stripped out of it'. This quote is a good example of the interplay between ideational and agential causal mechanisms, where an idea about something constitutes our way of talking and doing, in a specific context.

Respondent 5 also highlighted material causal mechanisms the most, with a score of 42. Structural mechanisms received a score of 24, while agential and ideational mechanisms scored 22 and 15, respectively (Fig. 3.38). Respondent 5 is a municipal official who works at the Buffelsdraai landfill site. Her responsibility is to oversee the implementation of alien invasive clearing and indigenous tree planting. In this regard, she said that her work is to enhance the work of the Wildlands Conservation Trust, which is why she highlighted material causal mechanisms the most. Respondent 5 also mentioned some of the other initiatives, for instance, rainwater harvesting, which the municipality is implementing at the land fill site. One of the structural causal mechanisms that she spoke about was local communities. She indicated that the municipality is educating local communities in terms of water security. She also said that the 'municipality sub-contact smaller SMME companies to actually do the clearing of alien invasive plants and to promote indigenous plants' (Respondent 5) (p. 128). Another structural causal mechanism that she noted was DRAP (p. 122), and the municipality also has a reforestation initiative that takes place at the Paris Valley Nature Reserve and Inanda Mountain. These initiatives are part of its ecology restoration programme (Respondent 5).

Respondent 6 was another municipal official and support engineer who highlighted material causal mechanisms the most, with a score of 45. Structural and agential causal mechanisms scored 38 and 32, respectively, while ideational causal mechanisms scored the least (16) (Fig. 3.39). One of the material causal mechanisms that he highlighted was SuDS, or the actual engineering infrastructure that manages the storm-water flow. Piacentini and Rossetto (2020) describe SuDS as water-related green infratsructures. An ideational mechanism is sensitive urban design, which looks at the whole water cycle, that mimicks natural hydrological processes (Piacentini and Rossetto 2020). It is more inclusive of the water cycle than the standard urban drainage systems, which only incorporate storm-water flow. Some of the structural causal mechanisms that Respondent 6 mentioned were D'MOSS and the partnership that the municipality has with Bremen to rehabilitate the River Horse

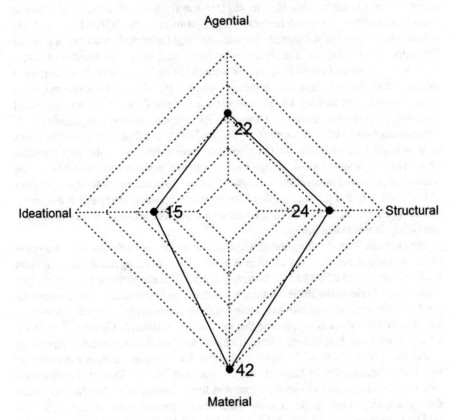

Fig. 3.38 Respondent 5's identified causal mechanisms (after Sil and Katzenstein 2010; Meissner 2017)

wetland (Respondent 6) (p. 125). The municipality is also looking at rehabilitating the Sundays River wetland. As has been done in the River Horse project, the municipality seeks to involve businesses (Fig. 3.17) that are situated near the wetland (p. 123). This initiative constitutes structural and material casual mechanisms, such as wetlands and the financial resources from businesses. To implement the Bremen partnership, eThekwini had to raise the financial resources, whereas the city of Bremen received funding from the German Federal government (Respondent 6).

The causal mechanism diagram of Respondent 7 has a similar shape to that of Respondent 6 (Fig. 3.40). Respondent 7 highlighted material causal mechanisms the most, with a score of 21, followed by structural causal mechanisms, with a score of 15, and agential and ideational mechanisms, with scores of 14 and 6, respectively. The high score of material and structural mechanisms is attributable to his role as a CSIR senior researcher in natural resources and governance structures. He noted that the municipality needs to communicate better on the issue of green infrastructures. It

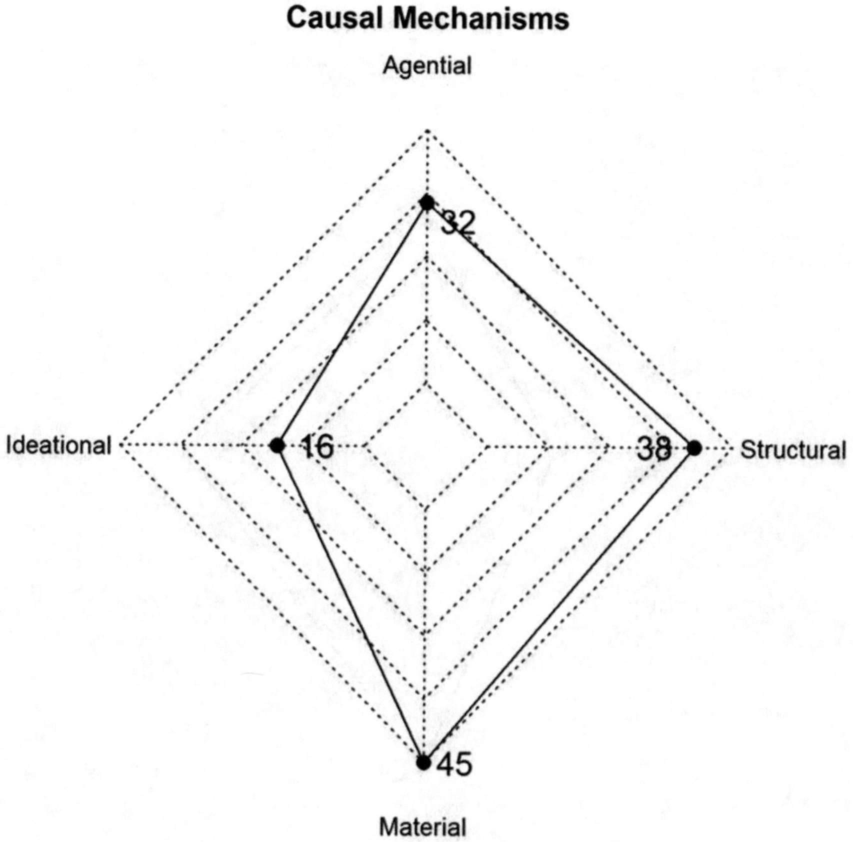

Causal Mechanisms

Agential

Fig. 3.39 Respondent 6's identified causal mechanisms (after Sil and Katzenstein 2010; Meissner 2017)

is here where the structural and agential causal mechanisms intersect. Furthermore, he spoke at length about the UEIP (p. 123), and hence the high score for material causal mechanisms, with natural infrastructure being the main focus within this structure. For instance, Respondent 7 mentioned the Msunduzi River, a tributary of the uMngeni River, which is a natural resource that supplies water to Durban. Furthermore, he indicated that there are some initiatives to clean-up the Msunduzi, to save water and to clear alien vegetation. These are all material mechanisms that are linked to agency and, to a certain extent, to the structure of the UEIP and to the actions of the actors who are involved in the UEIP.

Material causal mechanisms scored the most (37) in Respondent 8's interview analysis. This was followed by structural causal mechanisms, with a score of 27, followed by the agential and ideational mechanisms, with 16 and 12, respectively (Fig. 3.41). He spoke at length, and with authority, about natural resources linked to specific structures of rule, such as the D'MOSS and the Green Corridors initiatives,

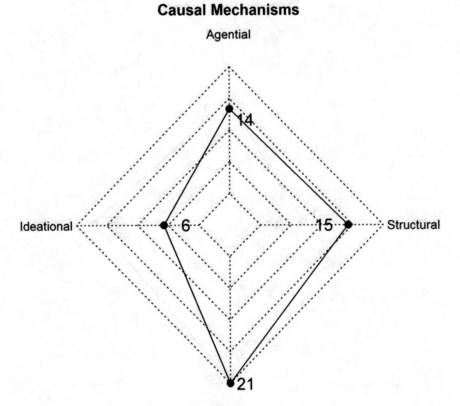

Causal Mechanisms

Fig. 3.40 Respondent 7's identified causal mechanisms (after Sil and Katzenstein 2010; Meissner 2017)

which were developed by DUCT and which are now independent entities funded by eThekwini (p. 124). He mentioned that the D'MOSS initiative is 25 years old and is linked to the 'idea' of having 'natural areas that could support water supply and cleaning water'. Moreover, he revealed information about the WfE enterprise, which is also an eThekwini enterprise. The idea behind the project is to 'look after the land as a whole…the land is cleared and they look after it for the benefit of the ecosystem and, more specifically, for [the] water or fire [project]' (Respondent 8). These are all material causal mechanisms that are contained within the specific structures of rule. Furthermore, he mentioned the clearing of alien vegetation and that Durban 'has realised that the uMngeni River is a bit like a sewer' (Respondent 8). Other material causal mechanisms that he spoke about were the technologies that navigated vehicles onto the N2 during the October 2017 flood (p. 111).

Material causal mechanisms received a score of 36 during Respondent 9's interview analysis, followed closely by the structural mechanisms (31) and the ideational

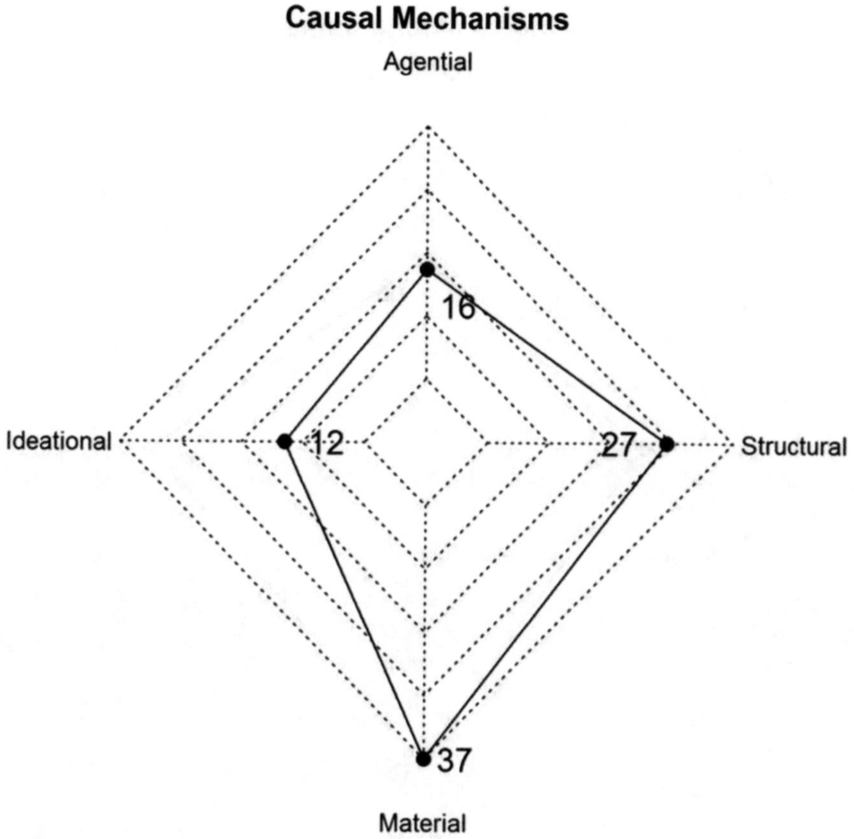

Causal Mechanisms
Agential

Fig. 3.41 Respondent 8's identified causal mechanisms (after Sil and Katzenstein 2010; Meissner 2017)

and agential mechanisms, which came in at 23 and 18, respectively (Fig. 3.42). She spoke more about ideational causal mechanisms than Respondents 6, 7 and 8, whose graphs are similar to hers. For instance, during our conversation about her under-standing of the concept 'green infrastructure', she indicated that 'there is quite a lot of confusion around the concept' and that it has various meanings. Because of this, her 'own understanding is quite a broad one' (Respondent 9) (p 86). In this case, we see that the use of ideational mechanisms by others has had an influence on her understanding (which is itself a causal mechanism) of the concept 'green infrastruc-ture'. Because she is a municipal official, she has good knowledge of its policies, plans, bylaws and standards regarding the implementation of green infrastructures. This explains the high scores that were received for structural causal mechanisms. Just like Respondent 8, she spoke at length about the D'MOSS, which she described as 'systematic conservation planning' and 'a scientific way of helping to identify key

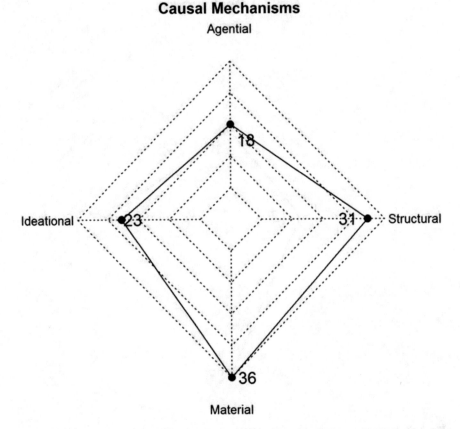

Causal Mechanisms

Agential

Ideational

Structural

Material

Fig. 3.42   Respondent 9's identified causal mechanisms (after Sil and Katzenstein 2010; Meissner 2017)

biodiversity or ecological assets and of working to protect them'. We see in this a nexus between material and structural causal mechanisms.

Respondent 10's causal mechanism diagram is similar in shape to those of Respondents 6, 7, 8 and 9. Material causal mechanisms received a score of 24, followed by structural mechanisms, with a score of 19, while agential and ideational mechanisms received scores of 13 and 8, respectively (Fig. 3.43). As a researcher at the INR, Respondent 10 had more knowledge about green and ecological infrastructures than structural mechanisms, such as the municipality's policies, plans, bylaws and standards. This explains her high score for material causal mechanisms. On this, Respondent 10 said that, '[m]y understanding is that their [SANBI's] definition is about natural systems, but I think you can view an ecological infrastructure as being a component of a green infrastructure, but not the other way around. She went on to say that a green infrastructure 'does not incorporate natural systems…' but '…refers

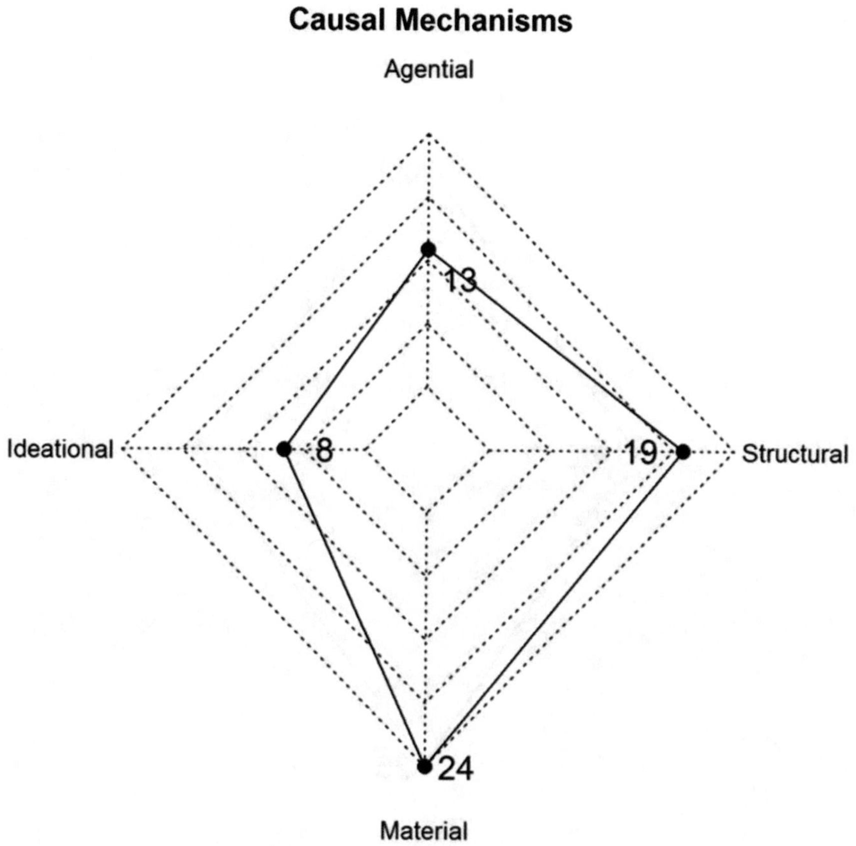

Causal Mechanisms

Agential

Ideational

Structural

Material

Fig. 3.43 Respondent 10's identified causal mechanisms (after Sil and Katzenstein 2010; Meissner 2017)

largely to a constructed natural system, like an artificial wetland' (Respondent 10) (p. 97).

The analysis of causal mechanisms in Respondent 11's interview shows that material and agential causal mechanisms received almost equal scores, with 25 and 22, respectively. We see the same for structural and ideational mechanisms, with scores of 17 and 15, respectively (Fig. 3.44). He indicated that the PRW is a voluntary group that was formed in 2013 based on the claim that 'ongoing and repeated pollution and habitat destruction has caused the aquatic creatures and plants that were in abundance a few years ago, to all but disappear'. This statement contains the philosophy of the PRW, which is an ideational mechanism that constitutes its actions. As an interest group representative, he was quite critical of eThekwini's structures of rule and he blamed these for the dismal state of the Palmiet River. Likewise, he showed that 'a parallel intervention to do a "green infrastructure" is required to reverse ongoing environmental degradation, namely "retro-engineering" storm-water and wastewater

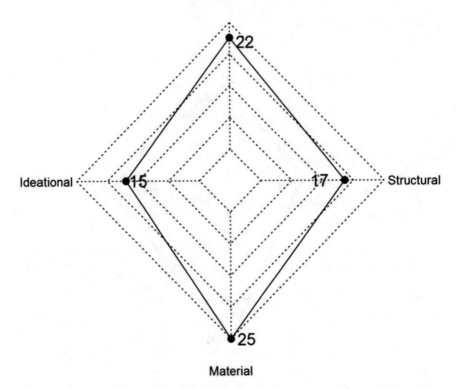

Fig. 3.44   Respondent 11's identified causal mechanisms (after Sil and Katzenstein 2010; Meissner 2017)

systems, so that they can function effectively and not destructively…'. This statement is an ideational causal mechanism that is aimed at changing the actions, policies and bylaws of the municipality. It is a good example of the interplay between ideational, agential and structural causal mechanisms, within the context of the Palmier River, and the perceived problems that the system experiences as a result of human activities.

Respondent 12's causal mechanism diagram reveals a score of 56 for material causal mechanisms, followed by structural mechanisms, with 47. Agential and ideational mechanisms received almost equal scores, namely, 34 and 30, respectively (Fig. 3.45). Respondent 12 is another municipal official who focuses his attention on the municipality's climate change responses. During the interview, he spoke at length about the solar initiatives that are being implemented by the municipality. This is the reason that the material and structural causal mechanisms scored the highest, which indicates that there is interplay between them. An example of this is the photovoltaic solar cells and the municipality's bylaws and policies that facilitate

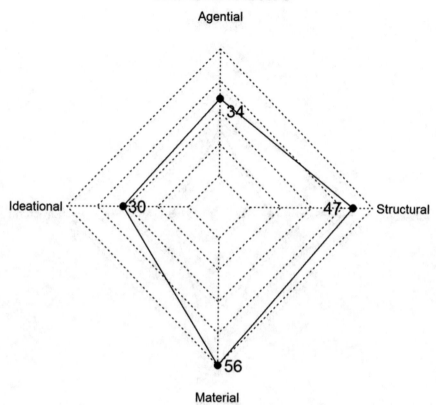

Causal Mechanisms

Fig. 3.45 Respondent 12's identified causal mechanisms (after Sil and Katzenstein 2010; Meissner 2017)

the implementation of such an infrastructure. With respect to structural causal mechanisms, his narrative revolved around the DCCS, which he described as a climate change response strategy (p. 126). He also referred to the C40 initiative (p. 122) by mentioning that the Mayor of Durban, Councillor Gumedi, is the Vice President of C40, and in line with C40, Durban is implementing a program for new buildings in the city, in a bid to reduce the its operational carbon to zero (Respondent 12).

Respondent 13's radar diagram indicates a score of 23 for material causal mechanisms, followed by agential mechanisms, with 10, and ideational and structural mechanisms lagging far behind, with scores of 4 and 3, respectively (Fig. 3.46). The high score for material causal mechanisms is no surprise, since he works at the Durban Botanical Gardens. During his interview, he spoke at length about capturing storm water to augment the Gardens' water resources and to make them less reliant on potable municipal water supplies. The municipality's Engineering department will implement this rainwater harvesting system, which indicates an interplay between the

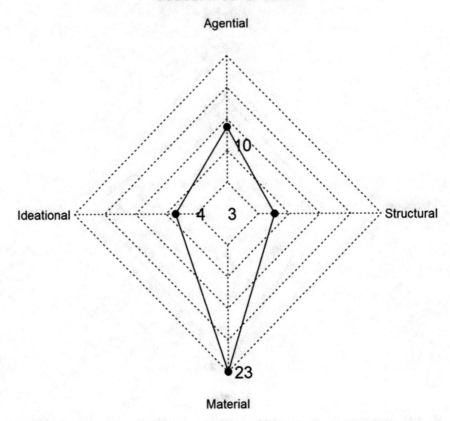

Fig. 3.46 Responding 13's identified causal mechanisms (after Sil and Katzenstein 2010; Meissner 2017)

material and agential causal mechanisms. Respondent 13 indicated that the Gardens will use the collected storm water for irrigation purposes (p. 112). The idea shows how the Botanical Gardens fit in with residential water management, which is an ideational causal mechanism. He also mentioned that litter in the storm water is a major issue, since it clogs the storm-water system during rain events and results in erosion from the overflow.

Respondent 14's analysis shows that material causal mechanisms received a score of 15, followed by structural causal mechanisms, with a score of 10, and then ideational and agential mechanisms, with scores of 7 and 4, respectively (Fig. 3.49). Just like the other respondents, Respondent 14 spoke about green and ecological infrastructures and gave examples, such as artificial wetlands, which she does not consider to be natural systems. For her, a natural wetland is a natural ecosystem, which equates to an ecological infrastructure (p. 93). With respect to structural causal mechanisms and water security, she mentioned that she does not think that the municipality has policies that are directly related to a green infrastructure and water security. Furthermore, she argued that: '[Municipal officials] do have a strategy that relates directly to ecological infrastructures, but that [the municipality] does not link an ecological infrastructure directly to water security'. She did, however, indicate that an ecological infrastructure is linked to flood management. Agential causal mechanisms were visible in her interview when she spoke about projects like the Sihlangzimvelo initiative, which focuses on rivers, their rehabilitation and the improvement of water security (p. 126).

What is interesting about Respondent 15's causal mechanism diagram is that the ideational and material causal mechanisms received almost equal scores, namely, 31 and 37, respectively, while agential and structural causal mechanisms received scores of 11 and 8, respectively (Fig. 3.50). The ratio between the material and ideational , and the agential and structural mechanisms is quite low for this respondent, compared to that of the others. This respondent is a Hydrology lecturer and researcher at

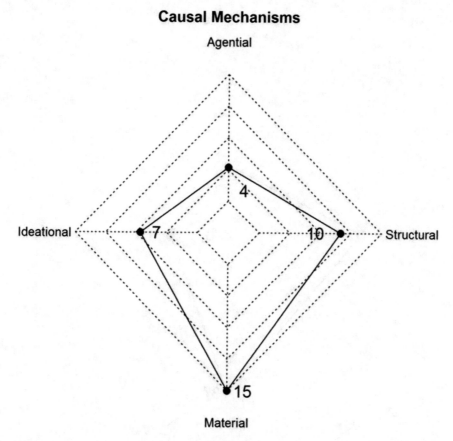

Causal Mechanisms

Agential

Ideational

Structural

Material

Fig. 3.47 Respondent 14's identified causal mechanisms (after Sil and Katzenstein 2010; Meissner 2017)

UKZN's Pietermaritzburg campus. Just like the other researchers (Respondents 2, 3, 4, 7 and 10), she gave the concepts of green and ecological infrastructures some thought during the interview (p. 93), which could explain the high score for ideational causal mechanisms in her interview analysis. Because she gave a lot of thought to the concept, in the context of catchments, it would explain why material and ideational causal mechanisms received such high scores. She also indicated that 'I am currently thinking about size, more than anything else. I know that someone might define a ridge line as an ecological infrastructure, but I think it is more about the elements in a landscape that play a role in ecosystem services' (Respondent 15).

Respondent 16 spoke more about material and structural causal mechanisms (64 and 38, respectively) than about agential and ideational mechanisms (36 and 20, respectively) (Fig. 3.50). This respondent is from the municipality's EPCPD and, according to him, the Climate Protection branch is an Environmental Planning department. For him, biodiversity is an intrinsic part of a green infrastructure, hence the

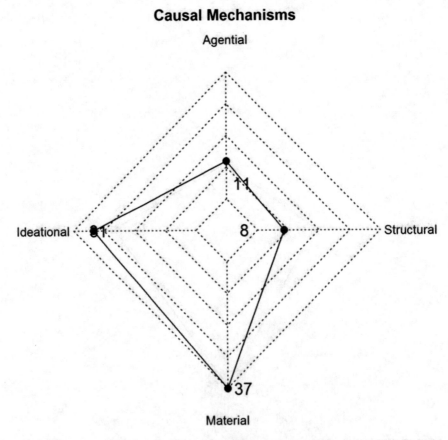

Causal Mechanisms

Fig. 3.48 Respondent 15's identified causal mechanisms (after Sil and Katzenstein 2010; Meissner 2017)

high count for material causal mechanisms (p. 87). He is also very knowledgeable about the municipality's green and ecological infrastructures, and he referred to the DCCS, which, according to him, outlines the policies regarding green infrastructures. Furthermore, he indicated that the municipality's Water and Sanitation Department was drafting its own climate change strategy at the time of the interview. He went on to speak about some of the projects and initiatives that have been introduced by the municipality with respect to green infrastructures, like the Sihlangzimvelo project (p. 126), and collaboration between the municipality and UKZN through the DRAP and the Sustainable and Healthy Food Systems projects. With respect to the link between structural and agential causal mechanisms, Respondent 16 specified that the municipality is planning to get together with the IPCC, and that he is the lead author of specific IPCC reports. During the interview, Respondent 16 also spoke about the C40 initiative and described it as a network of African cities that has been developed around the 'good principles' (ideational instruments) of climate adaptation.

In order to elaborate further on the identified causal mechanisms, a comprehensive list of the causal mechanisms that were identified during the PULSE[3] analysis will be presented in the Appendix. The list not only shows the causal mechanisms that each respondent emphasised, but also what case or occurrence it represented and whether the respondent flagged the occurrence as positive or negative. The Appendix demonstrates that the respondents view the presence or implementation of green and ecological infrastructures in a positive light and that their existence has favourable consequences for the environment and the goods and services that they provide. It is also significant to see that the interviewees perceive green and ecological infrastructures as being a service to the human population of eThekwini. This is in line with the dominance of the hydrosocial contract theory and the liberal institutionalism perspective, which emphasise the natural and economic function of natural resources and the systems from which they are derived.

Respondents 1, 2, 3, 4, 7, 8, 9 and 11 emphasised certain negative aspects with respect to the implementation of green and ecological infrastructure policies and initiatives, including flooding, littering, transparency, the lack of communicating the benefits of the initiatives to the public, rodents and venomous snakes, as well as the unharmonious governance of land in buffer zones.

Thus far, I have considered the causal mechanisms at a local government level and how they are linked to various processes that interact between, and within, the green and ecological infrastructure policy landscape and the natural environment. In the next section, I will elaborate on the visibility of the problem-solving and critical theory perspectives in the interviews, in order to show how extensively the respondents framed their narratives from the problem-solving and/or critical theory discourse.

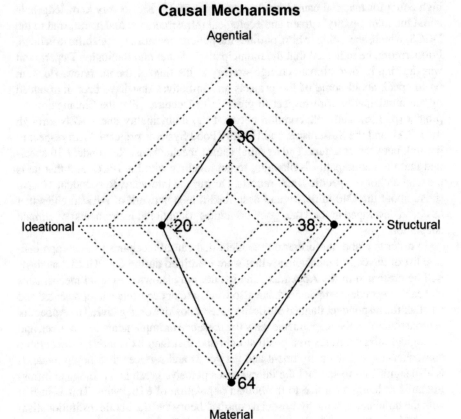

Fig. 3.49   Respondent 16's identified causal mechanisms (after Sil and Katzenstein 2010; Meissner 2017)

3.4.4   The Problem-Solving and Critical Theory Perspectives

The analysed interviews were dominated by problem-solving theories, with critical theories hardly being present (Fig. 3.50). This confirms my conclusion, thus far, that empiricism and positivism are the main paradigmatic foundation of the respondents' views; they see the policy arena from an overarching empirical perspective, as opposed to one that is normative.

The reality of long-term anthropogenic climate change informing their views is a certainty (Duvall and Varadarajan 2003). However, there is an abstraction to this reality, namely, that eThekwini's policy landscape is not devoid of theory(ies), albeit empirical theories. This dominance of a theoretical dispensation gives expression to the maxim that 'theory is always *for* someone and *for* some purpose' (emphasis in the original) (Cox 1996).

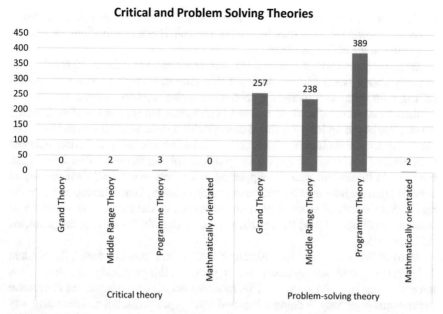

Fig. 3.50 Critical and problem-solving theory types (after Cox 1996; Mearsheimer and Walt 2013; Friedman 2001; Dixon-Woods et al. 2011; Coryn et al. 2011; Rogers et al. 2000; Meissner 2017)

In other words, Fig. 3.50 gives 'theoretical visibility' (Halliday 1991) to the policy arena. The respondents observe eThekwini's green and ecological infrastructures through a positivist epistemology and relate it to their work, and vice versa, through an empiricist methodology. This means that they view the policy landscape as common-sensical, which explains the high scores received for the programme theory (Fig. 3.50). Therefore, the perspectives that I identified during the analysis also envision a 'problem-solver agenda' at work (Duvall and Varadarajan 2003). The problem that needs resolving is the long-term anthropogenic climate change, and the way of solving it is by the implementation of green and ecological infrastructures that are supported by structural causal mechanisms, like D'MOSS, and the establishment of the water fund, among others. The municipality identified the systematic causes of climate change and the threats to water security, which were observed objectively by analysts, and this provided the policy-makers or municipal officials with the tools to manage these causes and threats.

Moreover, the respondents viewed the policy landscape as a natural feature of the existing local governance structures and as the acceptability of the predominant relations and power practices in eThekwini (Duvall and Varadarajan 2003). This means that some of the respondents who indicated the negativities of green and ecological infrastructures, attempt either to optimise the operation of these policies and initiatives, or minimise their deficiencies (Devetak and Walter 2016). This is the emphasis in Respondent 11's critique of the 'inability' of policies and initiatives to deliver the necessary goods and services to achieve water security on such a scale.

For him, the municipality must also address policy deficiencies by using systematic scientific knowledge, and thus he quotes Duncan Hay's observations of what is happening in the uMngeni River.

Since problem-solving theories are dominant, and the critical approaches are virtually absent, what can we do about this state of affairs? I agree with Brown (2013), who states that there is a need for a critical problem-solving theory or a '…theory that relates directly to real-world problems, but approaches them from the perspective of the underdog [critical theory]'. Therefore, when it comes to climate change, green and ecological infrastructures and water insecurity, there does not need to be a balance between the critical and problem-solving theories. What is essential, however, is to observe what is happening in the real world, and then to suggest solutions that are informed by critical theory approaches. I base this argument on the notion that the natural sciences offer an inadequate model for the practice of social disciplines (Skinner 1985; Brown 2013) and, by default, for solving the problems that face society.

Many of the respondents have already hinted at what 'could be done'. Respondent 1, for instance, said that engineers 'like to get rid of things quickly' and that 'this is not always the right thing to do'. This indicates that there is not only an alternative to the engineering way of doing things and solving problems, but an alternative way of thinking within the engineering fraternity. Respondent 7 indicated that the public does not understand the benefits of green and ecological infrastructure projects, due to inadequate communication. He noted that there must be more transparent communication about the benefits of the projects. Transparency is a normative and ethical matter that will not only help to raise awareness about the projects, but it will also give eThekwini's residents an idea of their existence and of who will benefit the most. This links back to the criticism in Chap. 1 of the misinterpreted multifunctionality of green infrastructures and the argument that such infrastructures often do not benefit the poor (p. 6). Respondent 8, just like Respondent 1, alluded to doing things differently to what they had previously been done, namely, in traditional and technical ways. He spoke about the Citizen Science project in Amanzimtoti that contributed to less flooding in October 2017; the storm-water system was not blocked and clogged by solid municipal waste because the people had not littered. Even so, we need to be cautious about interpreting this as the argumentation of a critical theory; the Citizen Science project is still a discursive and ideational causal mechanism that is based on empiricism and positivism. The predominant focus is on the way that science is conducted, instead of on emancipation through structural realignment. In this sense, it would be interesting to know how the Citizen Science project contributes to the day-to-day survival of residents living in Durban's poorer areas. Respondent 11 was quite critical of the municipality's green infrastructure initiatives and proposed ways of improving the way things are done, like introducing more grey engineering initiatives to retrofit storm-water systems and water treatment works, which would influence the ongoing environmental degradation. By invoking the same grey epistemology as in Chap. 1, he is also not arguing from a critical theory worldview, but from a problem-solving paradigm.

Please note that although I have mentioned practices throughout, I have not gone into detail. Before concluding this chapter, I will discuss the overarching practices by using the practice theory, and I will link these to thoughts on epistocracy that are manifested in routinised activities within an epistemic structure, and which characterise the policy landscape.

3.5 Practice Theory as a Direction-Finding Beacon

Focusing one's attention on practice theory is a productive way of producing real-world considerations (Pouliot and Thérien 2018) from the PULSE[3] analysis. As has already been mentioned above, a PULSE[3] analysis, as a practice, highlights the various research worldviews, theories, causal mechanisms and problem-solving and critical perspective types that are at play in a text that promotes a practice, or range of practices. The concept 'practice' is like an adhesive or 'ontological entity that cuts across paradigms under different names but with a related substance' (Adler and Pouliot 2011: 10). In other words, 'practice' is the 'ontological core concept that amalgamates the constitutive parts' of life, since practices link the material aspects with the meaningful elements of life, and the structures with processes (Adler and Pouliot 2011: 10). This conceptual glue does not only operate at an ontological level, but also in the epistemological realm. Having said that, research worldviews and perspectives, including causal mechanismsand problem-solving and critical theories, have a common and related substance—they all constitute various practices in academia and in the world of work.

Since practices are omnipresent in social life, various social scientific disciplines have theorised about them in isolation, with a theory highlighting a certain action, or group of practices, and another perspective doing likewise with dissimilar practices. For instance, in IR and Political Science, researchers have put certain practices, like global governance, in the realm of liberal institutionalism (Pouliot and Thérien 2018). This is because certain practices have specific traits (Adler and Pouliot 2011) or belong to a certain type of theorisation, based on particular characteristics. Practices that have material features generally fall under liberal institutionalism, while researchers associate practices that have a more symbolic appearance with social constructivism. Practices also have a psychological component, since they contain emotional states (Reckwitz 2002), and scholars could investigate them by using the political psychology approach. This shows that practices have either empirical or normative characteristics (Reus-Smit and Snidal 2008) that lend them to a certain type of theoretical analysis. Then, there is also the 'black boxing' of security issues that researchers do not investigate alongside economic matters or practices. For instance, rational choice theorists investigate practices that are aligned with cost-benefit calculations, while systemic theories explore structural constitutive practices (Adler and Pouliot 2011). Practice theory holds the 'promise' of '…a new dialogue of ideas, in order to better understand the pressing matters of…' social life (Adler and Pouliot 2011).

3.5.1 Conceptualising Practice

Reckwitz (2002: 250, cited in Bueger and Gadinger 2015) provided an elaborate definition of a practice, which highlights how process- and context-specific information relates to practices and practice theory. He notes that 'a practice is a routinized type of behaviour that consists of several elements, interconnected to one another: forms of bodily activities, forms of mental activities, "things" and their use, a background knowledge in the form of understanding, know-how, states of emotion and motivational knowledge'. When we perform a practice, all these elements are at play, which means that a practice is not like any of these elements (Reckwitz 2002; Bueger and Gadinger 2015), and it is because of this that practices produce the structural profiles of social life (Feldman and Orlikowski 2011). Pouliot and Thérien (2018: 164) spoke of practices less comprehensively as 'socially-meaningful and organized patterns of activities', while Jackson and Nexon (2019) asserted that '[p]ractices are obviously processes; they are the doings of agents', and they fill the voids between legal rules and procedures (Pouliot and Thérien 2018). In this way, Giddens (1984, in Feldman and Orlikowski 2011) defined practices as social actions that continuously produce and reproduce structures that enable actions, as well as constrain them. In a similar vein, Schatzki (2001, cited in Feldman and Orlikowski 2011: 1241) noted that 'the bundles of human activity that constitute practices enact social orders'. Practice theory, therefore, captures the consequentiality of practices (Feldman and Orlikowski 2011) and activities that beget other practices, which, in turn, bring about enabling and constraining societal structures.

3.5.2 The Practice Perspective

The purpose of this discussion is to deepen our understanding of eThekwini's green and ecological infrastructure policy landscape and the role of the individual therein. With this in mind, practice theory is not a single theoretical approach, like liberal institutionalism, but it consists of a broad range of approaches (Nicolini 2012; McCourt 2016), a 'broad intellectual landscape' (Feldman and Orlikowski 2011: 1241) or, in the words of Pouliot and Thérien (2018: 2), '…a large and diverse movement in social theory'. The extensive description of practice theory allows researchers to apply the theory in diverse settings; from individuals and organisations, to expansive geographical locations, and from municipalities and provinces, to the continental and global.

Practice theorists see social life as a consolidation of embodied and materially interwoven practices that are prearranged around a set of collective practical understandings of reality (Schatzki et al. 2001, cited in McCourt 2016). For Nicolini (2012), practice approaches contribute to awareness, since they uncover the actions, efforts and work of people that lie behind society's sturdy features, including institutions and their structures of rule. By focusing research on practices, it moves the attention away from scholarly work that is dedicated solely to structures and materiality, as well as their implications for governance and politics (Pouliot and Thérien 2018). Said differently, practice theory focuses more attention on the individual doings within the realm of governance and politics, which are constituted by structures and materiality. To reiterate, practice theory amalgamates the empiricism of agential and material causal instruments with the normativity of ideational and structural mechanisms, which are elements that are identified in the PULSE[3] analysis.

Practice theory indicates the dual and symbiotic dynamism between the processes and contexts in which individuals play a central role, and it is also relational and views the world as a combination of practices. For practice theory, the nature of reality (ontology) is relational, which means that it does not view reality as a clear distinction between the causation of agency and materiality, ideas and structure (Bueger and Gadinger 2015). This notion, or relationalism, resonates with the argument of Reus-Smit and Snidal (2008) that theory is simultaneously both empirical and normative and that it informs the activities of practitioners. The combination of actions, or the things that people do, constitutes the social context. The actions are interdependent to such an extent that one action becomes the foundation or well-spring for another performance, which constitutes the causal mechanisms within a certain activity realm. That said, practices beget practices via paradigms, theories, causal mechanisms and problem-solving and critical theory types. Should we become sensitive to practice theory, we could find '…remedy for a number of problems left unsolved by other traditions, especially the tendency of describing the world in terms of irreducible dualisms between actor/system, social/material, body/mind and theory/action' (Nicolini 2012: 2). Practices frame public policy processes when these actions create a baseline for political debates, as well as actor interactions (Pouliot and Thérien 2018). It is here where analytic eclecticism makes a valuable contribution (Sil and Katzenstein 2010; Meissner 2017) because of its ontological and epistemological pluralism, as it creates a cross-over between reducible dualisms. Therefore, practice theory rejects dichotomies as a method of theorising (Feldman and Orlikowski 2011).

In order to further enrich the discussion of the policy landscape in the following sections, I will discuss one element that stood out during the analysis, namely, activity, performance or work, and how it illuminates eThekwini's green and ecological infrastructures. This assumption emphasises the role of the individual in executing practices, as well as the knowledge that is necessary to perform activities or work. The municipality's initiatives are replete with practices and various practice-related activities or actions, like establishing partnerships and the advancement of both infrastructure types.

Activity, performance or work

The focal point of practice theory is the importance of activity, performance and work, which create and preserve social life and *all* its aspects, as has already been mentioned, and its social structures (Nicolini 2012). Practice is performance, the process of doing something, or the '…active social forces making and remaking the world' (Adler and Pouliot 2011, cited in Pouliot and Thérien 2018: 3). We cannot readily observe practices other than their unfolding or the processes that we observe in texts (Adler and Pouliot 2011). 'Family, authority, institutions, and organizations are all kept in existence through the *recurrent performance of material activities*, and to a large extent they only exist as long as *those activities are performed*' (emphasis added) (Nicolini 2012: 3).

This implies that practices, and particularly their performance, constitute and move with history (Adler and Pouliot 2011). In this way, actors perform activities that are not so much based on abstract forces, like interests, preferences and social norms, but more through practical requirements, habits and personified outlooks (McCourt 2016). This is not to say that these elements are absent from practices (Adler and Pouliot 2011). Practices express beliefs and preferences, but they are not like them, as Reckwitz's (2010) definition reminds us. Since this is the case, Kustermans (2016: 175) noted that '…the meaning of "practice" is a moving target. Sometimes it means process. Sometimes it refers to a type of knowledge and related actions. And sometimes it is used as a quasi-synonym for institution'. Herein, we see the empirical and normative elements locked up in activities, performances and work. With regard to the empirical and normative elements, practice theory shows that practices are open-ended doings and sayings that take place interdependently within a spatial-temporal dimension (Bueger and Gadinger 2015). Social structures are temporal outflows of performances, as they could always end, if people decide not to hold them together (Nicolini 2012). Even so, practices are not the same as institutions, although performance could provide a case for the establishment of institutions or supporting organisations (Adler and Pouliot 2011). Whatever the case may be, practices require activities. Because of this, we identify practices in texts when researchers use verbs like 'ordering', 'structuring' and 'knowing', instead of 'order', 'structure' or 'knowledge', which are nouns (Bueger and Gadinger 2015) that indicate the existence of the embodiment of a practice.

The Individual

The practice theory pinpoints the individual's role and the material things in the natural environment, as well as in social life. Practices would not be possible without material resources and individuals (Nicolini 2012). Individuals and their bodies are not mere instruments of the person because '…a practice *is* the routinized activity of the body' (emphasis in the original) (Nicolini 2012: 4). What is more, habits constitute our behaviour, with the social order inscribing routines into our bodies that, in turn, manifest themselves through discursive and bodily practices (Nicolini 2012). Kustermans (2016: 177) identified practices with 'doings' or with what actors say and do, which correspond to their discursive and bodily practices. He goes on to say that 'practice' is '*all of us doing all of our doings*, the myriad of human behaviours, which is forever going on, all at once' (emphasis in the original). Highlighting discursive and bodily practices bring to the fore agency as a form of discursive practice, in that the discourse of social practice is reflexive, since it adds to the constitution of the social realm and is created by the social (Davies 1990; Kettle 2005).

What this means is that every practice has a discursive element, since the discursive constitution of practices are practices in themselves. Furthermore, practices involve the use of language (Chouliaraki and Fairclough 1999; Kettle 2005) by recounting the essence and acquiring practices. Based on the assumptions of the practice theory, eThekwini's various green and ecological infrastructure enterprises are replete with activities and tasks that are performed by the involved stakeholders.

In what follows, I will focus on the role and function of epistocracies and how municipal officials shape the policy arena as an epistocracy.

Epistocracy

It appears as if municipal officials in the EPCPD, and other departments that are closely linked with green and ecological infrastructure initiatives, not only play the role of local government administrators, but also of scientists. This would be the case for several eThekwini officials who were interviewed. In Table 3.7, I identify numerous officials who have published books, book chapters and peer-reviewed articles. One of these, Debra Roberts, was unavailable for an interview during my fieldwork. However, as the former Manager of the municipality's EPCPD, and now the Head of eThekwini's Sustainable and Resilient City Initiatives Unit, I have included her publications. Debra is, after all, responsible for establishing D'MOSS and the EPCPD. This Department oversees the development and implementation of eThekwini's Municipal Climate Protection Programme, which includes the development and implementation of appropriate climate change adaptation and mitigation strategies and projects, of which D'MOSS is one (eThekwini 2020). When one looks at Table 3.7 and sees the number of publications that these officials have produced over the years, it would appear that an epistocracy is active within the EPCPD and other branches. To investigate this further, we need to ask the following: What is an epistocracy? What are its benefits and drawbacks? and What are the implications of these advantages and downsides?

The concept 'epistocracy' dates back to the time of Plato (428/427–348/347 B.C.E.), when he proposed this 'form' of government), instead of a democratic regime, for Kallispolis (Tremmel 2018; Lippert-Rasmussen 2012). The term 'epistocracy' literally means 'the rule of experts' (Min 2015; Tremmel 2018) or 'the rule of knowers' (Moraro 2018). Wissenburg (2019) sketched a more elaborate conceptualisation when he wrote that an epistocracy is rule that is exercised by individuals in society who have a superior knowledge of politics. Even so, in certain instances, epistocrats may not have a superior knowledge of politics but they could wield political power in a specific political arena, since they possess the relevant knowledge to make good policies (Mulligan 2015) concerning specific issues. This indicates that epistocracies are not only connected to the wider political context, but also to issue-specific policy areas, such as green and ecological infrastructures.

How does this notion fit in with eThekwini's green and ecological policy landscape? For one, as a local government, eThekwini is a political organ that has been established by the South African Constitution. It also means that the eThekwini Council needs to approve the implementation of green and ecological infrastructure projects, such as the River Horse project. Municipal Council approval involves a political function. In this regard, according to Lucky (2019), epistocrats claim that they can improve the outcomes of instrumental policy, since they possess a superior

Table 3.7 Scientific publications produced by eThekwini officials

Municipal official	Lead/Co-author	Title of publication	Year	Publication type
Andrew Mather	Co-author	Durban and its Port: An Investigation into the relationship between the City and port of Durban	No date	Book chapter
	Lead	City of Durban Sand Bypass Scheme: 20-year performance evaluation	2003	Proceedings
	Lead	Sea level rise and it's likely financial impacts	2008	Scientific journal article
	Lead	Projections and Modelling Scenarios for Sea-Level Rise at Durban, South Africa	2009	Report
	Lead	Adaptation in Practice: Durban, South Africa	2011	Book chapter
	Lead	Predicting extreme wave run-up on natural beaches for coastal planning and management	2011	Scientific journal article
	Lead	A Perspective on Sea Level Rise and Coastal Storm Surge from Southern and Eastern Africa: A Case Study Near Durban, South Africa	2012	Scientific journal article
	Co-author	Failed Coastal Stabilization: Examples from the KwaZulu-Natal Coast, South Africa	2012	Book chapter
	Co-author	KwaZulu-Natal coastal erosion events of 2006/2007 and 2011: A predictive tool?	2012	Scientific journal article
	Co-author	Durban Climate Change Strategy Water Theme Report: Draft for Public Comment	2014	Report
	Lead	Climate Change and the Coasts of Africa: Durban Case Study	2015	Book chapter
Geoff Tooley	Co-author	Adaptation in Practice: Durban, South Africa	2011	Book chapter
Joanne Douwes	Lead	Exploring transformation in local government in a time of environmental change and thresholds: a case study of eThekwini municipality	2018	Dissertation
	Lead	Innovative responses to biodiversity offsets in Durban	2018	Proceedings

(continued)

Table 3.7 (continued)

Municipal official	Lead/Co-author	Title of publication	Year	Publication type
Errol Douwes	Lead	Regression analyses of southern African ethnomedicinal plants: informing the targeted selection of bioprospecting and pharmacological screening subjects	2008	Scientific journal article
	Co-author	Exploring ecosystem-based adaptation in Durban, South Africa: "learning-by-doing" at the local government coal face	2012	Scientific Journal Article
	Lead	Buffelsdraai Landfill Site Community Reforestation Project: Leading the way in community-based adaptation to climate change	2015	Report
	Co-author	Evaluating the outcomes and processes of a research-action partnership: The need for continuous reflective evaluation	2016	Scientific journal article
	Co-author	How to build science-action partnerships for local land-use planning and management: Lessons from Durban, South Africa	2016	Scientific journal article
	Lead	Growing responsibility with growth	2017	Scientific journal article
Sean O'Donnoghue	Co-author	Economics of climate change adaptation at the local scale under conditions of uncertainty and resource constraints: the case of Durban, South Africa	2013	Scientific Journal Article
	Co-author	How to build science-action partnerships for local land-use planning and management: lessons from Durban, South Africa	2016	Scientific journal article
	Co-author	Evaluating the outcomes and processes of a research-action partnership: The need for continuous reflective evaluation	2016	Scientific journal article
	Co-author	Improving the management of threatened ecosystems in an urban biodiversity hotspot through the Durban Research Action Partnership	2016	Scientific Journal Article
	Co-author	City view: Durban, South Africa	2016	Book chapter
	Co-author	Mitigation and Adaptation: Barriers, Bridges, and Co-benefits	2018	Book chapter
Debra Roberts	Co-author	Urban open space planning in South Africa: a biogeographical perspective	1985	Scientific Journal Article
	Lead	Central and peripheral urban open spaces: need for a biological evaluation	1985	Scientific Journal Article

(continued)

Table 3.7 (continued)

Municipal official	Lead/Co-author	Title of publication	Year	Publication type
	Lead	An open space survey of municipal Durban	1990	Doctoral dissertation
	Lead	The vegetation ecology of municipal Durban, Natal. Floristic classification	1993	Scientific Journal Article
	Lead	The design of an urban open space network for the city of Durban (South Africa)	1994	Scientific Journal Article
	Lead	Urban environmental planning and management—the challenge of sustainable development	1996	Book chapter
	Lead	Durban's Local Agenda 21 programme: tackling sustainable development in a post-apartheid city	2002	Scientific Journal Article
	Lead	Durban's Local Agenda 21 Programme 1994-2001: Tackling Sustainable Development	2002	Book
	Lead	An urban challenge: Conserving biodiversity in the eThekwini Municipality, KwaZulu-Natal	2002	Book chapter
	Lead	Sustainability and equity: Reflections of a local government practitioner in Southern Africa	2003	Book chapter
	Lead	Using the development of an environmental management system to develop and promote a more holistic understanding of urban ecosystems in Durban, South Africa	2003	Book chapter
	Lead	Resource Economics as a tool for open space planning in Durban, South Africa	2005	Scientific Journal Article
	Lead	Durban's Local Agenda 21/Local Action 21 Programme: A vehicle for social inclusion?	2005	Book chapter
	Lead	Thinking globally, acting locally—institutionalizing climate change at the local government level in Durban, South Africa	2008	Scientific Journal Article
	Lead	Prioritizing climate change adaptation and local level resilience in Durban, South Africa	2010	Scientific Journal Article

(continued)

Table 3.7 (continued)

Municipal official	Lead/Co-author	Title of publication	Year	Publication type
	Co-author	Adaptation in Practice: Durban, South Africa	2011	Book chapter
	Co-author	Spatial justice and climate change: multiscale impacts and local development in Durban, South Africa	2011	Book chapter
	Co-author	Adaptation in practise: Durban, South Africa	2011	Scientific Journal Article
	Co-author	What car salesmen can teach environmentalists	2011	Scientific Journal Article
	Co-author	Environmental Inequalities Beyond Borders: Local Perspectives on Global Injustices	2011	Scientific Journal Article
	Co-author	Urban climate adaptation in the global south: planning in an emerging policy domain	2012	Scientific Journal Article
	Lead	Exploring ecosystem-based adaptation in Durban, South Africa: "learning-by-doing" at the local government coal face	2012	Scientific Journal Article
	Co-author	Preparing Cities for Climate Change: early lessons from early adaptors	2012	Scientific Journal Article
	Lead	Urban environmental challenges and climate change action in Durban, South Africa	2013	Scientific Journal Article
	Co-author	Economics of climate change adaptation at the local scale under conditions of uncertainty and resource constraints: the case of Durban, South Africa	2013	Scientific Journal Article
	Co-author	Experiences of Integrated Assessment Modelling in London and Durban	2013	Scientific Journal Article
	Lead	Cities OPT in while nations COP out: Reflections on COP18	2013	Scientific Journal Article

(continued)

Table 3.7 (continued)

Municipal official	Lead/Co-author	Title of publication	Year	Publication type
	Co-author	The future we really want	2013	Scientific Journal Article
	Co-author	Development: Time to leave GDP behind	2014	Scientific Journal Article
	Co-author	Urban areas	2014	Book chapter
	Co-author	Towards transformative adaptation in cities: the IPCC's Fifth Assessment	2014	Scientific Journal Article
	Co-author	Moving towards inclusive urban adaptation: approaches to integrating community-based adaptation to climate change at city and national scale	2014	Scientific Journal Article
	Co-author	What lies beneath: understanding the invisible aspects of municipal climate change governance	2015	Scientific Journal Article
	Co-author	Climate protection in mega-event greening: the 2010 FIFA™ World Cup and COP17/CMP7 experiences in Durban, South Africa	2015	Scientific Journal Article
	Co-author	Buffelsdraai Landfill Site Community Reforestation Project	2015	Scientific Journal Article
	Co-author	Climate change and the coasts of Africa: Durban case study	2015	Book chapter
	Lead	City action for global environmental change: assessment and case study of Durban, South Africa	2015	Book chapter
	Co-author	Roadmap towards justice in urban climate adaptation research	2016	Scientific Journal Article
	Co-author	Scientists must have a say in the future of cities	2016	Scientific Journal Article

(continued)

Table 3.7 (continued)

Municipal official	Lead/Co-author	Title of publication	Year	Publication type
	Co-author	How to build science-action partnerships for local land-use planning and management: lessons from Durban, South Africa	2016	Scientific Journal Article
	Lead	A global roadmap for climate change action: From COP17 in Durban to COP21 in Paris	2016	Scientific Journal Article
	Lead	Identifying ecosystem service hotspots for environmental management in Durban, South Africa	2016	Scientific Journal Article
	Co-author	Managing a threatened savanna ecosystem (KwaZulu-Natal Sandstone Sourveld) in an urban biodiversity hotspot: Durban, South Africa	2016	Scientific Journal Article
	Co-author	Re-imagining inclusive urban futures for transformation	2016	Scientific Journal Article
	Lead	Durban, South Africa	2016	Book chapter
	Lead	The New Climate Calculus: 1.5 °C = Paris Agreement, Cities, Local Government, Science and Champions (PLSC2)	2016	Scientific Journal Article
	Co-author	Durban's systematic conservation assessment	2016	Report
	Co-author	Cross-city analysis	2016	Book chapter
	Co-author	Improving the management of threatened ecosystems in an urban biodiversity hotspot through the Durban Research Action Partnership	2016	Scientific Journal Article
	Lead	City View: Durban, South Africa	2016	Book chapter
	Co-author	Evaluating the outcomes and processes of a research-action partnership: The need for continuous reflective evaluation	2016	Scientific Journal Article
	Co-author	Climate adaptation as strategic urbanism: assessing opportunities and uncertainties for equity and inclusive development in cities	2017	Scientific Journal Article

(continued)

Table 3.7 (continued)

Municipal official	Lead/Co-author	Title of publication	Year	Publication type
	Co-author	Planning for the future of urban biodiversity: a global review of city-scale initiatives	2017	Scientific Journal Article
	Co-author	Six research priorities for cities and climate change	2018	Scientific Journal Article
	Co-author	IPCC, 2018: Summary for policymakers	2018	Scientific Journal Article
	Co-author	Global warming of 1.5 °C	2018	Report
	Co-author	City transformations in a 1.5 °C warmer world	2018	Scientific Journal Article
	Co-author	Toward a Sustainable Wellbeing Economy	2018	Scientific Journal Article
	Co-author	New integrated urban knowledge for the cities we want	2018	Book chapter
	Co-author	Spatial analyses of threats to ecosystem service hotspots in Greater Durban, South Africa	2018	Scientific Journal Article
	Co-author	IPCC Expert Meeting on Assessing Climate Information for Regions	2018	Report
	Lead	Banksy and the Biologist: Redrawing the Twenty-First Century City	2018	Book chapter
	Co-author	Mitigation and Adaptation: Barriers, Bridges, and Co-benefits	2018	Book chapter
	Co-author	International Conference on Climate Risk Management, inputs for the Intergovernmental Panel on Climate Change's Sixth Assessment Report	2019	Scientific Journal Article

knowledge that informs governance (Min 2015; Lucky 2019) and politics. Based on this argument, the link between politics and eThekwini's green and ecological policy arena is not difficult to comprehend.

Nevertheless, epistocrats desire a certain shape of society and long for ends, rather than means (Wissenburg 2019). Tremmel (2018) and Wissenburg (2019) argue that we should make a distinction between an epistocracy and a technocracy. A technocracy is the rule of those that are most skilled in a practice, or those who use a means. In this section, I will only discuss the concept of epistocracy, as it is manifested in eThekwini's green and ecological infrastructure policy landscape. Even so, we cannot completely ignore the concept 'technocracy', since Tremmel (2018) and Wissenburg (2019) tie it explicitly to the practices that have been mentioned in the previous sections of this chapter.

In South Africa, the concept, forms and implications of epistocratic rule in society and in the global realm (Tremmel 2018 and Wissenburg 2019) have gained significant traction in recent years. This was especially evident after 2016 when the 35th International Geological Congress in Cape Town unofficially proclaimed a new geological epoch, namely, the Anthropocene, or the Epoch of Man (Wright and Simpson 2015; Tremmel 2018). Since the concept embodies the notion of human activity reshaping all parts of the earth's system, with its negative consequences manifesting themselves in anthropogenic climate change, scholars started asking whether society should overhaul democracy as a practice, since the Anthropocene redefines the place of humans in the world, for it speaks of humans in a different way (Tremmel 2018). We now see ourselves as a species with a geological force that is altering the earth's biosphere, and the Anthropocene, as a concept, encapsulates human transformation that is based on industrial, urban and technological change. Scientists have linked these developments to global climate change and the earth's sixth mass extinction event (Wright and Simpson 2015), which resonates with the idea of modernity as a juggernaut, as was mentioned during my discussion on a risk society (p. 146). We can link the influence of the Anthropocene concept with the rise of epistocracies in the following mantra: 'geological imagination influences political imagination' (Chakrabarty 2009; Wright and Simpson 2015). In other words, a scientific concept constitutes the rise of epistocracies as the subject matter in Philosophy and Politics.

Epistocratic forms

Anne Jeffrey (2018: 412) identified a form of epistocracy, which she terms a 'limited epistocracy' or the 'rule by institutions housing expertise in non-political areas that become politically relevant'. The doctrine of a limited epistocracy follows Rehg's (2011: 386, cited in Jeffrey 2018: 412) argument that 'the multidisciplinary complexity of the technical issues exceeds the technical expertise of one person'. Rehg (2011) used climate change as an example. Limited epistocracies exist where government institutions outsource political power to institutions or corresponding groups, like the IPCC, the World Health Organisation (WHO) and the Migration Policy Institute (MPI), which have the relevant expertise to solve urgent and complex problems, such as climate change, virus outbreaks and refugee crises. Governments do this because the problems outstrip democratic decision-making (Jeffrey 2018).

One objection might be that outsourcing problem-solving to an external organisation could bypass public involvement in the government decision-making process. Jeffrey (2018) counteracts this argument by saying that governments can arrange a limited epistocracy that is matched with robust political inclusion. She recommends that specialised institutions must issue directives that give citizens various working possibilities. In other words, scientists should not only produce recommendations based on their scientific investigations, but ideally, they should, in collaboration with government institutions, also incorporate working possibilities to allow citizens to make informed choices before, during and after policy implementation.

Based on an outline of Lippert-Rasmussen's (2012) moderate epistocratic views, we can identify three additional epistocratic forms that are compatible with Jeffrey's (2018) limited epistocratic formulation. Following Lippert-Rasmussen (2012), I will label these as issue-specific, temporal and small-group epistocracies.

An issue-specific epistocracy encompasses individuals who know the true, and the best, answers to questions surrounding certain issues. Such an epistocracy would ideally manage an issue-focused, political decision procedure. An issue-specific epistocracy consists of individuals who have authority relating to relevant issues. An example of this might be a research group that conducts research and publishes a report (List 2005) on a specific issue in which it has the relevant expertise. For instance, if guidance is required on educational policies, one might argue that the voices of teachers carry more weight (Lippert-Rasmussen 2012) than those of climate change experts.

The temporal epistocracy is characterised by a time-sensitive and privileged knowledge status. Because of its temporal nature, we are all, at some point, subject to the authority of others (Lippert-Rasmussen 2012), or they are subject to our authority, based on our expertise at a specific point in time. Not only is our exposure to influence temporal. Issues are also emergent and highly dependent (Danaher 2016) on society's needs and circumstances and, therefore, inherently dynamic. This means that individuals occupying this type of epistocracy may exert an influence for a limited period (Danaher 2016), when the issue is salient. The issue of xenophobia that emerges now and again in South Africa would be a case in point, as well as the Covid-19 pandemic at the time of writing.

A small-group epistocracy in society is one that has a better knowledge of the normative standards than any other group has. In other words, it is smaller than the entire political dispensation. The individuals within this epistocracy develop collective answers to an issue, but they do not include any of those individuals who know the best answer (Lippert-Rasmussen 2012; Kuljanin 2019). From a scientific perspective, the problem with this group is its identification (Kuljanin 2019). Lippert-Rasmussen (2012) uses the example of political factions in society. It would appear that a group-think (Janis, 1982) is the dominant form of the group's psychology.

Since the concept 'epistocracy' relates to the wider political society, observers might interpret political rule or political decision-making as law-making, which is a parliamentary function. If we also include opinion formulation, agenda setting and the construction of option menus in the conceptualisation of political decision-making, then we need to expand our focus to epistocratic forms. If we do so, we can include numerous actors, other than the organs of state, such as interest group- and counter-expertise, NGOs, corporations, the media (Holst 2012) and academics.

By spreading the actor net wider, we need to incorporate other classifications in the discussion. In this regard, Holst (2012) identifies a cultural epistocracy as a society that has a considerable respect for knowledge and knowers. Because of this admiration, many in such a society support the idea that decisions must be based on sound knowledge and that knowers must play a substantial role in the decision-making process. A cultural epistocracy comes in degrees and varies, depending on the type of knowers, as well as the type of knowledge that society recognises and values. According to Holst (2012), Norway is less of a cultural epistocracy than France and Germany, because the latter value academic degrees and merits more than does the Norwegian society. Different kinds of knowledge, such as moral knowledge, technical expertise and technocratic knowledge, exist within a cultural epistocracy (Holst 2012).

In addition, knowers can become governors, either through prescription or recognition. Epistocracies can also be formalised in relation to epistocratic policies and laws. For instance, a government system that prescribes recruitment requirements, such as a specific education level and cognitive aptitude for ruling positions in different institutional arenas, is more formally epistocratic that a system with more implicit staffing criteria. For example, certain government institutions have manuals pertaining to the recruitment of public administration staff, and these prescriptions could pertain to a specific educational background. Then, there are systems where these requirements act as an assumed framework that is less formalised and epistocratic than the former system. In this regard, the degree of formalisation may constitute the strengthening of a social epistocracy, since society reproduces certain knowledgeable leaders based on the recruitment procedures (Holst 2012).

We can also investigate epistocracies from a qualitative perspective. For instance, political units could recruit knowers based on the theoretical knowledge that they have acquired by means of an academic degree, or by being appointed as skilled practitioners. Organs of state can also appoint members of epistocracies because they have factual knowledge, as well as a specific technical competence or moral expertise, based on their academic training, or practical skills and experience. In this regard, it is also important to consider the content of the expert's expertise (Holst 2012). For example, a report that is authored by climate change experts on eThekwini's climate adaptation strategy will look different from a report authored by political scientists, because the paradigms and theories of their background knowledge differ.

Benefits, drawbacks and implications

As has already been mentioned, modern societies are large and complex, with the result that political problems, such as climate change, are multifaceted. According to Min (2015), it is for this reason that society should appeal to experts for policy development and evaluations. Efficient instrumental policy development happens through cost-benefit analyses that are based on positivism, since scientists exclude any reasonable doubt from the process (Wissenburg 2019). The numbers and equations speak for themselves. Estlund (2003) argued that certain political outcomes or policy consequences yield better results than others. If experts have a better knowledge than the general public for forecasting the outcomes of a specific issue, then an epistocracy could produce better results in governing certain policy processes. In other words, what makes an epistocracy valuable is not so much the efficiency of the decision-making process, but more the scientific and professional knowledge of the epistocrats (Holst 2012). If this is the case, the efficiency of policy decisions will increase, and the benefits for the rest of society will accrue.

Epistocracies play a compensatory role by correcting the relative unreliability of arguments, or the agential shortcomings, of authority actors (Viehoff 2016). Because of this, Pederson (2014) claimed that improving the use and impact of science-based policies and practices at various levels can enrich the quality and legitimacy of democratic governance systems. He added two specific qualifiers to this statement, namely, integrity and a trust in science. Society needs to develop and install proper measures to avoid dogmatism and a tendency towards irrational behaviour, and, at the same time, it needs to maintain its access to experts and scientific evidence. In this connection, without the relevant input of scientific facts from the experts, democratic decisions and public policies could be illegitimate or normatively defective (Pederson 2014). It is in the science-policy interface domain that epistocracies play a compensatory role in political processes.

In most forms of an epistocracy, it is possible that the more advantaged demographic groups, who have a superior education, would have more political representation than those that are less advantaged and educated. This is the so-called demographic objection that is levelled against an epistocracy, which argues, firstly, that such a form of representation is unjust and unfair. Secondly, it holds that some people will have more power than others, since epistocracies have the tendency to help the advantaged and harm the disadvantaged (Brennan 2018). This hints at a top-down decision-making process, with the experts at the top and society at the bottom, exemplified in who governs at the top benefits those at the bottom (p. 100). In this regard, Pederson (2014) noted that, as far as policy decisions that are based on normative questions are concerned, the wider democratic community should determine how decision-makers deal with normative questions. This implies that it has a balancing effect on the overbearing role of epistocrats and makes a compromise between epistocratic decision-making and the normative dimensions that affect society. This resonates with the notion of theory being both empirical and normative, where empiricism originates with scientists and normativity emanates from society. Pederson (2014) went on to say that to listen to the best-qualified scientist for policy advice does not necessarily guarantee that scientists conduct their research and development for the public good. Experts can suffer from public paralysis, and we cannot always avoid catastrophic events by listening to epistocrats. Therefore, it is necessary to filter and translate scientific expertise into policy-making, although science should not always speak truth to power under all circumstances (pp. 34, 101).

Based on this notion, epistocracies could suffer from several problems, such as a lack of popular acceptance and actual, or perceived, illegitimacy, which may lead to instability and resistance, since citizens will likely reject and be sceptical of any form of selection criteria or competence that determine political participation (Brennan

2019). As a result of their education, experts are not immune to biases, which could mean that their predispositions can offset the benefits of an epistocracy. It is for this reason that actors can distribute knowledge across society to offset the biases by perspectives from other parts of society (Min 2015).

Furthermore, experts cannot access the privileged information effecting citizens, and that carry the burden and consequences of expert decisions, especially when things go wrong. People have varying levels of knowledge, skills, intelligence and talents that are related to different areas of inquiry. Since experts are not epistemically privileged, they do not always know how their policies will affect individuals; only individuals know how the policies will influence them, either directly or indirectly. It is for this reason that deliberation is necessary to construct a more complete picture of the political problems and their solutions. Political problems are complex, which is reason enough for them to be framed from all the relevant perspectives, because what one person knows is often unknown to others (Meissner et al. 2017a, b), and vice versa. A missing perspective will therefore frame an incomplete picture of the problem, which is an argument that I have made when formulating the theoretical foundations of PULSE[3]. In this regard, I refer to the discussion on paradigms and theories in Chap. 1 in that perspectives are not merely knowledge claims about a proposition; rather, perspectives are the experiential source of an epistemic agent that informs a person's reasons, opinions, beliefs and worldviews. We listen to the perspectives of others, not only to critique them, but also to gain an open-ended understanding of an issue (Bohman and Richardson 2009; Min 2015).

Holst (2012: 53) stated that epistocracies are not inherently wrong. She went on to say that: '...knowledge-based decision-making is acceptable, and even a good thing if it is institutionalized adequately and legitimately, as well as democratically...whatever that might mean, more specifically'. In other words, science is not enough, which means that one cannot make decisions on science alone.

Over my many years as a researcher in water governance and transboundary water politics, I have witnessed many scientists being frustrated with practitioners, decision-makers and politicians; they have said that they 'do not listen to our policy recommendations that are based on robust science', or 'they ignore our advice completely'. These experts are usually, but not always, natural scientists who are often unaware of other disciplines, such as Political Science or International Relations, let alone Public Administration, and consequently, they often do not understand political and policy processes. They accept it as a forgone conclusion that their scientific research will save the day, but they ignore the wider governance, political, public and private administration processes that decision-makers need to consider, when developing and implementing policies. Policy development, adoption and implementation depend on various reflections, including the level of stakeholder participation, as well as the selection of either a top-down or bottom-up approach, which is only an insignificant consideration, compared to the complexities surrounding government, policy and political process. For instance, stakeholder participation, including that of scientists, will not automatically result in improved policy outcomes and effective legislation (Atisa 2020). Participation is just that; it is about participating in the policy process. Other steps in the process include agenda setting, as well as policy development, uptake, monitoring and evaluation. This speaks to the matter of specific academic expertise that may be lacking within an issue-specific epistocracy, which could lead to inefficiencies when communicating scientific evidence to decision-makers. In this respect, Wellstead et al. (2013), writing on climate change, called for an 'institutional revolution' and argued that:

> ...much systems-inspired modelling in the climate change adaptation area continues to rely upon structural-functionalist tautologies. That is, what are, in fact, some of the most determinate actors and variables in any policy-making process—politics and government—are simply assumed away in the belief that policy goals will just be set, and met, since they are "necessary" for the system to "function".

eThekwini's epistocracy

Based on the discussion on epistocracies, and returning to the EPCPD and other entities within the municipality, do the officials of the eThekwini departments constitute an epistocracy? Judging by the number of publications, we can argue that these individuals are experts in their respective fields. I would not say that they have a superior knowledge of the politics that is practised at a societal level, but they certainly do have an expert knowledge on political governance in the municipality. They are, after all, employees of the municipality, as a local government legislative and political structure, and because of this, they possess an inherent knowledge of this political arena. Therefore, they wield political power in this institution, as well as in transnational political settings like the IPCC, the municipality's international relations with Bremen and linkages to The Nature Conservancy.

The PULSE[3] analysis of their interviews indicates that they are capable of making 'good' policies, with regard to the River Horse project, the water fund and the UEIP, as well as collaborative activities with the CSIR, DUT and UKZN, and other non-governmental entities, such as DUCT, INR, ORI and WESSA. This indicates an aptitude on the side of the officials to inform green and ecological infrastructures with domestic and international epistemic structures. With regard to the green and ecological infrastructure projects, the EPCPD works towards improving the municipal policies by improving the biodiversity in the city, as well as climate change adaptation and mitigation. Their impressive list of publications shows that they utilise their superior knowledge to inform governance and politics in this issue-specific policy arena. This means that we are not dealing with a technocracy, although the officials are skilled in numerous practices, such as storm-water drainage, coastal and catchment management and climate change adaptation and mitigation.

We can link the work of the officials, and particularly those working in the EPCPD, to the conceptualisation of the Anthropocene, which encapsulates the influencing force of humans on the earth's biosphere. Durban is an urbanised environment, and its biosphere has been altered by human development since it was established in the nineteenth century. The fact that Durban's estuaries, open spaces, in-land forests and river catchments are being threatened by alien invasive flora was mentioned in the interviews. This means that the driving force behind the vision of the epistocratic officials is not only a matter of geological transformation through urbanisation, which constitutes the imagination, but also the alteration of the biodiversity, which influences governance and politics.

Moving on to the form of this epistocracy, I can say that we are dealing here with a limited epistocracy that concentrates on seemingly non-political issues that are politically relevant. Therefore, I would argue that the issue did not automatically become political, but that it became politically pertinent through the work of Debra Roberts and her colleagues. In other words, this limited epistocracy turned green and ecological infrastructures from a perceived non-political issue into a political issue, by the expert-infused practices in the political arena of the local government. However, it appears that they are treating it as apolitical in that they involve only a limited number of experts in the execution of green and ecological infrastructure policies. On the other hand, it is possible that eThekwini would not have had the EPCPD and green and ecological infrastructure initiatives if municipal officials had only been trained in public administration. Expertise in municipal management is not sufficient for dealing with the seemingly complex environmental issues confronting cities like Durban. When Roberts published her thesis in the mid-1980s on the impact of urbanisation on the natural environment, it resulted in a rising consciousness on this issue. This led to the implementation of a particular type of expertise that confronted environmental issues like biodiversity loss and climate change. Over the years, she also attracted other experts to the municipality, forming a limited epistocracy in the EPCPD. The PULSE[3] analysis also shows that EPCPD officials are not operating within a departmental vacuum, but that they interact with colleagues from other departments as they work towards achieving the ends of green and ecological infrastructure strategies and projects.

As already mentioned, it is also worth noting that municipal officials collaborate with experts from research institutions situated within, and outside of, eThekwini's boundaries. Many of the interviewed respondents indicated that the epistocratic officials had collaborated closely with them over the years, or at best, that they have intimately known what the municipality has been doing with regard to the promotion green and ecological infrastructures. The network of collaboration also includes NGOs, such as DUCT and WESSA, as well as other consultants. The individuals who work for these organisations are not merely campaigners who promote the cause of nature, but they are experts in various fields of natural science. Since this collaborative network exists, we can also say that there is not only an internal eThekwini epistocracy, but one that extends beyond the municipality's governance and geographic boundaries, even as far as Bremen, Fort Lauderdale and the IPCC. This is an indication that limited epistocracies operate not only within their direct structural sphere of influence (i.e. a municipality), but that they are also capable of playing a transnational and global governance role.

The eThekwini epistocrats collaborate with these external experts, as an intrinsic part of their work, to promote climate change adaptation and mitigation through green and ecological infrastructure enterprises. Therefore, the EPCPD is not only a limited epistocracy, but it is also issue-specific, since it focuses on matters relating to Durban's internal and external natural environment (i.e. the uMngeni River catchment).

Throughout my investigation, I realised that technical and technocratic knowledge play a pivotal role; however, what I did not notice was the presence of moral knowledge. That is not to say that the epistocrats are immoral, but it has more to do with their inability to incorporate values and norms into their epistocratic policies. What I have also found to be lacking, in this regard, is the incorporation of public opinions and critical voices. It is, therefore, not impossible to argue that the epistocrats have their eye on policy efficiency. They might argue that incorporating moral knowledge into Durban's green and ecological infrastructure policies might only add clutter to the research process and increase uncertainty regarding the instrumental policies. This is in line with the reasoning that epistocracies act as compensatory agents in policy processes. It appears as if they have ignored the lone critical voice of Respondent 11, since they view his arguments as being unreliable and as having inherent shortcomings. The epistocrats are, after all, backed by the IPCC's scientific processes, the results of which are more dependable and trustworthy than Respondent 11's normative arguments. Furthermore, the epistocrats enjoy more representation within municipal structures than in the Palmiet River Watch. The dominance of problem-solving theories in the analysis also hints at the top-down science-policy process of the epistocrats, for whom science speaks truth to power. This could explain why Respondent 3 mentioned the problematic relationship between the municipality and the *Amakhosi*. Van Klaveren (2018) reached a similar conclusion when he investigated eThekwini's green and ecological infrastructure projects from a constructivist perspective. When talking about the eThekwini epistocrats and their collaborating partners in other research entities and NGOs, he said that:

> The state and cooperative civil groups form an in-group, who embody expertise and a collective mandate, resulting in top-down expert-run projects. Projects end up revolving around community involvement through lobbying and physical labour programs.

Respondent 7 indicated that there is a lack of communication between the municipality and the public regarding green and ecological infrastructure projects. We see this in the establishment process of the water fund, where the EPCPD is the first to conduct a financial feasibility study, and then it communicates with the respective businesses to ascertain whether they are willing to contribute. The driving force behind the establishment of the water fund appears to be a best practice approach, with the epistocrats noticing that cities from around the world have established water funds to deal with water security issues in their jurisdictions. That said, a public participation process is absent, which means that there is no filter to translate the scientific expertise into policy-making via the voices of the wider public. The risk of this is that the water fund could be rejected by the businesses, since they might deem it to be illegitimate. Should eThekwini establish the fund, it could run into legitimacy problems in future, since there was no distribution of knowledge to offset the scientific biases during the feasibility phase of it's establishment. In other words, what the businesses know, might be unknown to the epistocrats (Meissner et al. 2017a, b). This implies that the epistocrats could be missing perspectives from the business community, should the municipality go ahead with the fund, which could adversely influence the feasibility and sustainability of the water fund. It is possible that the epistocrats may have actually closed off an open-ended understanding of the business community's perspectives, by not including them from the start.

Based on these arguments, the implementation of green and ecological infrastructures through whatever type of mechanism it may be, be it a water fund, a twinning agreement, DRAP or the UEIP, might succeed in a limited manner. In this sense, as Respondent 7 indicated, eThekwini's green and ecological infrastructures face the risk of becoming irrelevant, since the municipality does not adequately communicate their benefits to the public. There is also the possibility that the public might view such initiatives as illegitimate, since the epistocrats limit public participation, by closing off public perspectives and opinions, in certain instances. Although public participation is present when the municipality contracts community members to reforest certain areas, the community members act as a labour reserve (Van Klaveren 2018) and not as active and equal participants. In other words, green infrastructures act as an attraction for unifying environmental protection with economic growth (Lennon 2015; Van Klaveren 2018). The epistocrats could, therefore, be implementing a green and ecological infrastructure initiative by using structural hegemonic processes (Warner and Zawahri 2012; Van Klaveren 2018), as a vehicle of hegemonic control that hides the structural inequalities (Van Klaveren 2019), since the experts focus most of their empiricist practices on the environment and pay less attention to the normativity of the social structures.

3.6   Conclusion

The PULSE[3] analysis indicates that there is a predominance of the positivist paradigm in the interviews, with regard to both knowledge generation and the recommended actions. The dominance of positivism is a function of the subject matter at hand, namely, that green and ecological infrastructures relate to nature, and that the narrative around the infrastructure types is influenced by concepts that are 'measurable', namely, biodiversity, climate change and sustainable development goals, among other things. Municipal policies that relate directly to this, learn from these concepts. Since the early 1990s, the municipality has been implementing D'MOSS as its policy response to climate change adaptation and mitigation, especially with regard to adaptation. There is consensus about the types of infrastructure that fit neatly with empiricism and positivism, namely, that green and ecological infrastructures are something in 'reality' that humans can study in an objective manner, such as mountains and wetlands. This would also explain why the municipality promotes both infrastructure types to produce positive outcomes, since nature is controllable, to a certain extent, and everything that is 'natural' is good for human well-being. What is noteworthy is that several respondents used the SANBI definition to define an ecological infrastructure, which represents a positivist ontology. It is, therefore, not possible for interpretation and construction to take place within a dominant positivist narrative. This happens by borrowing from other concepts and applying them to one's own life-world or lived experience. However, the common denominator here is the positivist fit between the borrowed conceptualisation and the borrower.

The hydrosocial contract theory and liberal institutionalism dominate. What is noticeable about the hydrosocial contract theory is that the municipality and its epistemic partners utilise green and ecological infrastructures, together with grey infrastructures. Even so, the overall objective of this integration is to secure sustainable quantities of water for eThekwini, the economic hub of KwaZulu-Natal, with the port of Durban and its surrounding industrial areas, as well as tourism (p. 1), driving the city's economy.

Throughout the theory analysis, we see that liberal institutionalism incorporates assumptions from interactive governance and RIPE theories.

Considering social constructivism, the respondents widely share the concept of 'nature' to engage with the notions about green and ecological infrastructures. Since the respondents share these terms, my argument is that, along with agential constructivism (p. 145), systemic constructivism also characterises the policy landscape. The systemic description thereof lies within the empirically shared concepts, where nature and natural resources, such as alien invasive vegetation, estuaries, rivers, streams and wetlands, are central in defining the system of green and ecological infrastructure policies and initiatives. Therefore, the natural environment acts to inform and constitute policies and, most importantly, the various practices initiated by the network of stakeholders involved. This means that the natural environment constitutes an empirical ontology, which informs a positivist epistemology of, and for, the actors. In simple terms, nature, warts and all, represents a reality that *is*, for all involved, and as such, it leaves little room for interpretation, along with what it *ought* to be. For the respondents, nature is a wellspring of ecosystem services, as well as risks that can be measured empirically, so that humans can capitalise on its services and discover solutions for solving emerging problems.

The assumptions of a risk society are explained in the analysis by mentioning the various risks that are faced by the environment, and by default, eThekwini. Short-term temporal risks appear in the form of extreme weather events, such as the storms that Durban experienced in 2017 and 2019 (p. 2). A list of the longer term risks is not as explicit as that for the storms, but they range from alien invasive plants (Fig. 3.29), to deteriorating water quality and rodents, to venomous snakes. These risks are a consequence of modernity, within the context of an urban environment, and furthermore, they inform the practices of the various actors for their mitigation or solutions. We should read these hazards in terms of systemic constructivism and as being environmental risks that constitute the perceptions and norms held by the numerous stakeholders.

In the analysis, the environment appears as a material causal mechanism, with most respondents mentioning that it constitutes nature as a system for context-informing practices. The respondents described the issues through a predominantly problem-solving epistemology, with the programme theory type being revealed through the various green and ecological infrastructure projects. Linked to the manifestation of the programme theory, is the existence of a limited epistocracy, which has various links to other local and international epistemic actors and structures. What is noticeable within the context of the epistocracy is the limited role of normative understanding.

In the next and final chapter, I will continue to elaborate on the analysis and give my interpretation of eThekwini's green and ecological infrastructure policy landscape and how various stakeholders relate to the concepts, to the environment and to each other.

References

Adams P (2006) Exploring social constructivism: theories and practicalities. Education 34(3):243–257

Adler E (1997) Seizing the middle ground: constructivism in world politics. Eur J Int Relat 3(3):319–363

Adler E, Pouliot V (2011) International practices. Int Theor 3(1):1–36

African Conservation Trust (ACT) (2020) How it all began… African Conservation Trust, Durban. Accessed at: https://projectafrica.com/overview/. Accessed on: 30 Mar 2020

African Snakebite Institute (2019) Black Mamba (*Dendroaspis polylepis*). Available at: https://www.africansnakebiteinstitute.com/snake/black-mamba/. Accessed on: 10 Apr 2019

Albrecht TR, Crootof A, Scott CA (2018) The water-energy-food nexus: a systematic review of methods for nexus assessment. Environ Res Lett 13:1–26

Anguelovski I, Roberts D, Carmin J, Agyeman J (2011) Spatial justice and climate change: multi-scale impacts and local development in Durban, South Africa. In: Carmin J, Agyeman I (eds) Environmental inequalities beyond borders: local perspectives on global injustices. MIT Press, Cambridge, MA

Anscombe E, Geach P (eds) (1954) Descartes: philosophical writings. Nelson and Sons, Edinburgh

Archer D, Almansi F, DiGregorio M, Roberts D, Sharma D, Syam D (2014) Moving towards inclusive urban adaptation: approaches to integrating community-based adaptation to climate change at city and national scale. Clim Dev 6(4):345–356

Atisa G (2020) Policy adoption, legislative developments, and implementation: the resulting global differences among countries in the management of biological resources. In: International environmental agreements: politics, law and economics. https://doi.org/10.1007/s10784-020-09467-7

Bacon F (1620) The Novum Organon. In: Kitchen GW (ed) 1855. The Novum Organon. Oxford University Press, Oxford

Bai X, Dawson RJ, Ürge-Vorsatz D, Delgado GC, Barau AS, Dhakal S, Dodman D, Leonardsen L, Masson-Delmotte V, Roberts DC, Schultz S (2018a) Six research priorities for cities and climate change. Nature 555:23–25

Bai X, Elmqvist T, Frantzeskaki N, McPhearson T, Simon D, Maddox D, Watkins M, Romero-Lankao P, Parnell S, Griffith C, Roberts D (2018b) New integrated urban knowledge for the cities we want. In: Elmqvist T, Bai X, Frantzeskaki N, Griffith C, Maddox D, McPhearson T, Parnell S, Romero-Lankao P, Simon D, Watkins M (eds) Urban planet: knowledge towards sustainable cities. Cambridge University Press, Cambridge

Balza M, Radojicic D (2004) Corporate social responsibility and non-governmental organizations. Unpublished Master's thesis in Economics. Linköping University, Linköping

Beall J, Mkhize S, Vawda S (2004) Traditional authority, institutional municipality and political transition in KwaZulu-Natal, South Africa. Working Paper No. 48. Development Research Centre, London

Beck U (1995) Ecological politics in an age of risk. Polity Press, Cambridge

Beck U (1996) World risk society as cosmopolitan society? Ecological questions in a framework of manufactured uncertainties. Theor Cult Soc 13(4):1–32

Beck U (2009) Critical theory of world risk society: a cosmopolitan vision. Constellations 16(1):3–22

Beck U (2015) Emancipatory catastrophism: what does it mean to climate change and risk society? Curr Sociol 63(1):75–88

Beck U, Ritter M (1992) Risk society: towards a new modernity. Sage, Newbury Park, CA

Benedict MA, McMahon ET (2002) Green infrastructure: smart conservation for the 21st century. Sprawl Watch Clearinghouse, Washington D.C.

Bilderbeek SL, Gupta J, Ros-Tonen M (2016) Biodiversity conservation. SDG Policy Brief #2. Centre for Sustainable Development Studies, Amsterdam

Björkman M (1987) Time and risk in the cognitive space. In: Sjörberg L (ed) Risk and society. Studies of risk generation and reactions to risk. Allen and Unwin, London

Blowers A (1997) Environmental policy: ecological modernisation or the risk society? Urban Stud 34(5–6):845–871

Boehm A (2002) Corporate social responsibility: a complementary perspective of community and corporate leaders. Bus Soc Rev 107(2):171–194

Bohman J, Richardson H (2009) Liberalism, deliberative democracy, and "reasons that all can accept". J Polit Philos 17(3):253–274

Boon R, Cockburn J, Douwes E, Govender N, Ground L, Mclean C, Roberts D, Rouget M, Slotow R (2016) Managing a threatened savanna ecosystem (KwaZulu-Natal Sandstone Sourveld) in an urban biodiversity hotspot: Durban, South Africa. Bothalia-Afr Biodivers Conserv 46(2):1–12

Brand P (1999) The environment and postmodern spatial consciousness: a sociology of urban environmental agendas. J Environ Plan Manage 42(5):631–648

Brennan J (2018) Does the demographic objection to epistocracy succeed? Res Republica 24:53–71

Brennan J (2019) Giving epistocracy a fair hearing. Inquiry 1–15

Brown C (2013) The poverty of grand theory. Eur J Int Relat 19(3):483–497

Brown JS, Collins A, Duguid P (1989) Situated cognition and the culture of learning. Educ Res 18(1):32–42

Bueger C, Gadinger F (2015) The play of international practice. Int Stud Quart 59(3):449–460

Bulkeley H (2001) Governing climate change: the politics of risk society? Trans Inst Br Geogr 26(4):430–447

Callicot JH (1998) Structural functionalism as a heuristic device. Anthropol Educ Q 29(1):103–111

Carmin J, Anguelovski I, Roberts D (2012) Urban climate adaptation in the global south: planning in an emerging policy domain. J Plan Educ Res 32(1):18–32

Carpenter CR (2003) Stirring gender into the mainstream: constructivism, feminism and the uses of IR theory. Int Stud Rev 5(2):297–302

Cartwright A, Blignaut J, de Wit M, Goldberg K, Mander M, O'Donoghue S, Roberts D (2013) Economics of climate change adaptation at the local scale under conditions of uncertainty and resource constraints: the case of Durban, South Africa. Environ Urbanization 25(1):139–156

Chakrabarty D (2009) The climate of history: four these. Crit Inq 35(2):197–222

Chen AH, Warren J (2011) Sustainable growth for China: when capital markets and green infrastructure combine. Chin Econ 44(5):86–103

Chouliaraki L, Fairclough N (1990) Discourse in late modernity: rethinking critical discourse analysis. Edinburgh University Press, Edinburgh

Chu E, Anguelovski I, Roberts D (2017) Climate adaptation as strategic urbanism: assessing opportunities and uncertainties for equity and inclusive development in cities. Cities 60:378–387

C40 (2020) About C40. C40 Group, New York. Accessed at: https://www.c40.org/about. Accessed on: 30 Mar 2020

Cockburn J, Rouget M, Slotow R, Roberts D, Boon R, Douwes E, O'Donoghue S, Downs CT, Mukherjee S, Musakwa W, Mutanga O (2016) How to build science-action partnerships for local land-use planning and management: lessons from Durban, South Africa. Ecol Soc 21(1)

Cohen MJ (1997) Risk society and ecological modernisation. Futures 29(2):105–119

Corbella S, Stretch DD (2012) Coastal defences on the KwaZulu-Natal coast of South Africa: a review with particular reference to geotextiles. J S Afr Inst Civ Eng 54(2):55–64

Coryn CLS, Noakes LA, Westine CD, Schröter DC (2011) A systematic review of theory-driven evaluation practice from 1990 to 2009. Am J Eval 32(2):199–226

Costanza R, McGlade J, de Bonvoisin S, Farley J, Giovannini E, Kibizewski I, Lappe FM, Lovins H, Pickett K, Norris G, Prugh T (2013) The future we really want. Solutions 4:37–43

Costanza R, Kubiszewski I, Giovannini E, Lovins H, McGlade J, Pickett KE, Ragnarsdóttir KV, Roberts D, de Vogli R, Wilkinson R (2014) Development: time to leave GDP behind. Nat News 505(7483):283

Cox RW (1996) Social forces, states and world orders. In: Cox RW, Sinclair TJ (eds) Approaches to world order. Cambridge University Press, Cambridge

Danaher J (2016) The threat of algocracy: reality, resistance and accommodation. Philos Technol 29(3):245–268

Daniell KA, White I, Ferrand N, Ribarova IS, Coad P, Rougier JE, Hare M, Jones NA, Popova A, Rollin D, Perez P (2010) Co-engineering participatory water management processes: theory and insights from Australian and Bulgarian interventions. Ecol Soc 15(4)

Davids R, Rouget M, Boon R, Roberts D (2016) Identifying ecosystem service hotspots for environmental management in Durban, South Africa. Bothalia-Afr Biodivers Conserv 46(2):1–18

Davids R, Rouget M, Boon R, Roberts D (2018) Spatial analyses of threats to ecosystem service hotspots in Greater Durban, South Africa. Peer J 6:e5723

Davies B (1990) Agency as a form of discursive practice: a classroom scene observed. Br J Sociol Educ 11(3):341–361

Denzin NK, Lincoln YS (2018) Introduction: the discipline and practice of qualitative research. In: Denzin NK, Lincoln YS (eds) The SAGE handbook of qualitative research, 5th edn. Sage, Thousand Oaks

Development Bank of Southern Africa (DBSA) (2020) About us. Development Bank of Southern Africa, Midrand. Accessed at: https://www.dbsa.org/EN/About-Us/Pages/About-Us. aspx. Accessed on: 30 Mar 2020

Devetak R, Walter R (2016) The critical theorist's labour: empirical or philosophical historiography for international relations? Globalization 13(5):520–531

Diederichs N, Roberts D (2016) Climate protection in mega-event greening: the 2010 FIFA™ World Cup and COP17/CMP7 experiences in Durban, South Africa. Clim Dev 8(4):376–380

Dixon-Woods M, Bosk CL, Aveling EL, Goeschel CA, Pronovost PJ et al (2011) Explaining Michigan: developing an ex post theory of a quality improvement theory. Milbank Q 89(2):167–205

Doherty N, Coombs C, Loan-Clarke J (2006) A re-conceptualization of the interpretive flexibility of information technologies: redressing the balance between the social and the technical. Eur J Inf Syst 15(6):569–582

Douwes J (2018) Exploring transformation in local government in a time of environmental change and thresholds: a case study of eThekwini Municipality. Unpublished Master's Thesis, University of KwaZulu-Natal, Howard College Campus, Durban

Douwes E, Crouch N, Edwards T, Mulholland (2008) Regression analyses of southern African ethnomedicinal plants: informing the targeted selection of bioprospecting and pharmacological screening subjects. J Ethnopharmacol 119(3):356–364

Douwes E, Diederichs-Mander N, Mavundla K, Roberts D (2015) Buffelsdraai landfill site community restoration project: leading the way in community-based adaptation to climate change. eThekwini Municipality, Durban

Douwes E, Buthelezi N, Taylor C (2017) Growing responsibility with growth. Thola 19:64–68

Douwes J, Mullins G, Nkosi S, Roberts D, Ralfe K, Wilkinson R, Dlamini Z, Mungwe O, Macfarlane D (2018) Innovative responses to biodiversity offsets in Durban. In: IAIA18 conference proceedings, environmental justice in societies ibn transition, 38th annual conference of the international association for impact assessment, 16–19 May 2018, Durban, South Africa

Dray J, McGill A, Muller G, Muller K, Skinner D (2006) eThekwini municipality economic review 2006/2007. eThekwini Municipality, Economic Development Department, Durban

Du Plessis A (2000) Charting the course of the water discourse through the fog of international relations theory. In: Solomon H, Turton A (eds) Water wars: enduring myth or impending reality, Africa dialogue series, vol 2. ACCORD/Green Cross International and the African Water Issues Research Unit, Durham/Pretoria

Dunn AD (2010) Siting green infrastructure: legal and policy solutions to alleviate urban poverty and promote healthy communities. Environ Aff Law Rev 37:41–66

Durban Adaptation Charter (DAC) Secretariat (2011) Durban adaptation charter for local governments as adopted on the 4th December 2011 of the occasion of the "Durban Local Government Convention: adapting to a changing climate"—towards COP17/CMP7 and beyond. Durban Adaptation Charter Secretariat, Durban

Durban Adaptation Charter (DAC) Secretariat (2020) Introduction to the hub and compact approach. Durban Adaptation Charter Secretariat, Durban. Accessed at: http://www.durbanadaptationch arter.org/hub-and-compact-approach. Accessed on: 31 Mar 2020

Duvall R, Varadarajan L (2003) On the practical significance of critical international relations theory. Asian J Polit Sci 11(2):75–88

Duzi-Umngeni Conservation Trust (DUCT) (2020) Our vision. Duzi-Umngeni Conservation Trust, Pietermaritzburg. Accessed at: https://www.duct.org.za/. Accessed on: 31 Mar 2020

Eder K (2000) Taming risks through dialogues: the rationality and functionality of discursive institutions in risk society. In: Cohen MJ (ed) Risk in the modern age: social theory, science and environmental decision-making. Palgrave Macmillan, London

Eisner EW (2017) The enlightened eye: qualitative inquiry and the enhancement of educational practice. Teachers College Press, New York

Eriksson J (2014) On the policy relevance of grand theory. Int Stud Perspect 15:94–108

Estlund D (2003) Why not epistocracy? In: Reshotko N (ed) Desire, identity and existence: essays in honour of T.M. Penner. Academic Printing and Publishing, Canada

eThekwini Metropolitan Municipality (2020a) Climate protection branch. eThekwini Metropolitan Municipality, Durban. Accessed at: http://www.durban.gov.za/City_Services/development_pla nning_management/environmental_planning_climate_protection/About_Us/Pages/Climate-Pro tection.aspx. Accessed on: 20 Feb 2020

eThekwini Metropolitan Municipality (2020b) Working for ecosystems. eThekwini Metropolitan Municipality, Durban. Accessed at: http://www.durban.gov.za/City_Services/development_pla nning_management/environmental_planning_climate_protection/Projects/Pages/Working-for-Ecosystems.aspx. Accessed on: 30 Mar 2020

eThekwini Municipality (2007) Sustainability best practice portfolio. Special Edition: Water 2007/08. eThekwini Metropolitan Municipality Environmental Management Department. Development Planning Environment, Management Unit, Durban

Evans N (2019) Durban snake catcher captures 11 snakes in one week. Highway Mail, 13 Feb. https://highwaymail.co.za/312005/durban-snake-catcher-captures-11-snakes-one-week/

Feldman MS, Orlikowski WJ (2011) Theorizing practice and practicing theory. Organ Sci 22(5):1240–1253

Finnemore M, Sikkink K (2001) Taking stock: the constructivist research program in international relations and comparative politics. Annu Rev Polit Sci 4(1):391–416

Fitch HG (1976) Achieving corporate social responsibility. Acad Manag Rev 1(1):38–46

Friedman VJ (2001) Designed blindness: an action science perspective on program theory evaluation. Am J Eval 22(2):181–182

Friend RM, Anwar NH, Dixit A, Hutanuwatr K, Jayaraman T, McGregor JA, Menon MR, Moench M, Pelling M, Roberts D (2016) Re-imagining inclusive urban futures for transformation. Curr Opin Environ Sustain 20:67–72

Funke N, Meissner R (2011) Acid mine drainage and governance in South Africa: an introduction. Pretoria: Council for Scientific and Industrial Research (Publication number: CSIR/NRE/WR/IR/2011/0031/A)

Gerring J (2010) Causal mechanisms: yes, but… Comp Polit Stud 43(11):1499–1526

Ghosh R, Kansal A (2019) Anthropology of changing paradigms of urban water systems. Water Hist 11(1–2):59–73

Giddens A (1984) The constitution of society. Polity Press, Cambridge

Grafakos S, Pacteau C, Wilk D, Driscoll PA, O'Donoghue S, Roberts D, Landauer M (2018) Mitigation and adaptation: barriers, bridges, and co-benefits. In: Rosenzweig C, Solecki WD, Romero-Lankao P, Mehrotra S, Dhakal S, Ali Ibrahim S (eds) Climate change and cities: second assessment report of the urban climate change research network. Cambridge University Press, Cambridge. https://doi.org/10.1017/9781316563878

Grant JA (2018) Agential constructivism and change in the world. Int Stud Rev 20(2):255-263

Green Corridors (2020) Our journey: our foundation in the uMngeni valley, our aspirations transcending boundaries. Green Corridors, Durban. Accessed at: https://durbangreencorridor.co.za/our-journey. Accessed on: 31 Mar 2020

Guba EG, Lincoln YS (2005) Paradigmatic controversies, contradictions, and emerging confluences. In: Denzin NK, Lincoln YS (eds) The SAGE handbook of qualitative research, 3rd edn. Sage, Thousand Oaks

Guzzini S (2000) A reconstruction of constructivism in international relations. Eur J Int Relat 6(2):147–182

Gyawali D (2020) Can nexus avoid the fate of IWRM? Water Alternatives Forum, 15 Apr 2020. Accessed at: http://www.water-alternatives.org/index.php/blog/nexus. Accessed on: 16 Apr 2020

Halliday F (1991) International relations: is there a new agenda? Millennium J Int Stud 20(1):57–72

Hay D (2017) 'Our water, our future': securing the water resources of the uMngeni river basin, 5th edn. Institute of Natural Resources, Pietermaritzburg

Hobson JM (2000) The state and international relations. Cambridge University Press, Cambridge

Herod A (2007) The agency of labour in global change: reimagining the spaces and scales of trade union praxis within a global economy. In: Hobson JM, Seabrooke L (eds) Everyday politics of the world economy. Cambridge University Press, Cambridge

Heywood A (1997) Politics. Macmillan, London

Hobson JM, Seabrooke L (2007) Everyday IPE: revealing everyday forms of change in the world economy. In: Hobson JM, Seabrooke L (eds) Everyday politics of the world economy. Cambridge University Press, Cambridge

Hobson JM, Seabrooke L (2009) Everyday international political economy. In: Routledge handbook of international political economy (IPE): IPE as a global conversation, p 290

Hogg A (2016) Election final: Joburg completes ANC misery; 106 seats lost in big 6 metros. BizNews, 4 Aug 2016

Holst C (2012) What is epistocracy? Dimensions of knowledge-based rule. In: Øyen SA, Lund-Olsen T, Vaage NS (eds) Sacred science? On science and its interrelations with religious worldviews. Wageningen Academic Publishers, Wageningen, The Netherlands

Hopf T (2000) The promise of constructivism in international relations theory. In: Linklater A (ed) International relations: critical concepts in political science. Routledge, London and New York

Horwitz P, Finlayson M (2011) Wetlands as settings for human health: Incorporating ecosystem services and health impact assessment into water resource management. Bioscience 61(9):678–688

Huber LP (2010) Beautifully powerful: a LatCrit reflection on coming to an epistemological consciousness and the power of testimonio. J Gend, Soc Policy Law 18(3):839–851

Ikenberry GJ (2009) Liberal institutionalism 3.0. America and the dilemmas of liberal world order. Perspect Polit 7(1):71–87

Independent Electoral Commission (IEC) (2016) Results summary—all ballots: eThekwini. Independent Electoral Commission, Pretoria

Intergovernmental Panel on Climate Change (IPCC) (2019) Debra roberts: co-chair, Working Group II. Intergovernmental Panel on Climate Change, Geneva, Switzerland. Accessed at: https://www.ipcc.ch/people/debra-roberts/. Accessed on: 14 Oct 2019

Jackson PT, Nexon DH (2019) Reclaiming the social: relationalism in anglophone international studies. Cambridge Rev Int Aff 32(5):582–600

Jackson R, Sørensen G (2003) Introduction to international relations: theories and approaches, 2nd edn. Oxford University Press, Oxford

James A (2007) Constructivism about practical reason. Philos Phenomenol Res 74(2):302–325

Janis I (1982) Groupthink: psychological studies of policy decisions and fiascos, 2nd edn. Houghton Mifflin, Boston, MA

Jeffrey A (2018) Limited epistocracy and political inclusion. Episteme 15–4:412–432

Jessop B (2002) Liberalism, neoliberalism, and urban governance: a state–theoretical perspective. Antipode 34(3):452–472

Joenniemi P (2017) Others as selves, selves as others: theorizing city-twinning. J Borderlands Stud 32(4):429–442

Joenniemi P, Jańczak J (2017) Theorizing town twinning—towards a global perspective. J Borderlands Stud 32(4):423–428

Kettle M (2005) Agency as discursive practice: From "nobody" to "somebody" as an international student in Australia. Asia Pacific J Educ 25(1):45–60

Kinderman D (2012) 'Free us up so we can be responsible!' the co-evolution of corporate social responsibility and neo-liberalism in the UK, 1977–2010. Socio-Econ Rev 10(1):29–57

Knill C, Lehmkuhl D (2002) Private actors and the state: Internationalization and changing patterns of governance. Governance 15(1):41–63

Koch L (2020) Personal communication. University of Osnabrück, Osnabrück, Germany, 9 Apr 2020

Kooiman J, Bavinck M, Jentoft S, Pullin R (eds) (2005) Fish for life: interactive governance for fisheries. MARE Publication Series No. 3. University of Amsterdam Press, Amsterdam

Kooiman J, Bavinck M, Chuenpagdee R, Mahon R, Pullin R (2008) Interactive governance and governability: an introduction. J Transdisc Environ Stud 7(1):1–11

Kuljanin D (2019) Why not a philosopher king and other objections to epistocracy. Phenomenol Mind 16:80–89

Kustermans J (2016) Parsing the practice turn: practice, practical knowledge, practices. Millennium J Int Stud 44(2):175–196

Leck H, Roberts D (2015) What lies beneath: understanding the invisible aspects of municipal climate change governance. Curr Opin Environ Sustain 13:61–67

Lennon M (2015) Green infrastructure and planning policy: a critical assessment. Local Environ 20(8):957–980

Lincoln YS, Lynham SA, Guba EG (2011) Paradigmatic controversies, contradictions, and emerging confluences, revisited. In: Denzin NK, Lincoln YS (eds) The SAGE handbook of qualitative research, 4th edn. Sage, Thousand Oaks

Lincoln YS, Lynham SA, Guba EG (2018) Paradigmatic controversies, contradictions, and emerging confluences, revisited. In: Denzin NK, Lincoln YS (eds) The SAGE handbook of qualitative research, 5th edn. Sage, Thousand Oaks

Lippert-Rasmussen K (2012) Estlund on epistocracy: a critique. Res Republica 18:241–258

List C (2005) Group knowledge and group rationality: a judgement aggregation perspective. Episteme 2:25–38

Lovell ST, Taylor JR (2013) Supplying urban ecosystem services through multifunctional green infrastructure in the United States. Landscape Ecol 28(8):1447–1463

Lucky MC (2019) Flawed wisdom: the challenge of knowledge systems in epistocracy and democracy. Paper presented at WPSA, 20 Apr 2019

Mahon R (2008) Assessing governability of fisheries using the interactive governance approach: preliminary examples from the Caribbean. J Transdisc Environ Stud 7(1):1–12

Maharaj A, Mather AA (No date) Durban and its port: an investigation into the relationship between the City and port of Durban. Capsule Professionelle 2:127–141

Mander M, Roberts D, Diederichs N (2011) What car salesmen can teach environmentalists. Solutions 2(6):24–27

Masson-Delmotte V, Zhai P, Pörtner HO, Roberts D, Skea J, Shukla PR, Pirani A, Moufouma-Okia W, Péan C, Pidcock R, Connors S (2018) IPCC (2018) Global Warming of 1.5°C. An IPCC Special Report on the impacts of global warming of 1.5°C above pre-industrial levels and related global greenhouse gas emission pathways, in the context of strengthening the global response to the threat of climate change, sustainable development, and efforts to eradicate poverty (In press)

Masson-Delmotte V, Zhai P, Pörtner H-O, Roberts D, Skea J, Shukla PR, Pirani A, Moufouma-Okia W, Péan C, Pidcock R, Connors S, Matthews JBR, Chen Y, Zhou X, Gomis MI, Lonnoy E, Maycock T, Tignor M, Waterfield T (eds) IPCC (2018) Summary for policymakers. In: Global warming of 1.5°C. An IPCC special report on the impacts of global warming of 1.5°C above pre-industrial levels and related global greenhouse gas emission pathways, in the context of strengthening the global response to the threat of climate change, sustainable development, and efforts to eradicate poverty. World Meteorological Organization, Geneva, Switzerland

Mather A (2008) Sea level rise and it's likely financial impacts. IMFO: Off J Inst Municipal Fin Officers 8(3):8–9

Mather A (2009) Projections and modelling scenarios for sea-level rise at Durban, South Africa. eThekwini Municipality, Durban, South Africa

Mather A, Roberts D (2015) Climate change and the coasts of Africa: Durban case study. In: Glavovic B, Kelly M, Kay R, Tavers A (eds) Climate change and coast: building resilient communities. CRC Press, Boca Raton

Mather AA, Stretch DD (2012) A perspective on sea level rise and coastal storm surge from Southern and Eastern Africa: a case study near Durban, South Africa. Water 4:237–259

Mather A, Stretch D, Garland G (2011a) Predicting extreme wave run-up on natural beaches for coastal planning and management. Coast Eng J 53(2):87–109

Mayo Clinic (2020) E. coli: overview. Mayo Clinic, Scottsdale, AZ. Accessed at: https://www.mayoclinic.org/diseases-conditions/e-coli/symptoms-causes/syc-20372058. Accessed on: 31 Mar 2020

McCarthy J, Prudham S (2004) Neoliberal nature and the nature of neoliberalism. Geoforum 35:275–283

McCourt DM (2016) Practice theory and relationalism as the new constructivism. Int Stud Quart 60(3):475–485

McLean CT, Ground LE, Boon RGC, Roberts DC, Govender N, McInnes A (2016) Durban's systematic conservation assessment. eThekwini Municipality, Environmental Planning and Climate Protection Department, Durban

McPhearson T, Parnell S, Simon D, Gaffney O, Elmqvist T, Bai X, Roberts D, Revi A (2016) Scientists must have a say in the future of cities. Nature 538(7624):165–166

Mearsheimer JJ, Walt SM (2013) Leaving theory behind: why simplistic hypothesis testing is bad for international relations. Eur J Int Relat 19(3):427–457

Meissner R (2004) The transnational role and involvement of interest groups in water politics: a comparative analysis of selected Southern Africa case studies. D.Phil. Dissertation: University of Pretoria, Faculty of Humanities

Meissner R (2005) Interest groups and the proposed Epupa Dam: towards a theory of water politics. Politeia 24(3):354–370

Meissner R (2014) A critical analysis of research paradigms in a subset of marine and maritime scholarly thought. In: Funke N, Claassen M, Meissner R, Nortje K (eds) Reflections on the state of research and technology in South Africa's marine and maritime sectors. Pretoria, CSIR, p 303

Meissner R (2015a) The relevance of social theory in the practice of environmental management. Sci Eng Ethics 22(5):1–16

Meissner R (2015b) The governance of urban wastewater treatment infrastructure in the greater Sekhukhune District Municipality and the application of analytic eclecticism. Int J Water Gov 3(2):79–110

Meissner R (2016) Hydropolitics, interest groups and governance: the case of the proposed Epupa Dam. Springer, Heidelberg

Meissner R (2017) Paradigms and theories influencing policies in the South African and international water sectors. Springer, Cham, Switzerland

Meissner R (2020) WEF, paradigms and theory. Water Alternatives Forum, 16 Apr 2020. Accessed at: http://www.water-alternatives.org/index.php/blog/nexus. Accessed on: 16 Apr 2020

Meissner R, Turton AR (2003) The hydrosocial contract theory and the Lesotho highlands water project. Water Policy 5(2):115–126

Meissner R, Ramasar V (2015) Governance and politics in the upper Limpopo River basin, South Africa. Geo J 80(5):689–709

Meissner R, Jovanovic L, Petersen C (2017a) What one knows is unknown to others: a sediment transport study and its policy implications. Int J Water Gov 5(1):1–18

Meissner R, Stuart-Hill S, Nakhooda Z (2017b) The establishment of catchment management agencies in South Africa with reference to the *Flussgebietsgemeinschaft Elbe*: some practical considerations. In: Karar E (ed) Freshwater governance for the 21st century. Springer, Dordrecht

Meissner R, Steyn M, Moyo E, Shadung J, Masangane W, Nohayi N, Jacobs-Mata I (2018a) South African local government perceptions of the state of water security. Environ Sci Policy 87:112–127

Meissner R, Funke N, Nortje K, Jacobs-Mata I, Moyo E, Steyn M, Shadung J, Masangane W, Nohayi N (2018b) Water security at local government level in South Africa: a qualitative interview-based analysis. Lancet Planet Health 2(Supplement 1):S17

Meissner R, Steyn M, Jacobs-Mata I, Moyo E, Shadung J, Nohayi N, Mngadi T (2019) The perceived state of water security in the Sekhukhune District Municipality and the eThekwini Metropolitan Municipality. In: Meissner R, Funke N, Nortje K, Steyn M (eds) Understanding water security at local government level in South Africa. Palgrave Macmillan, London

Mell IC (2008) Green infrastructure: concepts and planning. FORUM: Int J Postgrad Stud Architect, Plan Landscape 8(1):69–80

Miles MB, Huberman AM (1984) Drawing valid meaning from qualitative data: toward a shared craft. Educ Res 13(5):20–30

Min JB (2015) Epistocracy and democratic epistemology. Polit Central Eur 11(1):91–112. https://doi.org/10.1515/pce-2015-0005

Moraro P (2018) Against epistocracy. Soc Theory Pract 44(2):199–216

Moufouma-Okia WM, Delmotte VM, Zhai P, Pörtner HO, Roberts D, Howden M, Pichs-Madruga R, Flato G, Vera C, Pirani A, Tignor M (2019) IPCC expert meeting on assessing climate information for regions. IPCC working group I technical support unit, Université Paris Saclay, Saint Aubin, France

Mulligan T (2015) On the compatibility of epistocracy and public reason. Soc Theor Pract 41(3):458–476

Mutombo K (2014) Port infrastructure: a holistic framework for adaptation to climate change. In: Funke Claassen N, Meissner M, Nortje K (eds) Reflections on the state of research and technology in South Africa's marine and maritime sectors. Council for Scientific and Industrial Research, Pretoria

Mythen G, Walklate S (2006) Criminology and terrorism: which thesis? Risk society or governmentality? Br J Criminol 46(3):379–398

Naidoo Y (2019) Former eThekwini mayor Zandile Gumede's home raided by Hawks, Asset Forfeiture Unit. Sunday Times, 10 Oct 2019

Nel P (1999) Theories of international relations. In: Nel P, McGowan PJ (eds) Power, wealth and global order: an international relations textbook for Africa. University of Cape Town Press, Rondebosch

Nel J, Colvin C, le Maitre D, Smith J, Haines I (2013) South Africa's strategic water source areas. Council for Scientific and Industrial Research, Pretoria. CSIR Report No: CSIR/NRE/ECOS/ER/2013/0031/A

Nicolini D (2012) Practice theory, work, and organization: an introduction. Oxford University Press, Oxford

Nilon CH, Aronson MF, Cilliers SS, Dobbs C, Frazee LJ, Goddard MA, O'Neill KM, Roberts D, Stander EK, Werner P, Winter M (2017) Planning for the future of urban biodiversity: a global review of city-scale initiatives. Bioscience 67(4):332–342

Nortje K, Funke N, Meissner R, Steyn M, Moyo E (2019) Beyond water quality and quantity: a typology towards understanding water security in South Africa. In: Meissner R, Funke N, Nortje K, Steyn M (eds) (2019) Understanding water security at local government level in South Africa. Palgrave Macmillan, London

Onuf N (1998) Constructivism: a user's manual. In: Kubálková V, Onuf N, Kowert P (eds) International relations in a constructed world. Armonk, M.E. Sharpe

Ownby T (2013) Critical visual methodology: photographs and narrative text as a visual autoethnography. Online J Commun Media Technol

Pederson DB (2014) The political epistemology of science-based policy making. Society 51(5):547–551

Piacentini SM, Rossetto R (2020) Attitude and actual behaviour towards water-related green infratsructures and sustainable drainage systems in four north-western Mediterranean regions of Italy and France. Water 12(5):1–17

Poff NL, Brown CM, Grantham TE, Matthews JH, Palmer MA, Spence CM, Wilby RL, Haasnoot M, Mendoza GF, Dominique KC, Baeza A (2016) Sustainable water management under future uncertainty with eco-engineering decision scaling. Nat Clim Change 6(1):25–34

Possingham HP, Wilson KA (2005) Turning up the heat on hotspots. Nature 436:919–920

Pouliot V, Thérien J-P (2018) Global governance in practice. Glob Policy 9(2):163–172

Poynton JC, Roberts DC (1985) Urban open space planning in South Africa: a biogeographical perspective. S Afr J Sci 81(1):33–37

Price R (1998) Reversing the gun sights: transnational civil society targets land mines. Int Org 52(3):613–644

Rannikko P (1996) Local environmental conflicts and the change in environmental consciousness. Acta Sociol 39(1):57–72

Reckwitz A (2002) Toward a theory of social practices: a development in culturalist theorizing. Eur J Soc Theor 5(2):243–263

Rehg W (2011) Evaluating complex collaborative expertise: the case of climate change. Argumentation 25:385–400

Republic of South Africa (1996) Constitution of the Republic of South Africa, Act 108 of 1996. Government Printer, Pretoria

Republic of South Africa (2003) Traditional leadership and governance framework act, No. 41 of 2003. Republic of South Africa, Pretoria

Reus-Smit C (2001) Constructivism. In: Burchill S, Devetak R, Linklater A, Patterson M, Reus-Smit C, True J (eds) Theories of international relations, 2nd edn. Palgrave, New York

Reus-Smit C, Snidal D (2008) Between utopia and reality: the practical discourse of international relations. In: Reus-Smit C, Snidal D (eds) The oxford handbook of international relations. Oxford University Press, Oxford

Revi A, Satterthwaite DE, Aragón-Durand F, Corfee-Morlot J, Kiunsi RBR, Pelling M, Roberts DC, Solecki W (2014) Urban areas. In: Field CB, Barros VR, Dokken DJ, Mach KJ, Mastrandrea MD, Bilir TE, Chatterjee M, Ebi KL, Estrada YO, Genova RC, Girma B, Kissel ES, Levy AN, MacCracken S, Mastrandrea PR, White LL (eds) Climate change: impacts, adaptation, and vulnerability. Part A: global and sectoral aspects. Contribution of working group II to the fifth assessment report of the intergovernmental panel of climate change. Cambridge University Press, Cambridge, United Kingdom and New York, NY, USA, pp 535–612

Revi A, Satterthwaite D, Aragón-Durand F, Corfee-Morlot J, Kiunsi RB, Pelling M, Roberts D, Solecki W, Gajjar SP, Sverdlik A (2014b) Towards transformative adaptation in cities: the IPCC's fifth assessment. Environ Urbanization 26(1):11–28

Rhodes RAW (1996) The new governance: governing without government. Polit Stud 44(4):652–667

Richter UH (2010) Liberal thought in reasoning on CSR. J Bus Ethics 97(4):625–649

Risse T, Wiener A (1999) 'Something rotten' and the social construction of social constructivism: a comment on comments. J Eur Public Policy 6(5):775–782

Ritzer G (2000) Sociological theory. McGraw-Hill, New York

Roberts DC (1993) The vegetation ecology of municipal Durban, Natal. Floristic classification. Bothalia 23(2):271–326

Roberts DC (1994) The design of an urban open-space network for the city of Durban (South Africa). Environ Conserv 21(1):11–17

Roberts D (1996) Urban environmental planning and management: the challenge of sustainable development. In: Reddy PS (ed) Readings in local government management and development. A Southern African perspective. Juta, Kenwyn

Roberts D (2003a) Sustainability and equity: reflections of a local government practitioner in Southern Africa. In: Bullard RD, Agyeman J, Evans B (eds) Just sustainabilities: development in an unequal world. Earthscan Publications Ltd, London and New York

Roberts DC (2003b) Using the development of an environmental management system to develop and promote a more holistic understanding of urban ecosystems in Durban, South Africa. In: Berkowitz AR, Nilon CH, Hollweg KS (eds) Understanding urban ecosystems. Springer, New York

Roberts D (2005) Durban's Local Agenda 21/Local Action 21 Programme: a vehicle for social inclusion? In: Herle P, and Walter U-J (eds) Socially inclusive cities: emerging concepts and practice. Lit Verlag, Munster

Roberts D (2008) Thinking globally, acting locally—institutionalizing climate change at the local government level in Durban, South Africa. Environ Urbanization 20(2):521–537

Roberts D (2010) Prioritizing climate change adaptation and local level resilience in Durban, South Africa. Environ Urbanization 22(2):397–413

Roberts D (2013) Cities OPT in while nations COP out: reflections on COP18. S Afr J Sci 109(5–6):1–3

Roberts D (2016a) A global roadmap for climate change action: from COP17 in Durban to COP21 in Paris. S Afr J Sci 112(5–6):1–3

Roberts D (2016) The new climate calculus: 1.5 C = Paris agreement, cities, local government, science and champions (PLSC2). Urbanisation 1(2):71–78

Roberts D (2016c) City action for global environmental change: assessment and case study of Durban, South Africa. In: Seto KCY, Solecki W, Griffith C (eds) The Routledge handbook of urbanization and global environmental change. Routledge, New York

Roberts D (2018) Banksy and the Biologist. In: Elmqvist T, Bai X, Frantzeskaki N, Griffith C, Maddox D, McPhearson T, Parnell S, Romero-Lankao P, Simon D, Watkins M (eds) Urban planet: knowledge towards sustainable cities. Cambridge University Press, Cambridge

Roberts DC (1990) An open space survey of Municipal Durban. Unpublished PhD thesis. University of Natal, Durban, South Africa

Roberts D, Diederichs N (2002) Durban's Local Agenda 21 programme: tackling sustainable development in a post-apartheid city. Environ Urbanization 14(1):189–201

Roberts D, O'Donoghue S (2013) Urban environmental challenges and climate change action in Durban, South Africa. Environ Urbanization 25(2):299–319

Roberts D, O'Donoghue S (2016) City view: Durban, South Africa. In: Masty L (ed) State of the world: can a city be sustainable?. Island Press, Washington, DC

Roberts DC, Poynton JC (1985) Central and peripheral urban open spaces: need for biological evaluation. S Afr J Sci 81(8):464–466

Roberts D, Mander M, Boon R (2002) An urban challenge: conserving biodiversity in the eThekwini Municipality, KwaZulu-Natal. In: Pierce SM, Cowlings RM, Sandwith T, MacKinnon K (eds) Mainstreaming biodiversity in development: case studies from South Africa. World Bank, Washington, DC

Roberts DC, Boon R, Croucamp P, Mander M (2005) Resource economics as a tool for open space planning Durban, South Africa. In: Trzyna T (ed) The Urban Imperative, urban outreach strategies for protected area agencies. IUCN-California Institute of Public Affairs, Sacramento

Roberts D, Boon R, Diederichs N, Douwes E, Govender N, McInnes A, McLean C, O'Donoghue S, Spires M (2012) Exploring ecosystem-based adaptation in Durban, South Africa: "learning-by-doing" at the local government coal face. Environ Urbanization 24(1):167–195

Roberts D, Morgan D, O'Donoghue S, Guastella L, Hlongwa N, Price P (2016) Durban, South Africa. In: Bartlett S, Satterthwaite D (eds) Cities on a finite planet. Routledge, London

Robinson ES (1934) Law. An unscientific science. Yale Law J 44(2):235–267

Roe M, Mell I (2013) Negotiating value and priorities: evaluating the demands of green infrastructure development. J Environ Plan Manage 56(5):650–673

Rogers PJ, Petrosino A, Huebner TA, Hacsi TA (2000) Program theory evaluation: practice, promise, and problems. New Dir Eval 87:5–14

Rosenau JN (2006) The study of world politics: theoretical and methodological challenges. Routledge, New York

Rosneu JN (2008) People count!: Networked individuals in global politics. Paradigm Publishers, London

Rouget M, O'Donoghue S, Taylor C, Roberts D, Slotow R (2016) Improving the management of threatened ecosystems in an urban biodiversity hotspot through the Durban research action partnership. Bothalia-Afr Biodiv Conserv 46(2):1–3

Rousseau J-J (1762) The social contract or principles of political right. Translated from the French by G. D. H. Cole 2001. University of Virginia Library, Charlottesville

Ruggie JG (1998) What makes the world hang together? Neo-utilitarianism and the social construction challenge. Int Org 52:855–887

Schalkwijk N, Tifflin W, Kohler K (2017) KwaZulu-Natal visitor book project report. Tourism KwaZulu-Natal, Durban

Schatzki TR (2001) Introduction: practice theory. In: Schatzki TR, Cetina KK, von Savigny E (eds) The practice turn in contemporary theory. Routledge, London

Schatzki TR, Cetina KK, von Savigny E (2001) The practice turn in contemporary theory. Routledge, London

Schnaiberg A (1994) The political economy of environmental problems and policies: consciousness, conflict, and control capacity. In: Freese L (ed) Advances in human ecology. Emerald Group Publishing, Bingley, UK

Scholes RJ, Biggs R (2005) A biodiversity inactness index. Nature 434:45–49

Schulze R, Mander ND, Hughes C, Mather A (2014) Durban climate change strategy water theme report: draft for public comment. eThekwini Municipality, Durban

Seabrooke L, Sending OJ (2006) Norms as doing things versus norms as things to do: analyzing everyday change in international politics. Norwegian Institute of International Affairs, mimeo, Ap 2006

Selby J (2018) Critical international relations and the impact agenda. Br Polit 13:332–347

Selin E (1987) Collective risks and the environment. In: Sjöberg L (ed) Risk and society: studies of risk generation and reactions to risk. Allen and Unwin, London

Shi L, Chu E, Anguelovski I, Aylett A, Debats J, Goh K, Schenk T, Seto KC, Dodman D, Roberts D, Roberts JT (2016) Roadmap towards justice in urban climate adaptation research. Nat Clim Change 6(2):131–137

Sil R, Katzenstein PJ (2010) Analytic eclecticism in the study of world politics: reconfiguring problems and mechanisms across research traditions. Perspect Polit 6(8):411–431

Singh O (2019) Zandile Gumede removed as mayor over 'performance of eThekwini Municipality'. Sunday Times, 3 Sept 2019

Singh RK, Arrighi J, Coughlan de Perez E, Warrick O, Suarez P, Koelle B, Jjemba E, van Aalst MK, Roberts DC, Pörtner HO, Jones RG (2019) International conference on climate risk management, inputs for the intergovernmental panel on climate change's sixth assessment report. Clim Dev 11(8):655–658

Skinner Q (1985) The return of grand theory in the human sciences. Cambridge University Press, Cambridge

Slootweg R (2005) Biodiversity assessment framework: making biodiversity part of corporate social responsibility. Impact Assess Proj Appraisal 23(1):37–46

Smith S (1997) New approaches to international theory. In: Baylis J, Smith S (eds) The globalization of world politics: an introduction to international relations. Oxford University Press, Oxford

Smith AM, Bundy SC, Mather AA (2012) Failed coastal stabilization: examples from the KwaZulu-Natal Coast, South Africa. In: Finkl CW (ed) Pitfalls of shoreline stabilization. Springer, Dordrecht

Smith A, Guastella LA, Mather AA, Bundy SC, Haigh ID (2013) KwaZulu-Natal coastal erosion events of 2006/2007 and 2011: a predictive tool? S Afr J Sci 109(3–4):1–4

Solecki W, Rosenzweig C, Dhakal S, Roberts D, Barau AS, Schultz S, Ürge-Vorsatz D (2018) City transformations in a 1.5 °C warmer world. Nat Clim Change 8(3):177–181

South African National Biodiversity Institute (SANBI) (2020) The uMngeni ecological infrastructure partnership. South African National Biodiversity Institute, Pietermaritzburg. Accessed at: http://biodiversityadvisor.sanbi.org/participation/umngeni-ecological-inf rastructure-partnership/. Accessed on: 31 Mar 2020

Spodarczyk E, Szelągowska-Rudzka K (2015) Is social responsibility required in the cooperation among universities, businesses and local government in the local environment? In: Rouco JCD (ed) Proceedings of the 11th European conference on management, leadership and governance. Academic Conferences International Limited, Reading, UK

StatsSA (2016) Community survey 2016: in brief. Statistics South Africa, Pretoria

Stone A (1994) What is a supranational constitution? An essay in international relations theory. Rev Polit 56(3):441–474

Tam M (2000) Constructivism, instructional design, and technology: implications for transforming distance learning. J Educ Technol Soc 3(2):50–60

Taylor J (2020) Personal communication. Wildlife and Environment Society of Southern Africa, Howick, South Africa, 8 Apr 2020

Taylor C, Cockburn J, Rouget M, Ray-Mukherjee J, Mukherjee S, Slotow R, Roberts D, Boon R, O'Donoghue S, Douwes E (2016) Evaluating the outcomes and processes of a research-action partnership: the need for continuous reflective evaluation. Bothalia-Afr Biodiv Conserv 46(2):1–16

The Nature Conservancy (TNC) (2019a) How we work. [Online]. Available at: https://www.nature.org/en-us/about-us/who-we-are/how-we-work/. Accessed 7 Mar 2019

The Nature Conservancy (TNC) (2019b) Water funds: a natural solution for water security in Sub-Saharan Africa. [Online]. Available at: https://www.nature.org/en-us/what-we-do/our-insights/perspectives/water-funds-a-natural-solution-for-water-security-in-sub-saharan-africa/. Accessed 7 Mar 2019

Tourism KwaZulu-Natal (TKZN) (2016) Statistics of our tourism sector 2016. Tourism KwaZulu-Natal, Durban

Tracy SJ (2013) Qualitative research methods: collecting evidence, crafting analysis, 2nd edn. Wiley, Hoboken, NJ

Tremmel J (2018) The Anthropocene concept as a wake-up call for reforming democracy. In: Hickmann T, Partzsch L, Pattberg HP, Weiland S (eds) The Anthropocene debate and political science. Routledge, London

Tuathail GÓ (1999) Understanding critical geopolitics: geopolitics and risk society. J Strateg Stud 22(2–3):107–124

Turton A, Meissner R (2002) The hydrosocial contract and its manifestation in society: a south african case study. In: Turton AR, Henwood R (eds) Hydropolitics in the developing world: a Southern African perspective. African Water Issues Research Unit, Pretoria

Turton A, Ohlsson L (1999) Water scarcity and social adaptive capacity: towards an understanding of the social dynamics of managing water scarcity in developing countries. Paper presented in the Workshop No. 4: water and social stability of the 9th Stockholm water symposium "urban stability through integrated water-related management," 9–12 Aug. MEWREW Occasional Paper No, 18. Stockholm Water Institute (SIWI), Stockholm, Sweden

University of KwaZulu-Natal (UKZN) (2020) Catherine Sutherland: Bio. University of KwaZulu-Natal, Durban. Accessed at: https://sobeds.ukzn.ac.za/staff-profile/development-studies/cather ine-sutherland/. Accessed on: 30 Mar 2020

Van Klaveren ES (2018) The allure of green infrastructure: a constructivist case analysis. Unpublished MSc. Thesis, Wageningen University, Wageningen, The Netherlands

Van Klaveren S (2019) Personal communication. Wageningen University, Wageningen, The Netherlands, 12 July 2019

Viehoff D (2016) Authority. And expertise. J Polit Philos 24(4):406–426

Vion A (2007) The institutionalization of international friendship. Crit Rev Int Soc Polit Philos 10(2):281–297

Viotti PR, Kauppi MV (1999) International relations theory: realism, pluralism, globalism, and beyond. Allyn and Bacon, Boston, MA

Von Korff Y, Daniell KA, Moellenkamp S, Bots P, Bijlsma RM (2012) Implementing participatory water management: recent advances in theory, practice, and evaluation. Ecol Soc 17(1)

Walsh CL, Roberts D, Dawson RJ, Hall JW, Nickson A, Hounsome R (2013) Experiences of integrated assessment of climate impacts, adaptation and mitigation modelling in London and Durban. Environ Urbanization 25(2):361–380

Warner J (2000a) Images of water security: a more integrated perspective. AWIRU Occasional Paper No. 4, African Water Issues Research Unit, Pretoria University. Available from Website http://www.up.ac.za/academic/libarts/polsci/awiru

Warner J (2000b) Integrated management requires an integrated society: towards a new hydrosocial contract for the 21st century. AWIRU Occasional Paper No. 24. African Water Issues Research Unit, Pretoria University. Available from Website http://www.up.ac.za/academic/libarts/polsci/ awiru

Warner JF (2004) Water, wine, vinegar, blood: on politics, participation, violence and conflict over the hydrosocial contract. In: Proceedings of the workshop on water and politics: understanding the role of politics in water management, Marseille 26–02-2004/27-02-2004

Warner J (2012) Three lenses on water war, peace and hegemonic struggle on the Nile. Int J Sustain Soc 4(1–2):173–193

Warner J, Zawahri N (2012) Hegemony and asymmetry: multiple chessboard games on trans-boundary rivers. Int Environ Agreements: Polit Law Econ 12(3):215–229

Weerts S (2019) Personal communication. Council for Scientific and Industrial Research, Durban, South Africa, 10 Apr 2019

Wellstead A, Howlett M, Rayner J (2013) The neglect of governance in forest sector vulnerability assessments: structural-functionalism and "black box" problems in climate change adaptation planning. Ecol Soc 18(3)

Wendt A (1995) Constructing international politics. Int Secur 20:71–81

Western Indian Ocean Marine Science Association (WIOMSA) (2020) History. Western Indian Ocean Marine Science Association, Zanzibar, Tanzania. Accessed at: https://www.wiomsa.org/ about-wiomsa/. Accessed on 31 Mar 2020

Wildlife and Environment Society of South Africa (WESSA) (2020a) About us: overview. Wildlife and Environment Society of South Africa, Bryanston, Johannesburg. Accessed at: http://wessa. org.za/about-us/overview/. Accessed on: 30 Mar 2020

Wildlife and Environment Society of South Africa (WESSA) (2020b) Working for ecosys-tems programme. Wildlife and Environment Society of South Africa, Bryanston, Johannes-burg. Accessed at: http://wessa.org.za/what-we-do/ecological-infrastructure-sustainability/wor king-for-ecosystems-programme/. Accessed on 30 Mar 2020

Wissenburg M (2019) The Anthropocene and the republic. Crit Rev Int Soc Polit Philos. https:// doi.org/10.1080/13698230.2019.1698152

Wright H (2011) Understanding green infrastructure: the development of a contested concept in England. Local Environ 16(10):1003–1019

Wright K, Simpson C (2015) Rethinking ecology in the Anthropocene: knowledge, practices, ethics and politics. Scan J Media Arts Cult 1:2–4

Chapter 4
Discussion

Abstract In this chapter, I will present my interpretation of the various epistemological and ontological dynamics that are present in eThekwini's green and ecological infrastructure policy landscape. Although these infrastructure types are situated within the natural landscape, human interaction within this setting, and the ontological views around it, have informed several policies and practices that have evolved into specific relations among various stakeholders. This means that the environmental landscape, with respect to green and ecological infrastructures, compliments the policy landscape, and vice versa. As I have shown in Chap. 3, the systemic nature of the natural landscape has an epistemological influence on the ontological character and expression of the policy landscape.

Keywords Environmental landscape · Policy landscape · Empirical science · Interpretive science

4.1 Introduction

This chapter contains a discussion of the PULSE[3] analysis and its findings. Academic conceptualisations and pronounced practices in the literature have influenced eThekwini's epistemological and ontological perceptions of green and ecological infrastructures. The municipality follows a networked approach for its green and ecological infrastructure policy development and projects. What is also noticeable is that the local government's approach to these infrastructure types has been influenced by engineering thinking. This is especially the case where eThekwini combines green and ecological infrastructures with civil engineering projects, like storm-water discharge systems and water purification. That said, the City Council of Durban has, through a deliberate policy measure called D'MOSS, elevated the environment to the same status as a technological infrastructure at various levels, from localised open spaces, to the uMngeni Catchment. D'MOSS also exemplifies a networked approach for the promotion of green infrastructures.

eThekwini's network of green and ecological infrastructures aspires to improve the well-being of its residents through a variety of motivated declarations, such as biodiversity improvement, climate change adaptation and mitigation, socio-economic and sustainable development, as well as improved water security, through the overarching notion of ecosystem service delivery. This means that water security is not the only focal point of the infrastructure types, but it is an indirect favourable consequence, together with other favourable outcomes, such as job creation.

I argue that a shift towards the benefits of ecosystem services should not come at the expense of normativity. The empirical benefits of ecosystem services in natural open areas are well-known in the natural sciences. Nonetheless, there is a paucity of normative aspects surrounding such services, because natural scientists view their advantages from an empirical cause-and-effect dogma. Therefore, there should not only be a balanced move away from the science- and data-driven epistemology of grey infrastructures (p. 8), towards the inclusion of green and ecological infrastructures that are based on a similar epistemology. The move should also include a normative ontology that focuses not only the concept of the material benefits, but also on the values and benefits for society. The PULSE[3] analysis reveals that eThekwini mixes and matches green and ecological infrastructures, but it does not mix and match empiricism and positivism with interpretivism and normativity, when developing and implementing the infrastructure types. The standard practice around green and ecological infrastructure policies and enterprises should, therefore, note how city dwellers' intersubjectively view and appraise the benefits of these infrastructure types. This means that the city planners and natural scientists who are involved in green and ecological infrastructure initiatives require an epistemological revolution towards a broader ontological realisation. This should help to relieve the ideological tension between the green and ecological infrastructures and the grey infrastructures, from an epistemological and ontological perspective. In other words, the technocratic thinking that informs the technocratic-infused governance system that I observed in eThekwini, should also have an intersubjectively-informed normative focus.

This means that an exclusive positivist viewpoint, which is infused with a liberal institutionalist and systemic constructivist perspective, is operating in the policy landscape. This creates an epistocracy in eThekwini that is epistemologically, methodologically and ontologically exclusive.

4.2 Self-Reflection and Exclusivity

At this point, I would also like to reflect on the development of my knowledge around green and ecological infrastructures. When I conceived this project, my plan was to investigate how green infrastructures influence eThekwini's water security, and my exclusive focus was to be on the interdependence that exists between them. During my open-ended (p. 54) data gathering and analysis activities, I realised that water security is not central in eThekwini's green infrastructure policy landscape; on the contrary, its central concern is the biodiversity and climate change adaptation within a sustainable development theory and ideology.

Later during my analysis, I realised that the ecological infrastructure, especially within the epistemic structure of the UEIP, focuses on water security as its main aspiration. Even so, policy-makers and scholars within the UEIP define water security as being relative to the water quantity and quality found in the uMngeni Catchment. This is another indication of the technocratic thinking that has influenced the implementation of the three ecological infrastructure pilot projects in the uMngeni River basin (p. 131).

I argue that the UEIP was established in 2013 (p. 5) by eThekwini and SANBI, as well as the other involved stakeholders, in order to place the uMngeni's water quantity and quality at a higher level on Durban's water governance agenda, as it relies on them for its survival. When the UEIP was established, we see that ecological infrastructures played an instrumental role in improving the uMngeni's water resources. However, what is unclear is who established the UEIP? SANBI played a central role in its formative stages, since the definition of the institute's ecological infrastructure was central in the motivation of the policy-makers to establish the partnership. Therefore, it is not impossible that SANBI promoted its definition as an ecological infrastructure and used it as an inroad to influence local government policy. That said, the empiricism located within the conceptualisation of an ecological infrastructure played an influential role in deciding who to involve in its formation. The purpose was to express a certain epistocratic vision of what constitutes water security and how to solve the uMngeni's quantity and quality problems through the ecological infrastructure. The unintended consequence of this is that the formation and functioning of the UEIP have, over the years, excluded normative structures that are imbedded within the catchment's civil society and general public structures including traditional authorities.

Between 2014 and 2018, I led a CSIR research team that investigated the views of a host of stakeholders on water security in the eThekwini Metropolitan and Sekhukhune District Municipalities. Our interviewees consisted of academics, community members, municipal officials and traditional leaders (Meissner et al. 2018a, b, 2019). For the eThekwini phase of the water security project, I interviewed some of the people with whom I had interacted during the green and ecological infrastructure research. I did not speak to community members and traditional leaders, since my focus was on eThekwini, as a local government structure, and its implementation of green and ecological infrastructures for the purpose of promoting water security.

In order to place the two research endeavours into context and to explain why I put so much effort into researching these matters in eThekwini, I will give an insight into my academic association with the city and the uMngeni River. I have been an Honorary Research Associate attached to the Centre for Water Resources Research (CWRR) at UKZN since 2014, when Graham Jewitt, who was then the Director of the Centre, invited me to accept this role. My informal association with the university started in 2011, when I met Professor Roland Schulze at a conference in Obergurgl, Austria. Based on the water governance research that I conducted at that time, he put me in touch with one of his students, Sabine Stuart-Hill, who I met at a conference later that year. Since then, we have established a collegial relationship and friendship and have also collaborated on a WRC-funded project investigating the lessons that were learned from the establishment of catchment management agencies in South Africa (Meissner et al. 2017 and Stuart-Hill and Meissner 2017).

Because of my formal association with UKZN and my professional partnership with Sabine, which have enabled an academic network, I have visited Durban and Pietermaritzburg on a regular basis to participate in conferences and workshops hosted by the university and the INR. During these numerous research activities, I have noticed that there has been an absence of community, public and traditional leaders' voices; not only have they not been physically present, but their voices have also not been expressed in the research of local researchers. The exception to this would be in the work conducted by Jim and Liz Taylor and their team from WESSA.

Since most researchers and other non-governmental representatives who attend these conferences and workshops are natural scientists from entities such as DUCT and WWF, I have concluded that their background knowledge does not consider the normative aspects expressed by the communities, the public and traditional leaders. In fact, most community activities that are spearheaded by DUCT and WESSA revolve around the practice of citizen science and education with regard to the water quality problems in the uMngeni Catchment. These activities are predominantly informed by empirical science, while, to a large extent, normativity falls by the wayside. This is also the case with eThekwini's green and ecological infrastructures, where organisations with an empirical background, like ACT, INR and SANBI, play a central role. These entities define and inform green and ecological infrastructures from an empirical and positivist ontology, within the framework of sustainable development, and as an ideology and theory of positive environmental change. This is linked to the existence of a limited epistocracy (p. 181) that operates within eThekwini's green and ecological infrastructure policy landscape.

Empirical and positivist scientific background knowledge is the function that informs the epistemological, methodological and ontological practices of this epistocracy. This means that the epistocracy does not deliberately exclude the normative and intersubjective voices expressed by communities, the public and traditional leaders, but that it does so subliminally, since empiricism and positivism practices inform a top-down objectivist notion of knowledge generation, where scientists govern, and the rest benefit.

By looking at the epistemological, methodological and ontological characteristics of the epistocracy, we see that 'science speaking truth to power' (pp. 33, 34, 100) operates in over-drive. The epistocracy is limited, since it is restricted to a group of people that operates in a specific environmental (the uMngeni Catchment) and epistemological (empiricist and positivist) intellectual setting. It is also not deliberately exclusionary, although this exclusionary feature is manifested in the way that it practices its science.

Therefore, how the various stakeholders view the reality of the uMngeni Catchment's problems and eThekwini's green infrastructure initiatives has a bearing on their knowledge and action-generating activities. Next, I will speak more about the concept of reality.

4.3 Reality

4.3.1 Empiricism

In the pages that follow, I will highlight the empirical and normative aspects that I have discovered during my analysis. An empirical or 'what is' perspective asks how stakeholders are currently acting in terms of eThekwini's green and ecological infrastructure policy landscape? Figure 3.3 shows that they are perceiving and acting from a predominantly empirical or 'what is' point of view, due to the dominance of positivism throughout the interviews. With respect to this empirical focus, the role of a green infrastructure in promoting biodiversity came out strongly. Not only did the respondents link green and ecological infrastructures with the concept of 'biodiversity', but they also linked them to the function of biodiversity, namely, to retain flood waters and clean water resources for human services, as well as to the function of natural ecosystems.

Therefore, if one follows Slootweg (2005) (p. 87), we see that the respondents frame their ontology, or the essence of their reality, around the concept of biodiversity, where 'biodiversity' becomes an 'uncomplicated concept', which means that it creates a sort of comfort zone in which to implement green and ecological infrastructures. In relation to this, Kaennel (1998) notes that biodiversity is a mental construct and, in my opinion, it is one that is widely accepted by society to promote the benefits of the infrastructure types. No-one would argue that the promotion of biodiversity, or sustainability for that matter, is bad because the opposite, namely, the loss of biodiversity and unsustainable practices, could be threatening to human existence (Bilderbeek et al. 2016); it would therefore be risky to promote biodiversity loss and unsustainable practices, both epistemologically and ontologically. These notions of biodiversity protection and sustainable development structure the ontology in which the respondents practice their research initiatives and various vocations, from implementing green infrastructures, to SuDS and the establishment of, and participation in, local and global governance network structures.

Having said this, it does not mean that biodiversity is creating an uncomplicated environment in which the practitioners can operate. I also uncovered a great deal of social constructivism around the concept of green and ecological infrastructures, when the respondents noted that, for them, these are from a certain perspective, like storm-water management or coastal protection, or something specific, like the black fly larvae, buffer zones, estuaries and wetlands. Therefore, an epistemology-informed objective reality exists that is beyond the human mind (Lincoln et al. 2011, 2018; Meissner 2017).

4.3.2 Systemic Constructivism

The social construction of the 'green and ecological infrastructure' concept falls within a positivist ontology and epistemology. Because of this, apart from agential constructivism, another social constructivist type is visible in the policy landscape, namely, systemic constructivism. Many of the respondents were clear about the meaning of the term 'green' and 'ecological' infrastructures, by structuring them in a specific way and for a specific reason. For instance, Respondent 3 said that she prefers the concept 'ecological infrastructure' over 'green infrastructure' and her answer was framed within certain terms that were based on the SANBI definition (p. 90). We also see this in how Respondents 2, 6, 9, 10, 11 and 16 frame the concepts. They coin them in such a way that their structured meaning avoids confusion. In other words, their advocacy around the meaning of the concept is to keep the research object (nature) at bay, for empirical objective purposes, so that other researchers can independently and objectively study it with the view to empirically informing the policy processes.

The pursuit of truth stands central here, with truth being viewed as a one-to-one mapping between reality and the respondents' statements (Lincoln et al. 2011, 2018; Meissner 2017). This means that objectivity informs the hegemonic position of the municipal officials and researchers in the framing of the concepts. However, I did not detect that the respondents took a domineering position when they talked about the conceptualisation of green and ecological infrastructures. In fact, the criticism levelled against the terms by Respondent 4 and, to a certain extent also by Respondent 11, indicates that the concepts are not 'holy cows', so to speak, but that they can be 'slaughtered' in the name of scientific progress and better policy formulation, in line with the empirically visible problems.

This is important, from a policy perspective. Since the respondents are not domineering in their views on how other actors should speak about and use the concepts, it is conducive for opening up a dialogue between a plurality of views. However, it could be detrimental when investigating it from a group-think perspective.

4.4 Group-Think

When returning to the exclusive epistocracy mentioned above, we see that there are several actors who direct the green and ecological infrastructure policy landscape. These actors are most notably the municipal officials, in their capacity as the implementers of the infrastructure types, and their epistemic NGO partners. Both actor types play an important role in buttressing the policy landscape through their respective research endeavours.

The danger with this is that a hegemonic 'in-group' could be operating, which could perceive itself to be immune to criticism because it is part of the group-think. A group-think is a term coined by Janis in 1982, who attributed it to 'flawed decision-making that values consensus over the best result' (Buchanan and O'Connell 2006: 40). According to Janis (1982), a group-think highlights the danger of conformity in group decision-making and the absence of vigorous debate (Waddell and Sohal 1998). A danger that could arise is bad decisions, along with cognitive inconsistency, improper assimilation of new data and information linked to old epistemological beliefs, a reliance on heuristics due to cognitive limitations and the framing of information (Mercer 2005) in a systematic manner. The wellspring of these attributes is the objective and rational framing of the green and ecological infrastructure concepts, and how these inform and promote the policies and programmes around such enterprises, which is also evident in the criticism of Respondents 4 and 11. However, what is hegemonic is an objective and rational epistemology that is akin to a 'grey epistemology', which is outlined in Chap. 1 by Finewood (2016) and Cousins (2017).

Should objectivity and rationality frame the green and ecological infrastructure approaches, it could impede democratic processes, and it could exclude marginalised groups, such as those living in informal settlements and traditionally governed areas. The criticism that has been levelled against eThekwini's green and ecological infrastructure initiatives is that the involved communities, such as those at the Buffelsdraai Landfill and Inanda Mountain, are acting as a labour reserve and that they do not play a meaningful role in the development of the initiatives (Van Klaveren 2019) and in epistemologically normative development. Respondent 8 referred to a community initiative in Amanzimtoti where the involvement of communities has had a positive influence on storm-water drainage systems. However, he did not elaborate on how the communities got involved, so I cannot draw any definite conclusion about this.

Furthermore, the presence of postpositivism, which is manifested as problem-solving theories in the interviews, indicates that not all the framings around eThekwini's green and ecological infrastructure landscape lend themselves to absolute objective rationality. There is also a measure of uncertainty, which is a further indication of a space that could promote democratic processes, which Cousins speaks of.

4.4.1 Ecological Dominance

Having looked at the empirical framing of the concepts, and following Finewood (2016) concept of a group-think, I would like to describe the prevailing ontology that informs the ecological dominance of the policy landscape. According to Jessop (2000: 328–329):

> The idea of ecological dominance emerged in work on plant and animal ecosystems, where it refers to the capacity of one species to exert an overriding influence on others in a given ecological community... I do want to suggest that the notion of "ecological dominance" can be usefully extended to social systems once allowance is made for their specificities as communicatively or discursively mediated systems, and for the capacity of social forces

to reflect and learn about their own evolution and engage in attempts (successful or not) to guide it. Thus, one could study social formations as bounded ecological orders formed by the co-presence of operationally autonomous systems and the lifeworld – with the structural coupling and co-evolution of these systems and the lifeworld mediated by various competitive, cooperative and exploitative mechanisms. Ecological dominance would then refer to the capacity of a given system in a self-organising ecology of self-organising systems to imprint its developmental logic on other systems' operations through structural coupling, strategic coordination and blind co-evolution to a greater extent than the latter can impose their respective logics on that system.

According to Jessop (2000, 2014), we need to take care not to apply the ecological dominance epistemology without considering its 'communicative' and 'discursive' mediated systems. We also need to consider that social forces (municipal officials and the epistemic community) reflect and learn about the evolutionary processes that guide their epistemology. Jessop (2000) goes on to say that, for the ecological dominance epistemology to be present, there need to be autonomously operated systems and an existing lifeworld. Both are present in eThekwini's green and ecological infrastructure policy landscape, for example, there is an interest group, municipal officials, NGOs, members of the epistemic community, as well as the municipality as a structural entity, which all exhibit autonomous preferences. Nevertheless, these independent inclinations are not free from influencing factors. Cost-benefit analyses play an important part in the policy landscape, especially when considering the establishment of a water fund and the role of liberal institutionalist practices, like attaching economic value to natural resources through structures like D'MOSS.

If we consider D'MOSS further, Roberts (1994) states that it frames a novel approach to public open space provisioning in Durban. Within these open spaces, the municipality has integrated their use as amenities and for conservation. D'MOSS envisions the creation of an open space network, which forms a permanent and inherent part of Durban's urban environment. The purpose of the policy is '...the preservation of maximum *sustainable* indigenous biotic diversity' (emphasis in the original) (Roberts 1994: 14). Therefore, the overarching theoretical framing of D'MOSS is sustainable development, with an apparent ecological dominance present.

From Robert's (1994) description of the vision and purpose of D'MOSS, we notice several normative concepts (Callicot et al. 1999), such as biotic diversity and conservation. The other normative concepts that interviewees identified, and which are linked to D'MOSS, include biodiversity (p. 87), ecosystem services, rehabilitation, adaptation and sustainable development. From a functionalist perspective (a sub-strand of neoliberalism) (p. 106), scholars comprehend '...nature primarily by means of ecosystem ecology and considers Homo sapiens a part of nature' (Callicot et al. 1999). Another school of conservation philosophy is compositionalism, which defines '...nature primarily by means of evolutionary ecology and considers Homo sapiens separate from nature' (Callicot et al. 1999: 22).

Biodiversity and biological integrity belong chiefly to the compositionalist vocabulary, while adaptation, conservation, ecosystem services, rehabilitation and sustainable development are part of the functionalist glossary (Callicot 1998).

The two philosophies are complementary, and not competitive, and they are mutually exclusive. Compositionalist norms are more appropriate for nature reserves (Callicot et al. 1999), such as those found in and around Durban, while functionalist norms are appropriate for areas that have been inhabited and exploited by humans (Callicot et al. 1999), such as Durban's beaches and recreational parks.

What is more, according to Freund (2001: 725), in the late 1990s, D'MOSS evolved from a document with a string of 'conservationist ethics' to a structure that promotes 'various environmental advantages', such as 'temperature control, as well as the cleansing of the water flow system'. D'MOSS packaged these into '[monetary] terms to give a business-like, cost-benefit analysis', which was not part of the original D'MOSS idea.

Therefore, eThekwini's green and ecological infrastructure policy landscape meets the criteria that are set out by Jessop (2000, 2014). The ecological dominance of the biodiversity discourse informs this underlying epistemology, which, in turn, informs the policy landscape. The ecological dominance of biodiversity imprints its positivist logic on social relations and institutions (MacKinnon and Derickson 2013), which is probably the reason why water security does not feature much, except by being an indirect benefit of D'MOSS. A salient feature of the EDCPD is its focus on anthropogenic global climate change adaptation through green infrastructures.

Therefore, D'MOSS incorporates compositionalist and functionalist norms that are applied to areas that have been inhabited and exploited by humans. It is functionalist, since humans are part of nature and exploit it for various purposes and values because it is a specific governance structure that serves an exact function, based on various norms and values, such as adaptation, preservation and sustainable development. D'MOSS is based on the notion that, for eThekwini to implement successful climate change adaptation and biodiversity conservation through green infrastructures, the policy is a necessity for the correct functioning of the system and for it to reach its normative goals.

Nevertheless, D'MOSS is not the only political mechanism that informs eThekwini's green infrastructure ontology. International relations that link eThekwini's climate adaptation and mitigation reality with other cities is an extension of its green and ecological infrastructure policy landscape across national borders.

4.4.2 International Relations of eThekwini as an Autonomous Actor

With regard to the municipality being an autonomous actor, Pieterse (2019) argues that South African municipalities have significant constitutional autonomy, as well as financial independence. We see these autonomous ideals manifesting themselves in the municipality's engagement with other cities (Bremen, Fort Lauderdale and African cities) through twinning agreements and the C40 initiative.

The South African Constitution informs the self-organising system in which the eThekwini Municipality operates as an autonomous actor that is capable of structurally coupling and strategically coordinating with a host of international actors to co-evolve the green and ecological infrastructure policy landscape.

We have seen this co-evolution, when several respondents mentioned that the policy landscape started in the mid-1980s with D'MOSS. Over the years, the policy landscape has evolved, with Durban being one of the hosting cities for events like the *Fédération Internationale de Football Association* (FIFA) 2010 Soccer World Cup and the 2011 COP17 conference (Respondent 12).

A practical effort informs these international relations that supply order and welfare maximisation by those who frame the green and ecological infrastructures (Hobson and Seabrooke 2007; Meissner 2017), as well as those who collaborate with the municipality, in the name of the environment. For instance, the Bremen twinning arrangement acts not only as an international relations conduit, but also as a mechanism for gaining access to scarce resources by certain municipal officials, in an environment that is characterised by actors who are competing for valuable financial resources (e.g. the different departments within the municipality) (Respondent 6).

At the time of writing, the municipality has also been in the process of establishing a water fund with the TNC. This translates into considerable autonomy for the municipality, because the water fund's establishment is not dependent on the national fiscus. Considerable rationality influences the initiative to get the water fund off the ground. In this sense, we see a structural coupling between the municipality, TNC and the idea of generating funding for green and ecological infrastructures, using private sector capital. Furthermore, this indicates the ecological dominance of financial innovation (Jessop 2009) in the name of biodiversity, in particular, and for the natural environment, in general.

4.5 Conclusion

What have we learned about eThekwini's green and ecological infrastructure policy landscape? What finer nuances can we discern that could help the relevant stakeholders to achieve a deeper understanding?

Although paradigms and theories are cognitive instruments, they highlight various unique aspects in the analysed data. As noted by Reus-Smit and Snidal (2008), theories and paradigms are simultaneously empirical and normative, at their most fundamental level, they instantaneously illuminate 'what is' and 'what ought' in an issue area. We now know 'what is'. It is a limited epistocracy that promotes green and ecological infrastructures, with the focus on biodiversity protection, climate change adaptation and mitigation, sustainable development and water security. Other aspirations that are linked to green and ecological infrastructure enterprises include socio-economic development through job creation. 'What ought' to be is that the eThekwini Municipality needs to include the public, as well as the critical voices, to add to its knowledge and understanding of the various initiatives.

What is more, paradigms and theories contain characteristics that assist the analyst to create a deeper understanding of what is practically relevant. The relationship between social theory and practice is not only based on determinism or linear causality, but also on a mutually enhancing relationship, with both of them forming part of an inseparable whole. In so doing, the municipality could enhance its knowledge of the implications of green and ecological infrastructures to generate a more inclusive and normative process, instead of a top-down hegemonic policy environment that is informed by a single paradigm.

That being said, as I argued in Chap. 2, paradigms and theories are not irrelevant to practice. Defining theories and paradigms in a specific way might create the impression that these cognitive instruments are irrelevant to practice, especially when they are conceptualised in an empirical and positivist manner. In our quest to build a firm foundation for knowledge, based on an exclusive scientific conceptualisation, we put the theory for practice on shaky ground. Using an empirical scientific meaning of theory, and focusing on theory's 'predictive capability', we make the development of practice straightforward by using regulatory cause-and-effect prognostications. These forecasts are usually in the form of hypotheses and this type of knowledge, on its own, is insufficient for practical knowledge purposes, since hypothesis-type statements are often devoid of process- and context-specific information. This is because the verification principle of positivism distinguishes valid knowledge from opinion, only when we can verify or confirm the truth of statements through hypothesis testing, which is often in the form of verification. Because of this characteristic, researchers often ignore the practical knowledge context, since they hold practical knowledge in the form of experience, which does not lend itself to hypothesis testing, let alone verification. With a hypothesis in hand, members of the epistemic community expect practitioners to know which variables they need to adapt to, in order to achieve the desired results.

The PULSE[3] analysis in Chap. 3 highlights the dominant practical knowledge that informs practices, as well as the 'neglected practical knowledge' that could enhance the practitioners' background knowledge. This overlooked practical knowledge is usually, but not always, in normative form, since this type of knowledge is often unknown to practitioners and researchers using empirical background knowledge.

PULSE[3], as a form of reflexive methodology, emphasises the dominant practical knowledge (e.g. paradigms, theories, causal mechanisms and problem-solving and critical theory types) that informs practice. PULSE[3] highlights how, by whom, and through which mechanisms, practitioners and researchers inform practices with the abovementioned practical knowledge types. For instance, whereas paradigms focus their attention on the worldviews that inform practices and/or methods, theories can, on the other hand, attenuate the role of the context and the actors. For causal mechanisms, it is about relations and processes, while the problem-solving and critical perspectives highlight various foundational principles that humans use, through explanation and understanding, when they are facing problems. Altogether, these cognitive instruments have one thing in common, namely, how to improve the human condition.

References

Bilderbeek SL, Gupta J, Ros-Tonen M (2016) Biodiversity conservation. SDG Policy Brief #2. Centre for Sustainable Development Studies, Amsterdam

Buchanan L, O'Connell A (2006) A brief history of decision making. Harvard Bus Rev 84(1):32–41

Callicot JH (1998) Structural functionalism as a heuristic device. Anthropol Educ Q 29(1):103–111

Callicot JB, Crowder LB Mumford K (1999) Current normative concepts in conservation. Conser Biol 3(1):22–35

Cousins JJ (2017) Structuring hydro-social relations in urban water governance. Ann Am Assoc Geogr 107(5):1144–1161

Finewood MH (2016) Green infrastructure, grey epistemologies, and the urban political ecology of Pittsburgh's water governance. Antipode 48(4):1000–1021

Freund B (2001) Brown and green in Durban: the evolution of environmental policy in a post-apartheid city. Int J Urban Reg Res 25(4):717–739

Hobson JM, Seabrooke L (2007) Everyday IPE: revealing everyday forms of change in the world economy. In: Hobson JM, Seabrooke L (eds) Everyday politics of the world economy. Cambridge University Press, Cambridge

Janis I (1982) Groupthink: psychological studies of policy decisions and fiascos, 2nd edn. Houghton Mifflin, Boston, MA

Jessop B (2000) The crisis of the national spatio-temporal fix and the tendential ecological dominance of globalizing capital. Int J Urban Reg Res 24(2):323–360

Jessop B (2009) The continuing ecological dominance of neoliberalism in the crisis. In: Saad-Filho A, Yalman GL (eds) Economic transitions to neoliberalism in middle-income countries: policy dilemmas, economic crises, forms of resistance. Routledge, London

Jessop B (2014) Capitalist diversity and variety: variegation, the world market, compossibility and ecological dominance. Capital and Class 38(1):45–58

Kaennel M (1998) Biodiversity: a diversity in definition. In: Bachmann P, Köhl M, Päivinen R (eds) Assessment of biodiversity for improved forest planning. Kluwer, Dordrecht

Lincoln YS, Lynham SA, Guba EG (2011) Paradigmatic controversies, contradictions, and emerging confluences, revisited. In: Denzin NK, Lincoln YS (eds) The sage handbook of qualitative research, 4th edn. Sage, Thousand Oaks

Lincoln YS, Lynham SA, Guba EG (2018) Paradigmatic controversies, contradictions, and emerging confluences, revisited. In: Denzin NK, Lincoln YS (eds) The SAGE handbook of qualitative research, 5th edn. Sage, Thousand Oaks

MacKinnon D, Derickson KD (2013) From resilience to resourcefulness: a critique of resilience policy and activism. Prog Hum Geogr 37(2):253–270

Meissner R (2017) Paradigms and theories influencing policies in the South African and international water sectors. Springer, Cham, Switzerland

Meissner R, Stuart-Hill S, Nakhooda Z (2017) The establishment of catchment management agencies in South Africa with reference to the *Flussgebietsgemeinschaft Elbe*: some practical considerations. In: Karar E (ed) Freshwater governance for the 21st Century. Springer, Dordrecht

Meissner R, Steyn M, Moyo E, Shadung J, Masangane W, Nohayi N, Jacobs-Mata I (2018a) South African local government perceptions of the state of water security. Environ Sci Policy 87:112–127

Meissner R, Funke N, Nortje K, Jacobs-Mata I, Moyo E, Steyn M, Shadung J, Masangane W, Nohayi N (2018b) Water security at local government level in South Africa: a qualitative interview-based analysis. The Lancet Planetary Health 2(Supplement 1):S17

Meissner R, Steyn M, Jacobs-Mata I, Moyo E, Shadung J, Nohayi N, Mngadi T (2019) The perceived state of water security in the Sekhukhune District Municipality and the eThekwini Metropolitan Municipality. In: Meissner R, Funke N, Nortje K, Steyn M (eds) Understanding water security at local government level in South Africa. Palgrave Macmillan, London

Mercer J (2005) Rationality and psychology in international politics. International Organization 59:77–106

Pieterse M (2019) A Year of Living Dangerously? Urban assertiveness, cooperative governance and the first year of three coalition-led metropolitan municipalities in South Africa. Politikon 46(1):51–70

Reus-Smit C, Snidal D (2008) Between utopia and reality: the practical discourse of international relations. In: Reus-Smit C, Snidal D (eds) The Oxford Handbook of international relations. Oxford University Press, Oxford

Roberts DC (1994) The design of an urban open-space network for the city of Durban (South Africa). Environ Conserv 21(1):11–17

Slootweg R (2005) Biodiversity assessment framework: making biodiversity part of corporate social responsibility. Impact Assessment and Project Appraisal 23(1):37–46

Stuart-Hill SI, Meissner R (2017) Lessons learnt from the establishment of Catchment Management Agencies in South Africa: Final Dissemination Report to the Water Research Commission. Water Research Commission, Pretoria. WRC Research Project Number: K5/2320

Van Klaveren S (2019) Personal communication. Wageningen University, Wageningen, The Netherlands, 12 July 2019

Waddell D, Sohal AS (1998) Resistance: a constructive tool for change management. Manag Decis 36(8):543–548

References

Appendix

Respondent	Stakeholder	Causal Mechanism Path	Respondent's motivation	Of what is it a case or an occurrence?	Long- or Short-term	Positive (+) or Negative (−)
1	Municipal official	Agential to Material	Creating artificial wetlands and detention and retention ponds	Flood mitigation	Long-term	+
		Material to Agential	Detention ponds are active during storms	Detention facilities detaining storm water	Short-term	+
		Material to Agential	Sports fields and parks mitigate downstream flooding	The built environment providing a service	Short-term	+
		Ideational to Agential	Engineers would like to get rid of everything quickly—and that is not the right thing	A paradigm constituting action	Short-term	−
		Ideational to Material to Ideational	Retention that is always wet containing fragmentise grasses with lots of biodiversity that the 'environmentalists' like	A socially constructed idea facilitating an eco-system service	Long-term	+
		Material to Agential	River systems, flood plains, and estuaries should not be built-up areas	Flood risk mitigation	Long-term	−
		Material to Agential	Estuarine protection and 'to get them to work properly'	Essential environmental functioning	Long-term	+
		Agential to Material	Pollution causing alien invasives to proliferate and chocking estuaries leading to killing 'all the animals and [the entire] productive system'	Consecutive negative environmental impacts	Long-term	−

(continued)

(continued)

Respondent	Stakeholder	Causal Mechanism Path	Respondent's motivation	Of what is it a case or an occurrence?	Long- or Short-term	Positive (+) or Negative (−)
2	Researcher at ORI	Material to Agential	Ecological infrastructure, such as estuaries, provide protection	Goods and service provisioning	Long-term	+
		Agential to Material incorporating Ideational to Agential	Green infrastructure provides protection of man-made infrastructure	Goods and service provisioning and essential environmental functioning for human benefit	Long-term	+
		Agential to Material	Ecological and green infrastructure fall 'under the umbrella of working with nature'	Using nature for human purposes	Long-term	+
		Ideational to Material	'People can relate to the need for the conservation of wetlands or the value they add'	Wetlands are valuable to humans	Long-term	+
		Material to Material	'The coast is still considered an economic resource…it is the biggest tourist attraction'	The financial value of a natural resource	Long-term	+
		Material to Agential	'The coast is extremely developed' and to implement green or ecological infrastructure is hard	Exploitation of natural resources	Long-term	−
		Agential to Materialto Ideational	Development has altered beaches to such an extent that people no longer consider them ecological infrastructure	Over exploitation of ecological infrastructure altering their conceptualisation	Long-term	−

(continued)

(continued)

Respondent	Stakeholder	Causal Mechanism Path	Respondent's motivation	Of what is it a case or an occurrence?	Long- or Short-term	Positive (+) or Negative (−)
		Material to Agential to Material	Estuarine functioning depends on the volume of water or sediments entering the system but has been upset in some way due to water abstraction	Essential environmental functioning for human advantage	Long-term	−
		Material to Agential	Dunes offer protection to infrastructure	Essential environmental functioning for human advantage	Long-term	+
		Agential to Material to Ideational	Durban Port is looking into working with nature in terms of green engineering that are sustainable initiatives because it offers better protection than engineered initiatives	Using nature to human's advantage	Long-term	+
3	Lecturer at UKZN	Ideational to Agential	People use the concept green infrastructure in a policy or applied field	Using a concept pragmatically	Long-term	+
		Material to Material	Green infrastructure provides ecosystem services to humans	Engineering the environment for human needs	Long-term	+
		Ideational to Agential to Material	When using the concept green infrastructure, you separate the environment and break the social ecological combination	The specific usage of a concept that does not related to the environment	Long-term	−
		Ideational to Agential	People use the concept green infrastructure in a political sense to defend the environment	A concept becoming an ideological tool	Long-term	−

(continued)

(continued)

Respondent	Stakeholder	Causal Mechanism Path	Respondent's motivation	Of what is it a case or an occurrence?	Long- or Short-term	Positive (+) or Negative (−)
		Ideational to agential	Researchers prefer to use the concept of ecological infrastructure	A concept becoming an ideological tool	Long-term	+
		Material to Material	Ecological infrastructure contains and produces ecosystem services	Nature producing services for the benefit of humans	Long-term	+
		Ideational to Material	Biodiversity is part of ecological infrastructure but also different to it	A concept becoming an explanatory tool	Long-term	+
		Material to Material	The concept green infrastructure goes with the concept green economy in a policy context	Using a concept pragmatically	Long-term	+
		Ideational to ideational	I am nervous framing something 'green', ecological infrastructure is the stronger discourse	Using a concept pragmatically	Long-term	+
		Agential to Ideational to Material	The Environmental Planning and Climate Protection Department uses ecosystem services to adapt to climate change	Using a concept to guide policy actions	Long-term	+
		Agential to Ideational to Material	The Environmental Planning and Climate Protection Department uses community-based ecosystem adaptation	Using a concept to guide policy actions	Long-term	+
		Agential to Ideationalto Material	The Environmental Planning and Climate Protection Department supports development and increase water security	Using a concept to guide policy actions	Long-term	+

(continued)

(continued)

Respondent	Stakeholder	Causal Mechanism Path	Respondent's motivation	Of what is it a case or an occurrence?	Long- or Short-term	Positive (+) or Negative (−)
		Structural to Agential to Material	D'MOSS defines the city's spatial development framework in terms of ecological infrastructure and ecosystem services	A structure of rule guiding policy actions	Long-term	+
		Structural to agential	D'MOSS is the pivot of other legislative structures	A structure of rule guiding policy actions	Long-term	+
		Structural to Agential to Material	Traditional authorities allocating land in coastal buffer zones	A structural of rule allocating resources	Long-term	−
		Structural to material	Policies and bylaws protecting coastal areas	A structural of rule allocating resources	Long-term	+
		Agential and Structural	The application of bibles, policies, and legislation in traditional authority areas is difficult	A structural of rule not being effective in allocating resources	Long-term	−
		Structural to Ideational to Agential to Material	The DRAP between the municipality and the University of KwaZulu-Natal Generated a lot of research and insight into protecting important spaces of ecological infrastructure in the city	A structure of rule generating specific knowledge for the policy environment	Long-term	+
		Agential to Structural	Management is hard in a dual governance context	Structures of rule are not always effective in certain contexts	Long-term	−

(continued)

(continued)

Respondent	Stakeholder	Causal Mechanism Path	Respondent's motivation	Of what is it a case or an occurrence?	Long- or Short-term	Positive (+) or Negative (−)
		Structural to Agential to Ideational	The KwaZulu-Natal Provincial Government has a lot of influence over the traditional authorities	Hierarchical structures of rule	Long-term	+
		Structural to Agential to Ideational to Material	The KwaZulu-Natal Provincial Government has guidelines for the Amakhosi to use to try and protect water buffers and wetlands	Hierarchical structures of rule	Long-term	+
		Structural to Ideational	The Amakhosi view these guidelines as imposing on the traditional authority and the right to make decisions over land	A clash between modern governance systems and traditional authorities	Long-term	−
4	Researcher at INR	Ideational to Ideational	Strong opinions around the concept green infrastructure and what people are trying to sell	Someone that does not agree with the nature of the concept and what it used for	Short term	+
		Ideational to Material	Ecological infrastructure is nature or naturally functioning ecosystems	A simplified way to explain the concept	Short term	+
		Ideational to Material	In a water security security context, it is naturally functioning ecosystems supporting water security objectives–clean water	A specific way in which one defines water security	Long-term	+
		Material to Agential to Material	Green infrastructure is created infrastructure or soft systems to improve water quality and quantity and holding soil together	A simplified way to explain the concept	Short term	+

(continued)

(continued)

Respondent	Stakeholder	Causal Mechanism Path	Respondent's motivation	Of what is it a case or an occurrence?	Long- or Short-term	Positive (+) or Negative (−)
		Ideational to Material to Ideational	With ecological infrastructure we are trying to reinvent the term nature to appeal to different audiences-economists and engineers	A statement critical of the concept ecological infrastructure	Short term	+
		Ideational to Ideational	Ecological infrastructure is like a sex doll	A critical analogy to emphasise the point	Short term	+
		Ideational to Structural	We are inventing terminology to suit disciplines	A statement critical of the concept ecological infrastructure	Short term	+
		Ideational to material	We have become quite mechanistic when we talk about nature	A statement critical of the concept ecological infrastructure	Short term	+
		Ideational to Agential to Ideational to Agential	Ecological infrastructure provides nature-based solutions for water security–nature-based solutions conserving wetlands and building of artificial wetlands	Paralleling nature with ecological infrastructure	Short term	+
		Structural to Agential to Material	D'MOSS is an explicit policy looking explicitly at open spaces that generate benefits to humans including flooded retention and flood mitigation	A specific policy for specific purposes	Long-term	+
		Materialto Material	A lot of open spaces are along rivers and streams	Aquatic environments are important in an open-space policy	Long-term	+

(continued)

(continued)

Respondent	Stakeholder	Causal Mechanism Path	Respondent's motivation	Of what is it a case or an occurrence?	Long- or Short-term	Positive (+) or Negative (−)
		Material to Ideational	Open spaces are important for the public	Nature is important to humans	Long-term	+
		Structural to Material to Agential	Durban's downstream geographic position bedevils its water supplysupplies policies because the water does not come from Durban	A downstream entity reliant upstream water supply	Long-term	+
		Agential to Structural	Durban negotiates with Umgeni Water from a powerful position	An entity's downstream position constituting its power position	Long-term	+
		Agential to Structural	Durbanand SANBI instrumental in establishing the UEIP	An entity's downstream position and interactive governance constituting its power position	Long-term	+
		Structuralto Agential	The UEIP is a strategy to engage upstream from Durban	An entity's downstream position and interactive governance constituting its power position	Long-term	+
		Ideational to Agential to Materialto Structural	It is with in Durban's material interest engaged upstream outside its jurisdiction, which is not illegal	A municipality acting based on its material interests	Long-term	+
		Material to Agential to Ideation	Durban explicitly engaging in nature-based efforts to improve its water security	Ecological infrastructure acting as a political structure	Long-term	+

(continued)

(continued)

Respondent	Stakeholder	Causal Mechanism Path	Respondent's motivation	Of what is it a case or an occurrence?	Long- or Short-term	Positive (+) or Negative (−)
		Ideational to Material	In the minds of the municipality it is cheaper and more effective to use rivers and streams to manage water than it is to use storm water systems (Sihlangzimvelo)	Ecological infrastructure as a cost-effective way to manage storm water	Long-term	+
		Structural to Material to agential	Sihlangzimvelo is a wonderful project, if institutionalised through policy and investment by the municipality and not so much through external funding	The independent management of a municipal project	Long-term	+
		Material to Agential to Material	Littering combined with alien invasives disrupts Durban Port operations during floods	The impact of pollution and alien invasives on the economy	Short-term	−
		Material to Agential	Places choked by alien invasive give criminals free reign because they can hide easily	The negative consequences of alien invasive plants	Short-term	−
		Material to Agential	Where sewer lines run along streams choked with alien invasives the municipality finds it difficult to clear sewer blockages in these areas	The negative consequences of alien invasive plants	Short-term	−
		Agential to Material	The municipality clear alien invasives and restore the topography so there is no huge erosion during floods	Mitigating potential risks emanating from alien invasives	Long-term	+
		Structural to Agential to Material	Policies prohibit people to develop properties close to the sea	Policies that mitigate against property damage during natural disasters	Long-term	+

(continued)

(continued)

Respondent	Stakeholder	Causal Mechanism Path	Respondent's motivation	Of what is it a case or an occurrence?	Long- or Short-term	Positive (+) or Negative (−)
		Material to Agential	Water is the epitome of an integrator—it brings together policiesand practices	A natural resource playing the role of connector	Long-term	+
		Material to Agential to Ideational	Water forces us to integrate our thinking	A natural resource constituting the nature and extent of human thought	Long-term	+
		Material to Agential to Material	Water forces us to integrate between freshwater and marine systems, between land and freshwater, between land and marine water, between social scienceand engineering and between ecology and hydrology—it forces us to bring everything together otherwise you are in trouble	A natural resource constituting the nature and extent of human thought and the practice of integration	Long-term	+
5	Municipal official	Material to Material	Plants and other natural resources clean water as it goes by	Natural resources supplying services to nature	Long-term	+
		Material to Material	Infrastructure to make more water available through rain water harvesting	Engineering solutions to augment water resources	Long-term	+
		Agentialto Ideational	Educating communities about water security	Providing an educational service to communities	Long-term	+
		Material to Agential to Material	Municipality has a spotter website that if someone sees an alien plant, they can report it on the website so the municipality can remove it	Using technology to increase green infrastructure and water security	Long-term	+

(continued)

(continued)

Respondent	Stakeholder	Causal Mechanism Path	Respondent's motivation	Of what is it a case or an occurrence?	Long- or Short-term	Positive (+) or Negative (−)
		Agential to Agential	The municipality subcontracts small companies to clear alien invasive plants	Job creation and wealth distribution of public funds	Long-term	+
		Agential to Ideational	Educate people around the importance of clearing alien invasives	Providing an educational service to communities	Long-term	+
		Agential to Material	Wildlands Conservation Trust gets communities to plant trees and the then buys the trees back from the communities—the communities then have a source of income	Wealth distribution through philanthropy using specific green infrastructure (trees)	Long-term	+
		Ideational to Agential to Structural	Students do research on green infrastructure and the municipality then use the research to enhance specific projects and adapt policiesand practices according to the research findings—DRAP	The science policy interface at work	Long-term	+
		Structural to Agential	The municipality encourages citizens to go green	A government structure influencing human behaviour	Long-term	+
6	Municipal official	Material to Material	Green infrastructure is natural areas and processes that provide us with services	A specific way of viewing the concept green infrastructure in line with predominant thinking	Long-term	+

(continued)

(continued)

Respondent	Stakeholder	Causal Mechanism Path	Respondent's motivation	Of what is it a case or an occurrence?	Long- or Short-term	Positive (+) or Negative (−)
		Material to Material	Green infrastructure also has a built environment component typically defined as engineered duplicates of natural systems eg green roofs, engineered wetlands, and rainwater harvesting systems	A typology of green infrastructure	Long-term	+
		Material to Agential	Water sensitive urban design that takes the entire water cycle into consideration	A practice highlighting water in urban design processes	Long-term	+
		Structural to Material to Agential	The storm water bylaws are going through a public participation process	The policy development process	Short-term	+
		Structural to Material to Agential	The storm waterpolicies deal with new developments and types of material and pipes and provide the storm water lens for new developments	Policies informing specific practices	Long-term	+
		Structural to Ideational to Agential	Storm waterpolicies do not incorporate water quality—Cape Town's bylaws do, and Durban would like to learn from them	Policy development and learning from other local governments	Long-term	+
		Material to Agential to Ideational	Monitoring of storm water quality is part of the Water Research Commission's research; producing assumed numbers if, for instance, a wetland or roof top garden are part of a storm water system, then the water quality would be a certain percentage	Science policy interface with green infrastructure as a component	Short-term	+

(continued)

(continued)

Respondent	Stakeholder	Causal Mechanism Path	Respondent's motivation	Of what is it a case or an occurrence?	Long- or Short-term	Positive (+) or Negative (−)
		Ideational to Material	Research conducted on the influence of engineered wetlands on the quality of sewerage discharges	Science policy interface with green infrastructure being a component	Short-term	+
		Agential to Material	The Green Corridor initiative by the municipality; clearing alien invasives and alien management on the uMngeni river	Getting rid of undesirable green infrastructure	Long-term	+
		Material to Structural to Material	River Horse wetland initiative with businesses starting a corporate social initiative to fund a long-term maintenance plan for the wetland	Private public partnership	Long-term	+
		Material to Agential	EU550 00000 to rehabilitate the River Horse wetland	Having a healthy natural environment costs money	Short-term	+
		Structural to Agential	Durban has a twinning agreement with the City of Bremen, Germany and both cities conducted fact finding missions to each other	Bilateral local government level international relations	Long-term	+

(continued)

(continued)

Respondent	Stakeholder	Causal Mechanism Path	Respondent's motivation	Of what is it a case or an occurrence?	Long- or Short-term	Positive (+) or Negative (−)
		Ideational to Agentialto Material	Durban is a new and expanding city while Bremen is an old city that wants to refurbish certain part So, both cities exchanged teams to learn from one another in terms of a new city that is expanding (Durban) with an old city that has challenges (Bremen) The River Horse Project showcased the issues Durban faces in terms of the environment	Mutual learning and to show the Germans Durban care about the environment like the Germans do	Short-term	+
		Structural to Agential	The Bremen partnership started with a lot of NGOs partnering with Bremen and then developed into a full partnership between the two cities	The development of international relations from one level to another	Long-term	+
		Agential to Material	Durban is looking at rehabilitating a wetland on the Sundays River, which is a similar initiative that the River Horse project	Private public partnership	Long-term	+
		Ideational to Material	The Sundays River and River Horse initiatives are about getting an understanding of what is happening in river catchments because it is about river catchment management	Catchment management for water security	Long-term	+

(continued)

(continued)

Respondent	Stakeholder	Causal Mechanism Path	Respondent's motivation	Of what is it a case or an occurrence?	Long- or Short-term	Positive (+) or Negative (−)
		Agential to Material	What you do in a catchment has an influence on water quality and water quality indicates what is happening in a catchment	Linear cause and effect relations between water quality and catchment-based activities	Long-term	+
7	Senior Researcher at the CSIR	Material to Material and structural	Green infrastructure is investment and development in projects into the green economy	Economic activities in the natural environment	Long-term	+
		Ideational to Material to Agential	Ecological infrastructure is like floating wetlands that are specific projects, technologies or infrastructure developed to assist ecological aspects like water functioning	Specific activities using natural resources to perform natural environmental functions	Long-term	+
		Ideational to Material	Green infrastructure is like a green building, roof top garden or solar panels on a roof	Modern technology that harnesses an element of the natural environment other than fossil fuels	Long-term	+
		Ideational to Ideational	Green infrastructure refers to the sustainable development debate	Green and ecological infrastructure is the manifestation of a certain ideology—sustainable development	Long-term	+

(continued)

(continued)

Respondent	Stakeholder	Causal Mechanism Path	Respondent's motivation	Of what is it a case or an occurrence?	Long- or Short-term	Positive (+) or Negative (−)
		Structural to Agential	Durban is a key partner in the UEIP	Durban playing a role in the uMngeni River catchment because it is the ultimate downstream user of the river's water	Short-term	+
		Ideational to Material	There is not a lot of information on how much projects that relate to the UEIP cost	Lack of financial transparency	Short-term	−
		Agential to Ideational	We don't fully understand the social costs and benefits of green and ecological infrastructure because researchers have not researched them	Lack of financial transparency	Short-term	−
		Agential to Material to Ideational	There is not enough communication to people that benefit from the green or ecological infrastructure projects	Lack of transparency	Short-term	−
8	Researcher at WESSA	Material to Material	Ecological infrastructure is natural infrastructure, non-man-made and non-built infrastructure supplying goods and services to people	The natural environment in the service of humans	Long-term	+
		Material to Structural	He linked ecological infrastructure to the energy nexus	The connection of two concepts that are currently in vogue	Long-term	+

(continued)

(continued)

Respondent	Stakeholder	Causal Mechanism Path	Respondent's motivation	Of what is it a case or an occurrence?	Long- or Short-term	Positive (+) or Negative (−)
		Ideational	He also uses the SANBI definition when talking about ecological infrastructure	Prevailing usage of established conceptualisations by an entity with perceived authority	Short term (Until the next conceptualisation comes along)	+
		Structural to Material	D'MOSS initiated 25 years ago to have natural areas supporting water supply and cleaning water	An established structure of rule providing water security	Long-term	+
		Structural to Material	The green corridor initiative for water supply to Durban directly linked to water quality	An initiative that highlights water quality in relation to water security	Long-term	+
		Material to Structural	The uMngeni River is the artery of the green corridor with walking trails along the bank	A natural resource being the centrepiece of an ecological infrastructure initiative	Long-term	+
		Agential to Structural to Material	DUCT started the green corridor initiative and not it is an independent entity funded by eThekwini	Interest group action leading to a change in local government policies	Long-term	+
		Ideational to Agential	Biodiversity, as the clearing of alien invasives, and entrepreneurship are the main emphases of the green corridor	An ideational element being the foundation of a local government initiative	Long-term	+

(continued)

(continued)

Respondent	Stakeholder	Causal Mechanism Path	Respondent's motivation	Of what is it a case or an occurrence?	Long- or Short-term	Positive (+) or Negative (−)
		Ideational to Agential	Water security is internal to ecological infrastructure initiatives—Durban realises that the uMngeni River is something like an open sewer, while also sorting out leaks and water wastage to protect the catchment supplying the water	Water security defined with water pollution and quantity in mind	Long-term	+
		Material to Material	With the October 2017 floods, heavy rains flooded the N2 highway because litter blocked the culverts and lots of people lost their lives	A calamity's origin from a specific source	Short term	−
		Agential to Material	Human behaviour (littering) made the disaster so much worse	The interface between natural disasters and human behaviour	Short term	−
		Agentialto Structural to Material	In Amanzimtoti community action over the past few years resulted in no flooding on the N2	Human behaviour preventing calamities	Long-term	+
		Material to Material	The flood cost insurance companies R600 million, while the community initiative was cost only R500 00000	Costs and benefits	Short term/Long-term	+/−
		Agential to Material	If you don't sort out the system to manage solid waste, you will pay a lot of money through insurance claims	The correct governance systems need to be in place to minimise risks to society	Long-term	+

(continued)

(continued)

Respondent	Stakeholder	Causal Mechanism Path	Respondent's motivation	Of what is it a case or an occurrence?	Long- or Short-term	Positive (+) or Negative (−)
		Material to Agential	Cellular navigational technology showed that the N2 had no traffic because all the cellular phones in the flooded cars stopped working and people's navigation systems steered them onto an assumed 'empty' road with no traffic	Technology is not always helping society to overcome problems	Short term	−
		Agential to Structuralto Material	WESSA and DUCT work a lot with SMME's that had been established through the working for ecosystems initiative, with eThekwini hiring them to do work	A private public partnership working for the good of the environment	Long-term	+
		Structural to Agential to Material	South African Pulp and Paper Industries (SAPPI) facilitates SMME involvement and entrepreneurship by using SMME's to clear solid waste and alien invasives, plant commercial timber and spraying the timber	Corporate social responsibility	Long-term	+
		Structural to Ideationalto Agential	Durban is a C40 city and has a twinning agreement with Bremen, meaning that sustainable development is on the city's front burner	Structures of rule/governance creating a certain impression	Long-term	+
9	Municipal official	Ideational	There is a lot of confusion around the concept green infrastructure because it means a lot of things to a lot of people	Concepts need to have specific meanings otherwise they create confusion	Long-term	−

(continued)

(continued)

Respondent	Stakeholder	Causal Mechanism Path	Respondent's motivation	Of what is it a case or an occurrence?	Long- or Short-term	Positive (+) or Negative (−)
		Ideational to Agential to Material	My understanding of the concept green infrastructure is quite broad—these are natural systems that provide a natural engineering function whether flood attenuation or purification	Natural resources that provide a service to humans	Long-term	+
		Ideational to Material	The broad conceptualisation of green infrastructure means that it can be anything from natural ecosystems to green roofs and certain technologies as well as the design of infrastructure like SuDS and green elements imposed on existing infrastructure like paving	A spectrum of green infrastructure	Long-term	+
		Ideationalto Material	Ecological infrastructure is at the end of the green infrastructure spectrum; and it is the natural systems and functions that these provide to support or add to green infrastructure's value or function on their own by performing engineering functions and providing services to people	A concept part of a class of concepts	Long-term	+
		Ideationalto Structural	Depending on how one defines green or ecological infrastructure will influence the type of policiesthe municipality has in place to promote green or ecological infrastructure for water security	The specific meaning of concepts constitutes specific structures of rule	Long-term	+

(continued)

(continued)

Respondent	Stakeholder	Causal Mechanism Path	Respondent's motivation	Of what is it a case or an occurrence?	Long- or Short-term	Positive (+) or Negative (−)
		Structural to Agential to Ideational	The department I work for protect biodiversity and biodiversity is our ecological infrastructure	Conceptualisation constituting practices and structures of rule	Long-term	+
		Ideational to Structural to Agential	D'MOSS is a scientific way to help key biodiversity or ecological assets and then working to protect those and to work specifically towards water security	The science policy interface at work	Long-term	+
		Structural to Agential to Material to Ideational	The municipality established the UEIP that aims to improve the state of the uMngeni's ecological infrastructure for water security	A structure of rule that establishes interactive governance in the interest of a specific governmental actor	Long-term	+
		Structural to Agential to Material	We are establishing a MOU with The Nature Conservancy to advance water security through a water fund	A structure of rule that establishes interactive governance through an economic instrument	Long-term	+
		Structural to Agential to Material	The Working for Ecosystem and Fire Programmes focus on invasive alien plants and landscape management for the purpose of water security and job creation	Specific structures of rule informing practices for human benefit and wealth creation and distribution	Long-term	+
10	Researcher at INR	Ideational to Ideational	Natural capital is a concept interchangeable to ecological infrastructure	The utilisation of concepts in differing contexts	Long-term	+

(continued)

(continued)

Respondent	Stakeholder	Causal Mechanism Path	Respondent's motivation	Of what is it a case or an occurrence?	Long- or Short-term	Positive (+) or Negative (−)
		Ideational to Ideational	Ecological infrastructure could include a component of green infrastructure, but ecological infrastructure is more an overarching concept whereas green infrastructure is a component thereof	The classification of things along certain philosophy of the natural sciences principles	Long-term	+
		Agential to Structural to Material	The eThekwini is working with The Nature Conservancy to establish a water fund	The economic way of dealing with water insecurity	Long-term	+
		Ideational to Agential to Structural	eThekwini is forward thinking in their consideration of incorporation of ecological infrastructure and that is why they are working with The Nature Conservancy	A concept incorporated into eThekwini's structures of rule	Long-term	+
		Material to Agential to Ideational	The water fund will enable the municipality to protect important water source areas outside the municipal boundary	The believe that a specific (economic) governance structure will have desirable outcomes	Long-term	+
		Structural to Agential to Material to Ideational	They have bylaws protecting natural resources, but the intent is not necessarily from a water security perspective, although water security is a benefit that the municipality gets	An observation that water security is not the main driver of governance structures	Long-term	+

(continued)

(continued)

Respondent	Stakeholder	Causal Mechanism Path	Respondent's motivation	Of what is it a case or an occurrence?	Long- or Short-term	Positive (+) or Negative (−)
		Ideational to Agential to Structural	The intent is, rather, around biodiversity conservation in terms of their mandate under Section 24 of the Constitution where people have the right to a clean environment	Biodiversity if the main driver of the municipality's Constitutional mandate	Long-term	+
11	Palmiet River Watch Interest Group	Ideational to Material to Agential	Green infrastructure is a 'buzzword' people use to get access to funding without delivering what is necessary	The politicsof policy influencing	Short term	−
		Ideationalto Material to Agential to Ideational	Green infrastructure is natural and man-made systems that enable and enhance nature's ability to provide goods and services; that allow nature to correct imbalances	A specific interpretation of what green infrastructure ought to be and do	Short term	+
		Material to Ideational to Agential	Natural forests, large bodies of water, the atmosphere and the earth's crust provide vital goods and services and humans are impairing all these resources in the name of economic and social development	An observation-based perception that humans are damaging the earth	Long-term	−
		Agential to Material to Agential	The introduction of green infrastructure at a vast scale would help to correct the imbalances	A believe that humans could reverse their impact on the earth through green infrastructure	Long-term	+

(continued)

(continued)

Respondent	Stakeholder	Causal Mechanism Path	Respondent's motivation	Of what is it a case or an occurrence?	Long- or Short-term	Positive (+) or Negative (−)
		Agential to Material	eThekwini has not committed to or implemented green and ecological infrastructure at a scale required to make significant improvements to water security	Protesting the perceived lack of green and ecological infrastructure investment	Long-term	−
		Structural to Agential to Material	D'MOSS does not protect eThekwini from the erosion of goods and services caused by development	Influencing a specific structure of rule	Long-term	−
		Material to Material	There is a need for effective city-wide storm water management projects to address the root causes of poor water quality and poor river health	Influencing public policy related to storm water drainage and water quality	Long-term	−
		Ideational to Agential	'Sexy projects' get votes and 'appear' to address service delivery but rather deal with the symptoms of environmental degradation	Influencing a specific structure of rule	Long-term	−
		Ideational to Agential	Respondent 12 used the research of Duncan Hay to illustrate the unhealthy state of the uMngeni river	Using scientific knowledge to influence policy	Long-term	+
		Ideational to Structural	There is no alignment between the Palmiet River Action Plan and the Take Back our Rivers Proposal	A perception that if structures of rule are synchronised, they will produce desired outcomes	Long-term	−

(continued)

(continued)

Respondent	Stakeholder	Causal Mechanism Path	Respondent's motivation	Of what is it a case or an occurrence?	Long- or Short-term	Positive (+) or Negative (−)
		Structural to Material	Proposals tend to produce financially sustainable outcomes for the business but not the natural environment	A critical assessment of public policy	Long-term	−
		Structural to Ideational to Agential to Material	The Palmiet River Rehabilitation Project (PRRP) brought new hope, but turned out to be dealing with the symptoms, wasting resources and funding instead of addressing the root causes of environmental degradation, violations and infractions in a conclusive and sustainable way, as well as hosting events and showcasing individuals	A critique of policy	Short term	−
		Agential to Structural to Material	The formation of the Palmiet River Valley Community (PRVC) indicates the need to ensure that resources and funding that are obtained and used effectively and not wasted particularly by ineffective models and approaches	A critique towards scientific approaches that ignores local community structures	Long-term	−
		Agential to Ideational	It will take 40 years to change behaviour	A believe that there is entrenched behaviour towards the environment that will take a long time to change	Long-term	−

(continued)

(continued)

Respondent	Stakeholder	Causal Mechanism Path	Respondent's motivation	Of what is it a case or an occurrence?	Long- or Short-term	Positive (+) or Negative (−)
		Structuralto Agential to Material	Government authorities have not yet made a significant and meaningfully measurable sustainable difference to water security, instead revenue collection, job creation, socio-economic goals, and research output requirements drive their behaviour	A critique towards state behaviour aligned to the dominant economic discourse	Long-term	−
12	Municipal official	Ideational	Green infrastructure could mean different things, like low carbon buildings, pavements made from recycled material s and building that generate their own electricity	A specific meaning of the concept 'green infrastructure'	Long-term	+
		Ideational to Agential to Material	Green infrastructure links to the work of climate adaptation dealing with wetlands, forests, and rivers (natural infrastructure)	A concept informing specific activities around climate change	Long-term	+
		Ideational	Green and ecological infrastructure mean one thing for professional and another for lay persons	An observation informed conclusion about the meaning of green and ecological infrastructure	Short-term	+
		Ideationalto Material	Ecological infrastructure is the natural environment	A specific meaning of the concept ecological infrastructure	Short-term	+

(continued)

(continued)

Respondent	Stakeholder	Causal Mechanism Path	Respondent's motivation	Of what is it a case or an occurrence?	Long- or Short-term	Positive (+) or Negative (−)
		Structural to Agential to Ideational	The DCCS is not a climate change adaptation or mitigation strategy but a climate change response strategy that includes adaptation, mitigation and over-arching thematic areas	A specific interpretation of a structure of rule	Long-term	+
		Structural to Agential to Ideational	The C40 initiative is a group of 90 cities that discuss and exchange information on climate change	A structure of rule designed specifically for cities to address climate change	Long-term	+
		Structural to Agential to Material	We should change our bylaws to say that 9 litres of water to flush toilets must come from grey water	An initiative to address water shortages at the household level	Long-term	+
13	Municipal official	Ideational	Green infrastructure is how green and open spaces like natural ecosystems like nature reserves and cultivated areas such as parks provide services and working to improve the quality of air and water and providing social benefits	A specific meaning linked to environmental goods and services as well as social benefits	Long-term	+
		Ideational to Ideational	Green infrastructure speaks to city resilience	Linking two concepts from environmental goods and services to humans with one another	Long-term	+

(continued)

Respondent	Stakeholder	Causal Mechanism Path	Respondent's motivation	Of what is it a case or an occurrence?	Long- or Short-term	Positive (+) or Negative (−)
		Ideationalto Material	The botanical gardens are very old green infrastructure that provides access to people so they can escape from normal city live	The botanical gardens fulfil a specific psychological function	Long-term	+
		Material to Material	We would like to utilise storm water for utilisation in the gardens	The use of a 'free' resource to water the gardens	Long-term	+
14	Researcher at INR	Ideational to Structural	The meaning of the concept green infrastructure depends on the context you are working in	The importance of context	Long-term	+
		Ideationalto Material	Green infrastructure is natural ecosystems in all their states and forms like artificial wetlands and not only natural wetlands	The notion that green infrastructure includes all things providing goods and services to people	Long-term	+
		Ideational to Material to Ideational	An artificial wetland is not a natural system and is green infrastructure but not necessarily ecological infrastructure because it is not natural	The classification of green and ecological infrastructure components based on the philosophy of the natural sciences	Long-term	+
		Structuralto Material	I don't think that the municipalityhas policies dedicated to water security, but strategies related to it	Municipal policies that incorporates water security but not dedicated to it	Long-term	+/−

(continued)

(continued)

(continued)

Respondent	Stakeholder	Causal Mechanism Path	Respondent's motivation	Of what is it a case or an occurrence?	Long- or Short-term	Positive (+) or Negative (−)
15	Lecturer at UKZN	Ideational to Material	When I think of green infrastructure, I think of the SANBI definition and about built, non-built and mimicked (ie, and artificial wetland) infrastructure	A specific classification of what constitutes green infrastructure	Long-term	+/−
		Ideational	I am familiar with the concept ecological infrastructure in terms of the UEIP and not so much with the concept green infrastructure	Familiarity with one concept as opposed to another based on the context the respondent is operating in	Short-term	+/−
		Ideational to Material	If green infrastructure is about environmental services from a river catchment, then it is about all elements of a catchment, be it terrestrial, atmospheric that provide services to people around water quantity and quality	Green infrastructure linked to environmental goods and services for human benefit	Long-term	+
		Ideational to Material	Could ecological infrastructure incorporate smaller elements like individual trees or a ridge line?	The construction of a concept based on specific observations	Short-term	+
		Ideational to Material	When thinking about ecological infrastructure we are always thinking 'big' and not in terms of 'individual' elements because bigger elements are easier to quantify around services	The construction of a concept based on specific observations	Short-term	+

(continued)

(continued)

Respondent	Stakeholder	Causal Mechanism Path	Respondent's motivation	Of what is it a case or an occurrence?	Long- or Short-term	Positive (+) or Negative (−)
		Agential to Agential	eThekwini engages with elements in the uMngeni River catchment with the climate change officials in the municipality playing a big role in this thinking	The linkage between water security and climate change at a catchment level	Long-term	+
16	Municipal official	Ideational to Material	Biodiversity is green infrastructure and the services it provides to humans	The notion that nature is in service of humans	Long-term	+
		Ideational to Material	Green infrastructure is all our natural ecosystems and created ecosystems, which is different to the biodiversity side	Grouping green and ecological infrastructure together with the common denominator being services to humans	Long-term	+
		Ideational to Material	Ecological infrastructure is more natural systems with ecological processes, with the botanical gardens not really being ecological infrastructure	A specific function attached to ecological infrastructure	Long-term	+
		Ideational to Material	The more artificial an infrastructure is the less I would consider it ecological infrastructure and greener infrastructure	A spectrum of green and ecological infrastructure	Long-term	+
		Structural to Ideational to Material	eThekwini's climate change strategy focuses on water security in terms of the provisioning of water and flood reduction that are important aspects of ecological infrastructure	A specific structure of rule to implicitly mitigate water insecurity	Long-term	+

(continued)

(continued)

Respondent	Stakeholder	Causal Mechanism Path	Respondent's motivation	Of what is it a case or an occurrence?	Long- or Short-term	Positive (+) or Negative (−)
		Structural to Agent to Material	The UEIP focuses attention on the uMngeni River and the river's water quality	A structure of rule geared towards a specific natural resource	Long-term	+
		Material to Agent to Structural	Black mambas bite people in the Quarry road informal settlement so we started a programme of snake monitoring and appointed people to remove the snakes	A specific reaction to a natural environmentalenvironment risk	Long-term	+

Index

A

Action, 15, 25, 28–35, 37, 38, 40–42, 44, 47, 49, 50, 52, 54–56, 62–65, 70, 71, 86, 107, 110–112, 117, 124, 126, 130, 132, 135, 136, 142, 143, 150, 158–160, 169, 176, 185, 189, 190, 200–202, 205–209, 219, 238, 249, 250, 252, 253, 265, 266

Active substantiation, 44, 56, 57, 72

Agency, 25, 48, 51, 54, 58, 63, 64, 109, 111, 117, 165, 185, 201, 202, 238

Agential, 51, 61, 62, 64, 130, 160, 176, 179–185, 187–195, 201, 214, 220, 240, 249–260, 262–267, 269–275, 278, 279

Agential, Ideational, Material and Structural (AIMS), 64, 65

Agential power, 63, 67, 115

Alien vegetation, 107, 108, 110, 143, 147, 170, 172, 174, 175, 185, 186

Amakhosi, 141, 148, 149, 154–157, 169, 218, 254

Ambiguity theory of leadership, 67

Analysis, 20, 31, 34, 40, 41, 47, 52, 60, 61, 64–67, 69, 72, 81, 82, 87, 114, 115, 129, 158, 185, 186, 189, 192, 194, 195, 197, 199, 201, 209, 216–221, 235–237, 239, 245

Analytic eclecticism, 41, 56–61, 67–70, 201

Andrew Mather, 204

Anthropogenic climate change, 19, 81, 196, 197, 211

Aristotle, 61–64

Axiology, 52

B

Biodiversity, 1, 5, 6, 8, 89–91, 93, 100, 101, 112, 123, 138, 140–142, 148, 150, 162, 164, 166, 171, 172, 188, 194, 204–206, 209, 210, 216, 217, 219, 236, 237, 239, 242–244, 249, 252, 265, 269, 271, 278

Black mamba, 126, 174, 177, 279

Blind-spots, 56, 60

Bremen, 138, 143, 151, 183, 184, 216, 217, 243, 244, 261, 262, 267

Buffelsdraai Landfill, 86, 124, 183, 208, 241

C

Cape Town, 82, 104, 106, 134, 139, 149, 174, 211, 260

Capex, 109, 110, 131

Case studies, 47, 66, 69, 141, 204, 208

Catchment management agency, 63

Causal, 25, 27, 30, 31, 34, 40, 41, 54, 57, 59–67, 69, 72, 82, 159, 176, 179–181, 183, 187, 192, 201

Causal mechanism, 25, 31, 32, 41, 54, 55, 59, 64–67, 71, 72, 87, 114, 115, 175, 176, 178–195, 197–199, 201, 220, 245, 249–279

Causal narrative, 31, 56, 62

Causation, 27, 31, 61–63, 65, 66, 176, 201

Citizen science, 124, 130, 147, 169, 172, 198, 238

City planners, 4, 13, 14, 19, 182, 236

Climate change, 19, 71, 118, 125, 126, 135–138, 140–144, 148, 150, 160, 163, 166, 171, 190, 191, 195, 197, 198,

R. Meissner, *eThekwini's Green and Ecological Infrastructure Policy Landscape*, https://doi.org/10.1007/978-3-030-53051-8

204–208, 210–213, 216, 217, 219,
    252, 274, 275, 278
Climate change adaptation, 1, 7, 10, 81, 114,
    137, 141, 143, 150, 151, 203, 205–
    207, 216, 217, 219, 236, 237, 243,
    244, 275
Coast, 82, 84, 123, 133, 137, 140, 152, 204,
    208, 250
Collaboration, 82, 108, 109, 132–137, 145,
    147, 157, 195, 212, 217
Commodification of nature, 133, 140, 151
Competition, 133, 141, 154
Complexity, 12, 15, 38, 42, 57–60, 68, 72,
    133, 154, 211
Complexity theory, 57, 59, 67
Complex systems, 42
Compositionalist, 242, 243
Confirmation bias, 44
Consciousness, 41, 130, 131, 158, 165, 217
Constitution, 17, 89, 171, 202, 203, 244, 271
Constitutive, 34, 165, 199
Context, 1, 2, 10, 17, 19, 26, 27, 29, 31, 33,
    34, 36, 38, 47, 48, 51, 60, 65–67, 69,
    71, 81, 85, 91–93, 96, 98, 108, 118,
    125, 126, 131, 141, 144, 145, 163,
    165, 169, 174, 176, 178–180, 182,
    183, 190, 194, 200, 201, 203, 220,
    221, 237, 245, 252–254, 269, 276,
    277
Control, 40, 42, 45, 46, 48, 52, 53, 63, 90,
    94, 109, 116, 117, 132, 142, 147, 166,
    174, 178, 219, 243
Cost-benefit analysis, 109, 213, 242, 243
Criminals, 143, 170, 257
Critical realism, 45, 104
Critical subjectivity, 46
Critical theory, 29, 33, 38, 41, 57, 70–72, 87,
    89, 195, 196, 198, 201, 245
Cultural theory of International Relations, 67

D
David Hume, 61, 62
Debra Roberts, 138, 143, 146, 161, 203, 205,
    217
Decision-makers, 7, 35, 40, 41, 51, 52, 57,
    59, 157, 214–216
Deon Geldenhuys, 39, 40
Dialogic/dialectical, 47
Dualist/objective, 46
Duncan Hay, 103, 109, 111, 119, 120, 198,
    272
Durban, 1–5, 9, 10, 19, 20, 81–87, 94,
    103, 108–112, 121, 123, 124, 128,
    130, 134–145, 147–151, 154, 169,
    170, 173–175, 182, 185, 186, 191,
    198, 204–210, 216–218, 220, 235,
    237, 238, 242–244, 251, 256, 257,
    260–262, 264–267
Durban Botanical Gardens, 86, 97, 99, 191
Dusi Canoe Marathon, 103

E
Ecological dominance, 241–244
Ecological infrastructure, 1–4, 6–15, 17–20,
    25, 26, 39, 54, 60, 72, 73, 81, 82,
    84, 86, 89–94, 96, 97, 99–101, 106–
    108, 110, 112, 114–116, 119, 121–
    123, 125, 126, 129–136, 138–140,
    142–144, 146–150, 152–154, 156–
    165, 167, 170, 172, 175, 176, 178–
    183, 188, 192, 194, 195, 197, 198,
    200, 201, 203, 211, 216–220, 235–
    245, 250, 252–257, 263–266, 268–
    270, 272, 274, 276–278
Ecological systems, 6, 12, 68
Ecosystem, 1, 5, 8–11, 16, 58, 69, 84, 91, 93,
    97–99, 101, 123–125, 129, 137, 143,
    148, 150, 164, 168, 171–174, 186,
    192, 205–207, 209, 239, 241, 242,
    252, 254, 267–269, 275, 276, 278
Ecosystem services, 5, 6, 9–12, 16, 19, 20,
    89, 92, 93, 100, 138, 162, 182, 194,
    209, 210, 220, 236, 242, 251–253
Education, 10, 28, 37, 52, 107, 111, 134, 142,
    147, 167, 213–215, 238
Emotions, 36, 38, 39, 44, 62, 200
Empirical, 2, 14, 16, 31, 33–35, 39, 40, 43,
    47, 55, 57, 60, 61, 69, 101, 106, 109,
    115, 150, 154, 160, 196, 199, 201,
    202, 214, 220, 236, 238–241, 244,
    245
Empirical science, 115, 238
Empiricism, 11, 16, 35, 36, 39, 68, 69, 85,
    90, 92, 94, 100, 158, 160, 196, 198,
    201, 214, 219, 236–239
Engineer, 3, 5, 7, 9, 13, 14, 17, 19, 33, 89,
    92, 93, 101, 117, 118, 122–124, 127,
    129, 130, 134, 139, 159, 178, 179,
    181, 183, 189, 191, 198, 235, 249,
    251, 255, 258, 268
Environmental functions, 6, 249–251, 263
Environmental landscape, 235
Epistemic community, 14, 20, 40, 50, 81, 97,
    111, 112, 130, 136, 144, 148, 160,
    161, 163, 165, 167, 242, 245

Epistemology, 28, 30, 37, 44, 46, 66, 91, 146, 149, 197, 220, 236, 239, 240, 242, 243
Epistocracy, 81, 199, 203, 211–218, 220, 221, 236, 238, 240, 244
Epistocratic rule, 211
Epistocrats, 203, 211, 214, 217–219
Errol Douwes, 205
Estuary, 121, 125, 128, 139, 154–156, 163, 168, 176
EThekwini, 3, 9, 10, 14, 19, 20, 26, 27, 39, 54, 60, 67, 73, 81, 82, 84–86, 89, 90, 92, 94, 96, 97, 101, 104, 106, 108– 112, 122, 123, 125, 127, 129, 131– 137, 141–144, 146–151, 153, 154, 158, 160–163, 165, 168, 170–174, 181, 184, 186, 189, 195–198, 200, 201, 203, 204, 206, 211, 213, 216– 221, 235–239, 241–244, 265, 267, 270, 272, 278
EThekwini Metropolitan Municipality, 1–3, 9, 19, 81, 82, 84, 85, 104, 122, 132, 143, 144, 147, 157, 178
Ethics, 52, 58, 94, 111, 243
Ethnographic studies, 47
Ethnomethodological, 47
Everyday international political economy, 67
Experimental/manipulative, 13, 46, 47, 117
Expert, 14, 18, 39, 81, 90, 108, 132, 133, 142, 162, 167, 203, 210, 212–219

F
Feminism, 29, 45, 67
First transition, 116, 117, 122, 123, 130
Flooding, 2, 3, 16, 123, 124, 126, 127, 140, 168–170, 179, 181, 195, 198, 249, 266
Foreign policy, 39, 40, 151
Forest, 9, 96, 108, 170, 172, 173, 216, 271, 274
Fort Lauderdale, 150, 151, 217, 243
France, 213
Francis Bacon, 116
Functionalist, 148, 216, 242, 243

G
Gauteng, 82
Geoff Tooley, 123, 124, 204
Germany, 138, 143, 213, 261
Global warming, 109, 171, 210

Government, 1, 2, 10–12, 14, 17–19, 33, 35, 39, 44, 63, 67, 90, 108, 116–118, 121, 122, 130–133, 135–139, 141– 143, 145–147, 149–155, 157, 159, 160, 167, 169, 184, 195, 203–207, 209, 211–213, 215–217, 235, 237, 259–261, 265, 274
Graham Jewitt, 136, 182, 237
Grand theories, 70
Green infrastructure, 3–8, 10–14, 19, 81, 86, 89, 91–94, 96–99, 101, 107, 109, 111, 112, 123–127, 129–131, 140, 150, 152, 153, 160–164, 168, 171, 172, 179–184, 187–189, 193–195, 198, 219, 235, 236, 238–240, 243, 250, 251, 254, 258, 259, 261, 263, 267, 268, 270, 271, 274, 276–278
Green open spaces, 92, 101, 182
Grey epistemology, 13, 25, 131, 152, 198, 241
Group-think, 212, 240, 241

H
Harbour, 2, 170
Hard infrastructure, 8, 9, 141, 152, 170
Hegemony, 52, 111
Hermeneutical/dialectical, 47
Historical realism, 45
Hobbesian, 116, 117, 122, 126, 129, 130
Household refuse, 170
Hydraulic mission, 116, 117, 122, 126, 129, 165
Hydro-hegemony, 43
Hydro-social contract theory, 67, 112, 116, 118, 121, 141, 220
Hypothesis, 47, 57, 60, 69, 72, 245

I
Ibandla, 155
Ideational, 26, 27, 42, 50, 59, 61–64, 132, 158, 159, 161, 165, 176, 178–195, 198, 201, 249–252, 254–256, 258– 260, 262, 263, 265, 267–278
Ideology, 2, 26, 29, 63, 64, 155, 237, 238, 263
Inanda Dam, 104, 117, 125, 173
Individual, 16, 17, 29, 32, 35–37, 41, 44, 45, 50, 58, 59, 70, 82, 86, 100, 110– 112, 129–131, 157–159, 165–175, 200–203, 212, 215–217, 273, 277
Interactive governance theory (Govern-ability), 67

Interest group, 20, 43, 44, 94, 108, 117, 118,
        121, 127, 130–132, 136, 137, 141,
        159, 167, 189, 213, 242, 271
Interest group corporatism, 67
Interest group pluralism, 43, 67, 133, 141
International relations, 27, 36, 39, 70, 131–
        134, 151, 215, 216, 243, 244, 261,
        262
Interpretive science, 236
Interpretivism/constructivism, 29, 42, 44,
        45, 56, 59, 87, 90, 92, 107
Inter-subjective, 2, 158, 236, 238

J
Java, 54
Jean Jacques Rousseau, 121
Jeroen Warner, 116
Jim Taylor, vii
Joanne Douwes, 204
Johannesburg, 11, 39
John Locke, 118

K
Kate Pringle, vii
Knowledge, 12, 13, 25, 29, 31–38, 40, 41,
        44–46, 48, 49, 51–56, 58, 65, 67, 68,
        72, 82, 86, 87, 101, 111, 112, 115,
        117, 129, 130, 133, 135, 137, 139,
        143, 144, 147, 150, 156–160, 165,
        167, 181, 187, 188, 198, 200–203,
        210–216, 218, 236, 238, 244, 245,
        253, 272
Knowledge generation, 37, 45, 89, 147, 219,
        238
Knowledge system, 30, 159
KwaZulu-Natal, 2, 4, 19, 82, 83, 124, 136,
        137, 139, 141, 147, 169, 204, 206,
        209, 220
KwaZulu-Natal Provincial Government,
        154, 254

L
Level of analysis, 51, 64
Liberal institutionalism, 71, 81, 129, 131–
        134, 145, 149, 153, 157–159, 195,
        199, 200, 220
Littering, 20, 127, 170, 173, 195, 257, 266
Lived experience, 46–48, 86, 220
Lockean, 116–118, 121, 122, 129

M
Marc Pienaar, 54, 87
Market function, 152
Marxism, 29, 67, 86
Material, 31, 51, 59–64, 69, 90, 93, 104, 110,
        132, 139, 149, 158, 159, 165, 168,
        176, 178–186, 188–195, 199, 201,
        202, 220, 236, 249–279
Mathematically oriented theories, 70, 71
Meta-theoretical assumptions, 25, 28, 55
Methodology, 28, 42, 47, 49, 53, 54, 56, 65,
        66, 82, 85, 86, 148, 158, 197, 245
Middle-range theory, 70, 71
Midmar Dam, 102, 120, 147
Milja Kurki, 61
Modernity, 67, 112, 165–167, 211, 220
Modified, 46, 47, 70
Modified experimental/manipulative, 47
Mozambique, 137, 144
Msunduzi River, 103, 105, 107, 147, 185
M theory, 71
Multi-functionality, 10, 11
Municipality, 1–3, 10, 12, 17–20, 67, 81,
        82, 84, 86, 89, 97, 101, 106–108,
        110–112, 122, 123, 126, 129–131,
        133–139, 141–144, 147, 149, 151,
        154, 156, 160, 170, 171, 173, 178,
        179, 182–184, 188, 190, 191, 193–
        195, 197, 198, 200, 201, 203, 204,
        206, 216–220, 235, 242–245, 253,
        256–259, 261, 268–271, 276, 278

N
Natural environment, 15, 16, 19, 25, 28, 30,
        33, 42, 44, 59, 68, 82, 84, 94, 117,
        145, 149, 162, 168, 170, 172–175,
        195, 202, 217, 220, 244, 261, 263,
        264, 273, 274, 279
Natural resources, 81, 91, 101, 116, 152, 167,
        168, 171, 176, 180, 184, 185, 195,
        220, 242, 250, 258, 263, 265, 268,
        270, 279
Natural science, 7, 36, 70, 89, 117, 198, 217,
        236, 270, 276
Nature, 1–3, 5, 6, 8, 9, 20, 26, 29, 32, 34,
        36–39, 41, 45, 50, 52, 54, 56, 57,
        65, 67, 84, 85, 90, 92, 93, 96, 97,
        100, 101, 108, 110, 111, 115–117,
        123, 127, 130, 136, 137, 139, 148,
        149, 153, 157, 159–161, 165–167,
        175, 183, 201, 212, 216, 217, 219,
        220, 235, 240, 242, 243, 250–252,
        254–256, 258, 269–271, 275, 278

Neo-liberalism (Liberal pluralism), 67, 112, 131, 145
Neo-realism, 67, 149
Networking, 133, 134
Norm, 37, 58, 62, 71, 92, 116, 117, 132, 149, 151, 158–160, 162, 167, 202, 218, 220, 243
Normative, 1, 2, 11, 13, 14, 16, 20, 26, 29, 31, 33–35, 44, 55, 60, 85, 90, 94, 101, 109, 116, 118, 122, 129, 131, 151, 153, 155, 157, 164, 196, 198, 199, 201, 202, 212, 214, 218, 221, 236–239, 241–245
Normative commensalism, 67

O

Objective, 1, 7, 13, 26, 30, 35, 36, 38, 45–48, 52, 54, 68, 72, 135, 138, 147, 148, 171, 219, 220, 239–241
Objective truth, 36
Ontology, 10, 28, 30, 45, 51, 57, 64, 66, 90, 91, 104, 111, 131, 146, 149, 152, 153, 160, 201, 219, 220, 236, 238–241, 243
Open-ended analysis, 60, 72, 73, 112
Organising question, 50, 110, 149

P

Palmiet River, 82, 86, 94, 95, 114, 125, 127, 129, 130, 147, 171, 173, 189, 218, 271–273
Paradigm, 11, 13, 25–30, 32–42, 44, 45, 51, 52, 54–56, 58–63, 70–72, 82, 85, 87, 90, 92, 104, 107, 114, 116, 129, 198, 199, 201, 213, 215, 219, 244, 245, 249
Paradigm assessment, 41, 44, 54, 55, 72, 87, 90
Paradigm assessment index, 45
Paradigm peace, 54
Paradigm war, 54
Participative reality, 30, 45
Participatory, 45–47, 49, 57, 63, 86, 129
Participatory paradigm, 29, 30, 38, 42, 44, 59
Phenomenographic studies, 47
Philosophy, 25, 29, 61, 64, 68, 117, 142–144, 146, 148, 149, 158, 189, 211, 242, 243, 270, 276
Policies, 2, 7, 10–12, 14, 15, 17, 18, 25–27, 29, 35, 37–44, 51, 52, 54–56, 60, 63, 64, 68–71, 81, 85, 86, 89, 90, 92, 101,
104, 108, 111, 122, 123, 125, 127, 129, 130, 132–134, 138–142, 144, 146–148, 151, 155, 159, 161, 166, 170, 171, 176, 181, 182, 187, 188, 190, 193, 195–198, 201, 203, 207, 211–220, 235, 236, 240, 241, 243, 245, 251, 253, 255–261, 265, 268, 269, 271–273, 276
Policy landscape, 14, 20, 26, 27, 39, 54, 60, 67, 73, 81, 112, 131, 133, 134, 146, 158, 160, 161, 165, 176, 195–197, 199–201, 203, 211, 220, 221, 235, 236, 238–244
Policy-makers, 11, 25, 26, 112, 143, 180, 197, 210, 237
Policy paradigm, 26, 27, 29, 112
Political, 1, 13, 15, 16, 18–20, 26, 34, 36, 43, 45, 47, 48, 52, 55, 58, 60, 64, 67, 81, 84–86, 92, 109, 112, 116, 118, 122, 123, 130–132, 138, 147–149, 151, 154, 157–161, 166, 167, 173, 199, 201, 203, 211–217, 243, 251, 256
Political ecology or Green politics, 67
Political landscape, 12, 15
Political participation, 47, 214
Political science, 36, 199, 215
Politics, 15, 27, 38, 39, 43, 64, 68, 86, 118, 131–134, 144, 149, 151, 157–159, 166, 167, 201, 203, 211, 215–217, 271
Pollution, 2, 3, 5, 11, 16, 66, 87, 91, 105, 108, 111, 116, 117, 121–123, 126, 127, 129, 139, 145, 147, 166, 168, 171, 173, 174, 178, 189, 249, 257, 266
Port of Durban, 2, 6–8, 82, 83, 122, 204, 220
Positivism, 25, 29, 34, 36, 38–41, 44, 45, 54, 56, 57, 59, 61, 81, 85, 89, 90, 94, 104, 107, 109, 148, 158, 159, 196, 198, 213, 219, 236, 238, 239, 245
Positivist, 30, 36, 38, 39, 42, 47, 51, 52, 54, 56, 59, 61, 66, 68, 70, 87, 94, 107–109, 111, 114, 116, 148, 158, 197, 219, 220, 236, 238, 243, 245
Post-positivism, 29, 39, 40, 42, 44, 45, 56, 59, 87, 104, 107, 241
Practical, 18, 19, 32–35, 38, 44, 46–49, 52, 55–61, 72, 85, 90, 114, 117, 200, 202, 213, 244, 245
Practice, 1, 5, 12, 13, 15–17, 25–33, 35–38, 40, 42–44, 52, 54, 58–63, 68–72, 81, 85, 106–108, 116, 117, 127, 129, 130, 137–139, 147, 151, 155–160, 165, 166, 197–202, 204, 207,

211, 214, 216–220, 235, 236, 238, 239, 242, 245, 258–260, 269
Practice theory, 199–203
Practitioners, 17, 25, 26, 28, 32, 33, 37, 38, 41–44, 50, 51, 54, 55, 57, 58, 60, 61, 65, 67–71, 73, 85, 129, 201, 206, 213, 215, 239, 245
Prime empirical focus, 50, 111
Programmes, 16, 25, 42, 44, 51, 54, 55, 69–71, 87, 104, 118, 134, 135, 142–144, 172, 203, 206, 241, 269, 279
Programme theory, 70, 197, 220
Public, 11–14, 17–20, 29, 33, 40, 60, 63, 68, 69, 107, 108, 110, 112, 116, 117, 132, 133, 135, 139, 149, 155, 159, 160, 162, 166, 179, 195, 198, 201, 204, 212–215, 217–219, 237, 238, 242, 244, 256, 259–262, 267
PULSE³, 20, 25, 29, 41–44, 54, 57, 61, 64, 69, 72, 73, 81, 82, 85, 87, 90, 114–116, 158, 195, 199, 201, 215–217, 219, 235, 236, 245

Q
Qualitative, 47, 50, 54, 66, 85, 86, 115, 213
Quantitative, 47, 50, 54, 85, 94

R
Rat, 126, 174, 177
Rational choice, 39, 158, 160, 199
Raw realism, 45, 90, 91
Realism, 43, 67, 70
Reality, 10, 28–31, 33, 36, 39, 41, 45–49, 52, 54, 57, 58, 62, 70, 89–91, 99, 100, 104, 107, 109, 150, 158, 160, 196, 200, 201, 219, 220, 238–240, 243
Reductionism, 60
Reductionist, 58, 69, 70
Regular determinism, 61
Relativism, 29, 45, 99
Reliability, 30, 49
René Descartes, 116
Research object, 46, 48, 66, 240
Research paradigm, 20, 25–27, 29, 41, 54, 62, 87
Risk, 8–10, 12, 13, 16, 40, 57, 129, 143, 150, 152, 165–176, 210, 218–220, 249, 257, 266, 279
Risk society, 67, 112, 165–174, 211, 220
River health, 127, 173, 272

River Horse, 86, 106, 109, 111, 118, 130, 136, 145, 151, 153, 183, 184, 203, 216, 261, 262
Rural areas, 3, 5, 6, 8, 11, 19, 81, 174

S
Sabine Stuart-Hill, 237
Scale, 6, 8, 10, 16, 48, 68, 96, 109, 110, 121, 127, 137, 141, 166, 167, 171, 172, 175, 197, 205, 207, 208, 210, 271, 272
Scholars, 2, 4, 5, 14, 19, 26, 28, 29, 31, 32, 34, 39, 40, 44, 58, 61, 64, 69, 70, 72, 85, 86, 92, 133, 159, 199, 211, 237, 242
Science, 13, 26, 28, 29, 32, 34, 38–40, 45, 57–59, 61, 71, 72, 90, 92, 108, 116, 117, 135, 138, 139, 143, 144, 182, 198, 205, 209, 214, 215, 218, 236, 238
Science speaking truth to power, 39, 40, 111, 238
Scientific, 2, 4, 9, 16, 20, 28, 29, 32–34, 37, 40, 44, 57, 59, 64, 65, 71, 72, 81, 85, 86, 92, 93, 101, 108, 114, 138, 143, 144, 146, 148, 165, 172, 182, 187, 198, 199, 204–212, 214–216, 218, 238, 240, 245, 269, 273
Scientific method, 35, 36, 47, 50
Sean O'Donnoghue, 205
Second transition, 116–118, 122, 127, 130
Sekhukhune District Municipality, 67, 178, 237
Service delivery, 12, 18, 109, 143, 173, 272
Sewage pollution, 127, 144, 171
Sex doll, 93, 159, 161, 178, 255
Shared belief, 158, 160, 161
Social constructivism, 67, 112, 115, 157–159, 161, 165, 199, 220, 239
Social science, 25, 31, 48, 50, 58, 61, 64, 70, 94, 115, 142, 158, 258
Social theory, 27, 56, 157, 200, 245
Socio-economic development, 8, 10, 11, 16, 17, 84, 244
Software, 54, 55, 71
Solid waste, 2, 4, 7, 8, 20, 110, 119, 126, 143, 144, 147, 148, 169, 174, 266, 267
South African economy, 82, 83
South African National Biodiversity Institute (SANBI), 8, 9, 93, 96, 99–101, 107, 122, 123, 146–148, 160, 161, 165, 182, 188, 219, 237, 238, 240, 256, 265, 277

Storm, 2, 5, 20, 87, 89, 91, 119, 124, 144, 169, 204, 220, 249

Storm-water, 1, 6, 10, 13, 17, 19, 91, 104, 106, 123, 124, 127, 129, 139, 141, 144, 147, 152, 169, 170, 173, 179, 183, 189, 191, 192, 198, 216, 235, 239, 241, 249, 257, 260, 272, 276

Strategic adaptive management or adaptive management, 67

Structural, 18, 34, 63, 64, 91, 93, 94, 97, 100, 116, 131, 145, 148, 150, 151, 176, 180–195, 197–201, 216, 217, 219, 242, 244, 253–257, 259–276, 278, 279

Subjective, 30, 38, 45, 48, 58, 86

Substantive, 43, 50, 60, 132, 155

Sustainable development, 1, 36, 90, 93, 100, 117, 118, 138, 142–144, 152, 206, 219, 236–238, 242–244, 263, 267

Systemic, 38, 51, 131, 175, 199, 220, 235, 236

Systemic constructivism, 100, 220, 240

T

Temporal, 48, 166, 168–175, 180, 202, 212, 220

Theoretical synthesis, 59

Theories for practice, 67, 245

Theory, 11–13, 15, 20, 25–28, 30–46, 48, 50, 52, 54–63, 66–72, 81, 82, 85–87, 93, 112, 114–116, 123–127, 131–133, 145, 149, 158, 159, 161–165, 195, 196, 198, 199, 201, 215, 218, 220, 237, 238, 241, 244, 245

Theory of social learning and policy paradigms, 67

Theory of truth, 30, 47, 48

Thomas Hobbes, 116

Time, 4, 37, 38, 42, 44, 45, 48, 49, 60, 73, 110, 116, 132, 143, 145, 146, 148, 149, 162, 166, 167, 170, 173, 175, 195, 203, 204, 208, 212, 214, 237, 244, 273

Traditional Leadership and Governance Framework Act, 157

Training, 35, 50, 52, 66, 144, 213

Transactional/subjective, 46

Transdisciplinarity, 58, 59

Transparency, 14, 195, 198, 264

Twinning, 137, 138, 150, 151, 178, 219, 243, 244, 261, 267

Typology, 29, 41, 61, 62, 64, 66, 72, 260

U

UMgungundlovu District Municipality, 81, 136, 147

UMngeni Ecological Infrastructure Partnership, 9, 96, 100, 101, 108, 122, 123, 125, 126, 129, 136, 146–149, 160, 163, 173, 185, 216, 219, 237, 256, 264, 269, 277, 279

UMngeni River, 4, 9, 10, 19, 81, 101, 102, 105, 109–111, 122, 124, 129, 130, 134, 136, 146–148, 165, 168, 170, 177, 185, 186, 198, 217, 237, 261, 264–266, 272, 278, 279

UMngeni Vlei, 119, 120

Understanding, 11, 25–27, 30–33, 35, 36, 38, 40–42, 45, 52, 55–58, 60, 61, 65, 67–73, 85, 86, 89, 91–94, 97–100, 104, 106, 107, 114, 115, 123, 124, 137, 140, 146, 159, 160, 162, 165, 180, 181, 183, 187, 188, 200, 206, 208, 215, 219, 221, 244, 245, 262, 268

Unit of analysis, 50

University of KwaZulu-Natal, 10, 82, 89, 92, 99, 111, 130, 131, 134–136, 142, 144, 147, 148, 194, 195, 216, 237, 238, 251, 253, 277

UThukela District Municipality, 81

V

Validity, 30, 49

Values, 9, 11, 14, 19, 26, 28, 34–37, 42, 44–46, 49, 50, 52, 58, 60, 63, 68, 84, 85, 94, 111, 112, 129, 140, 147, 152, 154, 156, 164, 213, 218, 236, 241–243, 250, 268

Voice, 51, 94, 111, 115, 116, 121, 212, 218, 238, 244

W

Water, 1–4, 6–13, 15–19, 26, 43, 63, 64, 67, 69, 81, 84, 86, 89, 90, 92–94, 96, 97, 101, 108–112, 116–118, 121–127, 129–131, 136, 137, 139–141, 145, 147, 152, 163, 164, 168–174, 183, 185, 186, 191, 192, 195, 198, 204, 215, 220, 235, 237, 239, 243, 251, 254, 256, 258, 260, 263–266, 269–271, 275, 276, 278

Water fund, 108–111, 130, 134, 137, 148–150, 153, 154, 197, 216, 218, 219, 242, 244, 269, 270

Water quality, 9, 10, 85, 103, 104, 108, 109, 111, 122, 127, 129, 147, 171, 173, 220, 238, 254, 260, 263, 265, 272, 279
Water quantity, 16, 99, 108, 126, 140, 237, 277
Water resources management, 1, 2, 17–19, 64
Water security, 1, 2, 6, 9, 10, 15–20, 26, 67, 86, 107, 108, 111, 112, 114, 122, 124–126, 129, 135, 136, 138, 140, 144, 146–150, 161, 170–172, 178, 183, 192, 193, 197, 218, 236, 237, 243, 244, 252, 254–256, 258, 262, 265, 266, 268–270, 272, 274, 276, 278
Water supplies, 90, 123, 129, 149, 170, 172, 186, 191, 256, 265

Z
Zandile Gumede, 143

Printed in the United States
by Baker & Taylor Publisher Services